S. Niwa · S.M. Perren
T. Hattori (Eds.)

Biomechanics in Orthopedics

With 210 Figures

Springer-Verlag
Tokyo Berlin Heidelberg
New York London Paris
Hong Kong Barcelona
Budapest

Professor SHIGEO NIWA, M.D.
Chief eng. TOMOKAZU HATTORI, BSC.
Department of Orthopedic Surgery, Aichi Medical University, 21 Yazako Karimata,
Nagakute, Aichi, 480-11 Japan

Professor STEPHAN M. PERREN, DR. MED., DR. SC.(H.C.)
AO/ASIF Research Institute, Clavadelerstrasse, CH-7270
Davos Platz, Switzerland

ISBN-13:978-4-431-68218-9   e-ISBN-13:978-4-431-68216-5
DOI:10.1007/978-4-431-68216-5

Printed on acid-free paper

Library of Congress Cataloging-in-Publication Data
Biomechanics in orthopedics / S. Niwa, S.M. Perren, T. Hattori (eds.). p. cm. Includes biblio-
graphical references and index. ISBN-13:978-4-431-68218-9
1. Musculoskeletal system — Mechanical properties — Congresses. 2. Orthopedic surgery —
Congresses. 3. Biomechanics — Congresses. I. Niwa, S. (Shigeo), 1932– . II. Perren, S.M.
(Stephen M.), 1932– . III. Hattori, T. (Tomokazu), 1957– . [DNLM: 1. Biomechanics —
congresses. 2. Orthopedics — methods — congresses. WE 170 B615] RD732.B58 1992, 617.3
— dc20, DNLM/DLC, for Library of Congress, 92-49957

Preface

A major part of orthopedics is the treatment of musculoskeletal diseases caused by structural disorders and mechanical breakdown of living tissue. Therefore, biomechanical consideration of static structures and dynamic mechanisms is compulsory for both diagnosis and treatment of orthopedic diseases.

Previous biomechanical studies have enabled great advances in orthopedic implant technology, such as artificial joint replacement and instrumentation for spinal fusion. Consequently the importance of biomechanics is increasing more and more in daily clinical practice and development. In addition, biomaterial research into mechanical properties and tissue reactions of implant materials is certainly an important area of related study.

This book is comprised of 22 papers presented at the International Seminar on Biomechanics in Orthopedics and the 17th Annual Meeting of the Japanese Society for Orthopedic Biomechanics, held in Nagoya in 1990. The volume contains full descriptions of both conventional and updated knowledge of the spine, ligaments, artificial joint replacement in the hip and knee, fracture treatment, and gait analysis, as well as biomaterials. I earnestly hope that this book will be of benefit to readers in daily clinical work and research.

To close, I would like to thank profoundly the two coeditors, Prof. S.M. Perren and Mr. T. Hattori, and also a quiet supporter Mrs. J. Buchanan in Davos, for their cooperation in producing this book.

SHIGEO NIWA

Preface

In recent years orthopedic biomechanics has become of increasing interest worldwide. The International Seminar on Biomechanics in Orthopedics in Nagoya, Japan, covered a wide spectrum of topics. The study of the mechanics of living things has become inseparable from the manual and instrumental measures involved in the assessment and corrective treatment of musculo-skeletal deformities and diseases. The diversity of this field of investigation encompasses progress in the area of implants and prostheses and aims to expand our knowledge of the biomechanical properties of molecular, cellular, and organ material. This knowledge facilitates the selection of optimal forms and ideal materials which will harmonize with the conditions in the human body. The important evolution of acetabular prostheses and artificial limbs in the last few years and the efforts to find a nickel-free, non-allergenic material for their construction has been addressed in this volume. The role of materials and coatings as part of the wider issue of biocompatibility has been recognized as an essential factor in the consideration of the biomechanical properties of tissues, in particular, in the stages of healing after fracture.

Stimulating material was presented on topics relating to the locomotor system. This includes information on the kinematics of the musculoskeletal system and the quantification of forces, the knowledge of which provides the basis for optimal engineering design of implants.

The heterogeneous nature of the presentations at this seminar underlines the interdisciplinary nature of orthopaedic biomechanics.

The seminar itself and the keen interest of the participants has certainly contributed to stimulating thought and further experimentation in this area, in particular, by the advancement of international cooperation. The contribution of the Japanese Society of Orthopaedic Biomechanics and Prof. Niwa to these fields of research should be acknowledged here. I feel sure that the readers of this volume will profit greatly and be encouraged to pursue their research with renewed zest.

STEPHAN M. PERREN

Preface

As an engineer studying biomechanics in orthopedics, it was a great opportunity to undertake co-editorship of this book, the proceedings of the International Seminar on Biomechanics in Orthopedics with several papers from the 17th annual meeting of the Japanese Society for Orthopedic Biomechanics, both held in 1990.

It goes without saying that orthopedic biomechanics is an interdisciplinary field consisting of orthopedics and engineering. Recently there has been remarkable progress in research and development technology, including the Finite Element Method for stress analysis, computerized control and measurement systems, various fine sensors, and CT and NMR imaging systems. So we have received a great deal of benefit from systems providing more accurate information on living tissue as well as fine prototype products. However, we are also faced with new complicated problems requiring broader and deeper knowledge of both the living body and engineering. Under these circumstances, a stronger link between orthopedics and engineering is surely essential in future research.

I hope that this book will be a milestone in our interaction, and that previous annoying problems will be solved by collaboration of the authors here and readers who have a strong interest in biomechanics. To end, I would like to sincerely thank Prof. S. Niwa for assigning me to this worthwhile work, and also to thank all authors for their contributions to this book.

TOMOKAZU HATTORI

Contents

New Aspects of Orthopedic Biomechanics

Fracture

Ligament

Spine

Hip Joint

Knee Joint

List of Contributors

New Aspects of Orthopedic Biomechanics

Fracture

Stability and Bone Healing

STEPHAN M. PERREN[1]

Summary. In fracture healing, the mechanical properties of biological tissue play an important role. This paper discusses the mechanical aspects of fracture healing and their contribution to it. The response of biological tissue to a variety of forces is a function of their mechanical properties. The type, magnitude, direction, and point of application of a force as well as whether or not it is applied slowly, rapidly, repetitively, or for a long duration will influence this response. These mechanical properties depend on the type of tissue under consideration and the conditions under which they are determined. There is a need for further detailed study of these parameters.

Key words. Stability — Bone healing — Bone fracture — Biomechanics — Temporary porosis

Introduction

There is a strong interrelation between nature's way of healing a fracture and the results of surgical intervention. An essential question is why does the surgeon intervene in the first place. It is generally assumed that every fracture requires operation if healing is to be achieved. As part of our research we carried out animal experiments to investigate the effects of readapting the fracture and stably fixing the fragments to prevent their displacement in relation to each other. We observed that an animal in which an osteotomized bone has been internally fixed will use the treated limb freely, weight-bear and regain functioning immediately after the operation. That is to say, the aim of surgical intervention is the recovery of normal functioning as soon as possible. The time required to achieve this is understandably longer in clinical cases. The goal of treatment is early and complete recovery of limb function. Early means

---

[1] AO Research Institute, Clavadelerstr., 7270 Davos Platz, Switzerland

that the limb should be mobile immediately after the operation, and complete means that correct anatomical shape, in particular of the articular ends of the limb, should be regained.

Fracture

What exactly happens when a bone breaks? When a bone breaks it is very sudden. At a speed of 10,000 frames per second it is possible to film a fracture as it occurs. As the fragments move apart a vacuum develops and then a powerful implosion [1], a phenomenon that can be compared to cavitation. This process damages the soft tissue immensely.

Untreated fractures can be observed in wild animals and our findings show that they usually unite solidly. The problem is that of malalignment. Angulation, rotation, and shortening of the limb are often present.

Stabilization

The concepts behind stabilizing a fracture are important with regard to how and why we stabilize it. The first major concept is that of compression. Fragments under compression should be incapable of displacement. This can be demonstrated by applying axial dynamic loading, bending, and torsion to a fractured bone in which the fracture has been placed under compression. Although the bone itself will deform, the fracture surfaces will not move in relation to each other.

Preload

If you preload the fracture and twist the bone, then it is not the preload which is active but friction. Friction prevents movement, but movement will occur if the load is too great. If the compression force is greater than the tensile force, there is no movement. Friction is, therefore, essential to maintain torsional stabilization. Previously, torsion was considered detrimental to fracture healing. However, this was probably because the extent of torsion could not be controlled.

Pressure Necrosis

When compression was first introduced as a treatment method, there was a certain amount of anxiety that compression would produce necrosis of bone. Experimental experience has shown that the compression applied to a plate decreases very slowly [2]. After 10–12 weeks it has reduced by about 50%. If there was any amount of pressure-induced necrosis and resorption, the compression would fall off very rapidly (Fig. 1).

Fig. 1a,b. a Compression applied to cortical bone in vivo. **b** The initial value of compression of 1800 N force decreases very slowly. This pattern of change in compression proved that pressure necrosis with surface resorption in the compressed area did not occur

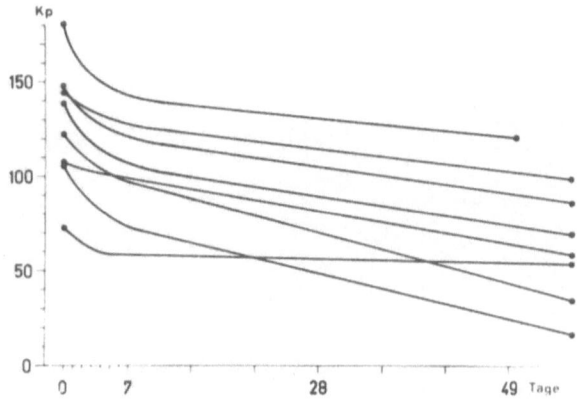

The situation for screws is somewhat different. When a screw is used to produce compression, there is a similar initial decrease as for the plate; the compression then remains relatively constant for about 4 months. We can reasonably assume that compression under stable conditions does not result in resorption and does not lead to necrosis. Therefore, it makes sense to stabilize the bone using compression exerted by lag screws. This procedure results in direct or primary healing.

Direct Healing

In the process of direct healing, Haversian osteons can be seen crossing the fracture gap. The question to be dealt with is whether this kind of healing is essential or whether it is merely an adjunct to absolutely stable fixation.

It is important to realize that rigid implants are not actually rigid. A minimal force will cause the fracture gap to open or close more tightly even if the fracture has been fixed with a plate. The plate is fifty times more flexible than the bone for geometric reasons. The same is true for the medullary nail which is rotationally unstable and which relies on an interlocking of the fracture surfaces. These must be pressed firmly together and/or interlocking screws

Fig. 2a–c. Experimental method used to study bone reaction at unstable interfaces **a** Hydraulically operated bellows with plunger mounted within a U-shaped jig. **b** Cross-section of the tibia: the histological section shows the marked bone surface resorption in the vicinity of an intermittently contacting plunger surface. **c** Cross section of the tibia: no resorption in the case of a plunger which does not contact the bone surface

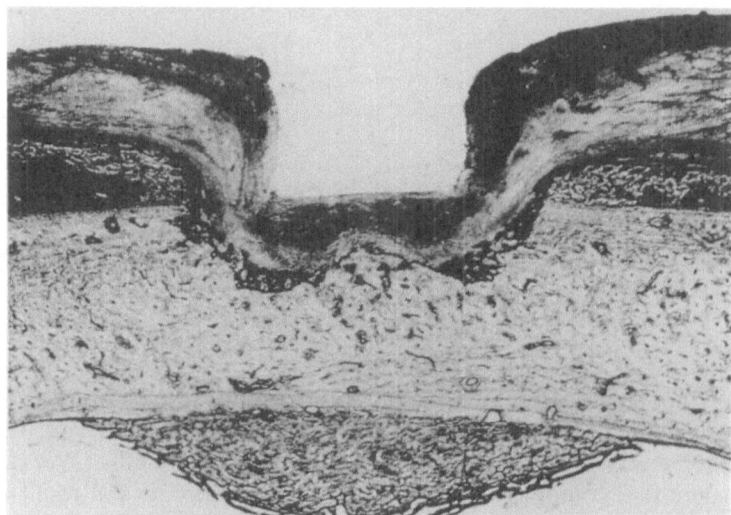

must be inserted in order to adequately stabilize the fracture. The same is true for the external fixation frames which have a similar function, namely to allow only a minimal amount of movement. This is the case unless compression is used as well.

Micromotion

To study bone reaction at an unstable interface, Stadler et al. [3] devised an experimental method using hydraulically operated bellows with a plunger mounted on a U-shaped jig to simulate the amount of micromotion which occurs in fracture fixation without compression (Fig. 2). The plunger was then applied to the bone intermittently. If the plunger hits the bone surface, even with almost negligible force, then resorption occurs. If the plunger does not impinge on the bone surface there is no resorption.

To further investigate motion-induced bone resorption, a straightforward experiment can be performed. A screw is fixed to the bone under both stable and unstable conditions. If a screw is intermittently loaded, even though it remains in constant contact with the bone, it will not lead to bone resorption. If the loading forces permit the fracture gap to open only a few micrometers, bone resorption immediately occurs. This is the case for internal fixation and hip prosthesis alike.

A further study of bone resorption was performed [4–6] using a wedge of bone with an external fixator constructed to permit movement of the wedge around a fulcrum. There were two possible conditions: At rest and with hydraulic loading. If there is too much motion, bone cannot form. However, at 10 cycles a day within a period of 12 h followed by 12 h of rest, bone regeneration is very marked. At 1000 cycles in 12 h, the regeneration is less but still pronounced. This may explain the mechanism of indirect or secondary healing.

Micromotion at the fracture site may be crucial to the healing process. If a plate fixation is unstable, the inevitable micromotion can lead to inadequate healing (angulation) and the need for a second operation may eventually arise. The displacement distance may be smaller than the size of an individual cell but produces a deformation which is very important for the cell. The wider the gap, the less deformation is caused by the same amount of motion and the amount of strain is less. The critical issue is that if there is micromotion it may cause elongation and rupture of cellulous material and the damage may be done before the micromotion is detected. The data from Yamada and Evans [7] clarifies the stage at which repair tissue will be critically affected.

Early Temporary Porosis and Stress Protection

One of the complications related to internal fixation is not altogether obvious. Many cases of refracture after plate removal are ascribed to Wolff's law. It is argued that the plate takes part of the load away from the bone and thus a

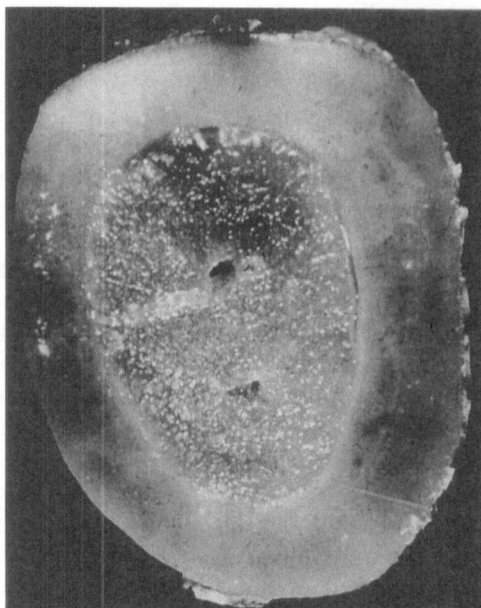

a

Fig. 3a,b. Blood supply, remodelling, and porosis beneath the plate
in **a** areas with disturbed blood
supply and **b** areas of normal bone
and remodeled, temporary porotic
bone (20 weeks post operative)

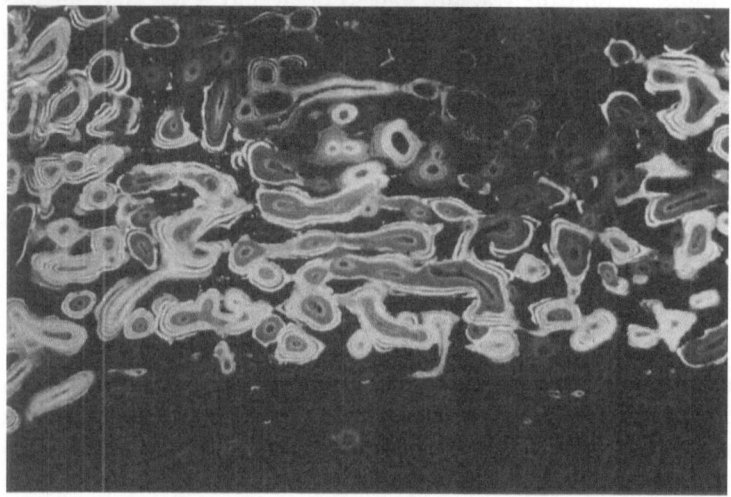

b

weak bone structure is formed. This explanation is not necessarily correct.
Wolff's law states that bone adapts its structure to suit its function. However, it
has been shown that bone does not adapt its structure as often as is usually
assumed. In an experiment carried out on sheep by Bereiter et al. [8], an
acetabular component was fitted closely onto the surrounding bone which grew
onto the component, providing a very stable fixation. We see a clear indication
of structural change of the cancellous bone. The question of whether unloading
results in porosis and/or bone loss was investigated by Gautier et al. [9] using
steel and plastic plates (Fig. 3). He found that remodeling and porosis for the
plastic plate in the proximal part of the plate was significantly different from

Fig. 4a,b. Vascular damage produced by different plates. **a** Using disulphine blue as a vital marker of blood supply the cross-section of the sheep tibiae displayed a markedly better blood supply beneath a grooved plate. **b** At 12 weeks a distinct difference in disturbed circulation is seen. Type of plate: *1*, DCP; *2*, DCP with transverse grooves between the screw holes; *3*, DCP with transverse and longitudinal grooves between the screw holes

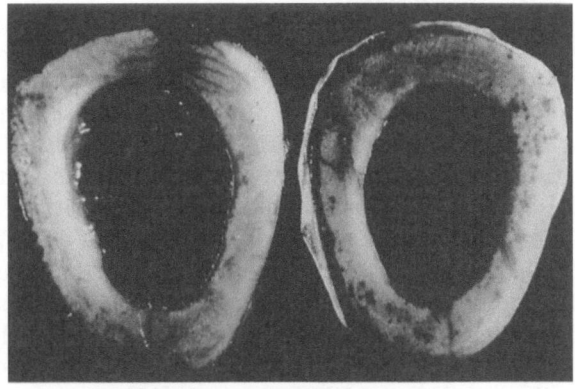

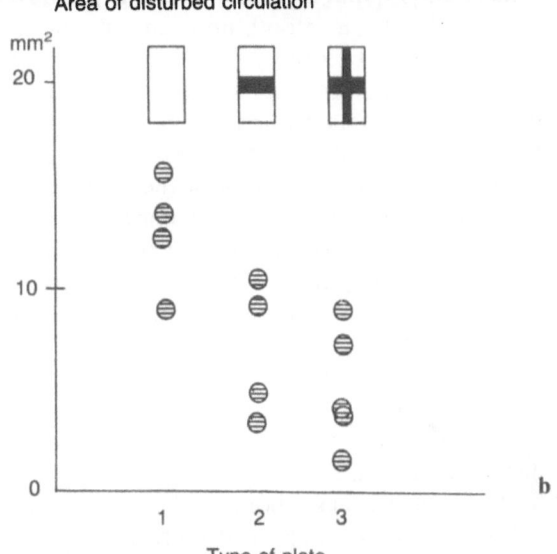

Area of disturbed circulation

Type of plate

that for steel. Distally, the results were similar for both plates. The soft plate resulted in more bone loss than the rigid plate.

One explanation for the non-uniform results could be that the contact area is larger proximally. Gautier et al. [9] investigated the relation between the contact surface and its effect on porosis. The porosis under the plastic plate is markedly less than for the steel plate. The undersurface of the plate may play an important role [10–12].

Staining shows that the area beneath the smooth plate is not well vascularized. Therefore, if porosis is related to blood supply, remodeling which is too intensive may be a reason for refracture. Sequestration or infection may result from porosis. If the induction of porosis can be avoided, sequestration may be avoided and thus the risk of refracture reduced.

An experiment was carried out by Jörger [13] using plates with grooves in the undersurface and plates with a smooth undersurface. He found that using grooved plates resulted in an improved blood supply (Fig. 4).

Vattolo [14] investigated the long-term effect of the difference in plate surfaces for a period of 5–10 weeks. He found that the bonè structure at the end of this period definitely improved less under a smooth surface plate than under a plate with grooves. The area of porosis can be determined over 15–19 weeks under a smooth plate. The magnitude of porosis is similar for the grooved plate initially, however, it disappears more rapidly. Therefore, both the shape of the plate and the nature of its surface must be considered.

The implant-to-bone interface is a crucial area in fracture healing. For this reason a plate has been introduced into the clinic which has a completely changed undersurface with grooves. This alteration in the form of the plate also affects its stiffness. The plate is least stiff at the screw hole and very stiff between the holes. The grooves improve the blood supply under the plate. Tepic et al. [15] and Lippuner et al. [16] removed the plate after 10 weeks and observed that for a smooth undersurface, bone healing and bone strength are less as compared to the grooved plate. The smooth plate prevents the area from healing and produces a stress concentration. This may lead to refracture, not along the original fracture line but just beside it through the screw hole. More evenly distributed stiffness of the plate is important to avoid bending under load. Even distribution of the screw holes allows the plate to be moved during the operation and to be compressed from both sides.

Another problem is allergic reaction to the implant, for example, to nickel. Yamage [17] investigated the migration of blood cells, and noted that a systemic reaction occurs in the circulatory system. The migration test can be used to follow the changes on implant removal but not to predict hyper-sensitivity. Titanium was then considered as an alternative implant metal to steel. More blood vessels are seen close to titanium implants than to steel ones.

Titanium is relatively inert, and does not corrode. Titanium does not evoke a reaction in human tissue. Titanium is strong and ductile. Disadvantages include the fact that it breaks more easily when bent or twisted. It also frets, wears, and leaves an inert deposit if there is instability. Its price and availability are two other essential issues.

Porosis is seen regularly in the case of medullary nailing. The nail changes the blood supply of the bone in this area. The outer blood supply is good, the inner is poor and between is an area of porosis which moves towards the implant. It may be the same for prosthetic stems. The porosis can only start where there is blood supply. It starts in the area between necrotic bone and well-vascularized bone, and moves outward towards the plate or inwards towards the nail. These observations show that there is a clear relationship between temporary avascularity and remodeling which may weaken the bone. Dagrenat et al. [18] demonstrated that the blood supply after medullary nailing in sheep leaves large areas not perfused. This must be avoided. The implant or the method of application should be changed. Klein [19] studied the method of application. He used medullary nails in dogs in a preterminal experiment with and without reaming [20]. Results showed that with reaming there is much less perfusion than without reaming. An unreamed procedure can only be carried out if a small nail is used.

One problem is that of resorption around the pin track and infection. Movement of the pins should be avoided [21]. A study was carried out using a hydraulic element to move the pin back and forth [22]. If you have no preload and the pin is exactly the same size as the hole, there is resorption along the pin track. If the pin is properly preloaded there is either no or very little resorption. The situation of pins in external fixation can be improved through analysis, no preload, and changes to the amount of cortical contact surface, bending preload, and radial preload. Prebending leads to an area of instability.

In one experiment, the effect of low preload and low load was investigated to see if resorption could be completely abolished. However, there was still a definite resorption around the pins with bending preload compared to radial preload. High load is not important but micromotion is. At zero preload there is resorption, at 0.5 mm preload there is complete destruction of the bone due to mechanical overload. 0.1 mm seems to be the ideal amount of preload. At 0.3 mm, the bone starts to fragment again [21].

In conclusion, the mechanical conditions at the interface are the most critical elements in successful bone healing. Mechanical stability and adequate vascular supply to the bone are the two most important considerations in the fixation of fractures.

Acknowledgement. My thanks go to Mrs. Joy Buchanan for her contribution to this paper.

References

1. Moor R, Tepic S, Perren SM (1989) Hochgeschwindigkeits-Film-Analyse des Knochenbruchs. Z Unfallchir Versicherungsmed Berufskr 82(2):128–132
2. Ganz R, Perren SM, Rüter A (1975) Mechanische Induktion der Knochenresorption. Fortschr Kiefer Gesichtschir 19:45–49
3. Stadler J, Brennwald J, Frigg R, Perren SM (1982) Induction of bone surface resorption by motion. An in vivo study using passive and active implants. Second International Symposium on Internal Fixation of Fractures, Lyon, September 16–18, pp 62–64
4. Hente R, Cheal EJ, Perren SM (1990) Tissue response to controlled interfragmentary strain gradients in fracture healing. In: Proceedings of the 2nd Conference of the International Society for Fracture Repair, Stockholm, September 6–8, p 148
5. Hente R, Cheal EJ, Hagerty T, Perren SM (1991) Differentiation of repair tissue under controlled strain gradients. 37th Annual Meeting of the Orthopaedics Research Society, Anaheim, California, March 4–7, vol 16(2):479
6. Hagerty TA, Hente R, Cheal EJ, Perren SM (1990) Fracture healing under controlled cyclic displacement in the ovine model. In: Proceedings of the 2nd Conference of the International Society for Fracture Repair, Stockholm, September 6–8, pp 49–50
7. Yamada H, Evans FG (1970) Strength of biological materials. Williams and Wilkins, Baltimore

8. Bereiter H, Buergi M, Rahn BA, Morscher E (1990) Primary mechanical and secondary biological stability with the press fit cup (PEC). SICOT 90/XVIII Congres International, Montreal, 9–14 September, 1990. No. 436:244
9. Gautier E, Rahn BA, Perren SM (1986) Effect of steel versus composite plastic plates on internal and external remodeling of intact long bones. Orthop Trans 10:391
10. Cordey J, Schneider M, Belendez C, Ziegler WJ, Rahn BA, Perren SM (1991) Osteoporosis and mechanics: In vitro densitometrical and mechanical study of the proximal femur. Calcif Tissue Int 48 Suppl. A71:261
11. Cordey J, Schneider M, Belendez C, Ziegler WJ, Rahn BA, Perren SM (in press) Effect of the bone size and not the density on the stiffness of the proximal part of normal and osteoporotic femora
12. Predieri M, Gautier E, Sutter F, Tepic S (1990) Vermeidung der Porose unter Osteosyntheseplatten. Acta Med Austriaca [Suppl] No. 40:48–49
13. Jörger KA (1987) Akute intrakortikale Durchblutungsstörung unter Osteosyntheseplatten mit unterschiedlichen Auflageflächen. Inaugural dissertation, Bern
14. Vattolo M (1986) Der Einfluss von Rillen in Osteosyntheseplatten auf den Umbau der Kortikalis. Dissertation, Bern
15. Tepic S, Bresina S, Perren SM (1988) Distribution of bone density and stiffness by ultrasonic microscopy. Proceedings of the 6th Meeting of the European Society of Biomechanics, Butterworths, London
16. Lippuner K, Vogel R, Tepic S, Rahn BA, Cordey J, Perren SM (in press) The effect of animal species and age on plate-induced vascular damage in cortical bone.
17. Yamage M (1987) In vitro tests on patients with metal allergies. Unfallheilkunde 189:890–898
18. Dagrenat D, Moncade N, Cordey J, Rahn BA, Kempf I, Perren SM (in press) An experimental study of dynamization following static medullary nailing in comminuted diaphyseal fractures
19. Klein MPM (1990) Aufbohren oder nicht Aufbohren? Zirkulationsstörung durch Marknagelung an der Hundetibia. Dissertation, Basel
20. Klein MPM, Rahn BA, Frigg R, Kessler S, Perren SM (1990) Reaming versus non-reaming in medullary nailing: Interference with cortical circulation of the canine tibia. Arch Orthop Trauma Surg 109(6):314–316
21. Hyldahl C, Pearson S, Tepic S, Perren SM (in press) Induction and prevention of pin loosening in external fixation: An in vivo study on sheep tibiae
22. Perren SM, Pearson S, Hyldahl K, Biliouris T (1991) Entwicklung auf dem Gebiet des Fixateur externe: Radiale Vorspannung, Pinless. 6. Deutsch-Österr.-Schweiz Unfalltagung, Wien, p 284

Ligament

The Human Anterior Cruciate Ligament and Its Replacement: Biomechanical Considerations

Savio L-Y. Woo, Douglas J. Adams, and Shinro Takai

Summary. The inability of a mid-substance tear of the anterior cruciate ligament (ACL) to heal often leads to surgical intervention to restore knee stability, but there is much debate on graft selection and reconstruction techniques. We believe that increased knowledge of the function of the ACL, such as its tensile properties and its role in maintaining knee kinematics, will provide the baseline data needed to improve surgical treatment regimens.

Tensile testing of the femur-ACL-tibia complex (FATC) from 27 donors (22 to 97 years) demonstrated significant reductions in stiffness and strength ($p < 0.05$) with age. Using a kinematic linkage system, the anteromedial (AM) portion of the ACL was found to lengthen and sustain a majority of the applied load with knee flexion, while the posterolateral (PL) portion lengthened and sustained applied load only during extension.

During ACL reconstruction, graft placement and initial tension both significantly affect joint kinematics and graft force. An increase in initial tension from 10N to 40N resulted in substantial increases in measured graft force. Quadriceps muscle force further increased the graft force, while conactivation of the hamstrings served to decrease graft force.

Key words. Knee — ACL — Ligament — Kinematics — ACL reconstruction — Quadriceps — Hamstrings

Introduction

The human knee is a primary weight-bearing joint that is known to have a complicated kinematic signature as determined by the geometry of the femoral and tibial articulating surfaces as well as by its four major ligaments. The ACL,

Connective Tissue and Musculoskeletal Research Laboratory Department of Orthopaedic Surgery, University of Pittsburgh and Orthopaedic Bioengineering Laboratory, University of California, San Diego

named to reflect its anterior insertion into the tibia, lies within the capsule of the knee joint and is structurally complex, consisting of intertwining bands of different lengths that twist as they traverse from the femoral to tibial insertions. The ACL is a major stabilizer of knee motion, aiding in the control of the six degrees of freedom (df) of knee motion. For reasons that are still debated, the ACL does not heal once torn. Its absence can lead to severe joint instability and potential disruptions of other tissues around the knee, and finally progress to degenerative joint diseases. Thus, many have advocated surgical intervention to replace a torn ACL in order to partially restore stability to the knee.

Until recently, however, only limited information was available on the properties of the ACL and its function in the knee on a quantitative basis. The design of replacement materials or surgical techniques to reconstruct this ligament have been done empirically with only limited knowledge of the tensile properties of the human ACL. Moreover, these data are needed for kinematic analyses of the human knee in order to understand the roles that the ACL plays in normal knee motion. These topics are of recent interest as investigators strive to quantitatively describe the function of the ACL.

Clinically, the patellar tendon (PT), a thick fibrous structure connecting the patella to the anterior aspect of the tibia, has been one of the most widely used autologous tissue grafts for ACL reconstruction since 1963 [1]. Noyes et al. [2] reported that the central or medial third of the human patellar tendon (14 mm wide) had an ultimate load approximately 1.6 times that of the native anterior cruciate ligament, while the energy absorbing capability was similar. The linear stiffness, however, was four times that of the normal ACL. The PT autograft has also been the most extensively studied in experimental animal models. However, the tensile properties of the grafts reached only one-third of the intact ACL value after as long as 24 months [1, 3, 4].

All too frequently, criteria for ACL replacement have focused on reproducing or surpassing the natural strength of the ACL with little attention given to other structural parameters. Because of the differences in biomechanical properties between the graft materials and the ACL, recent investigations have begun to examine the entire load-elongation curve in tension and use these characteristics for graft selection.

In this study, we will describe a series of in vitro studies designed to investigate the structural properties of the ACL and its role in maintaining the normal kinematic function of the knee. First, we established baseline data for the load-elongation behavior of the human FATC, including ultimate load, linear stiffness, and energy absorbed to failure as a function of age. Second, we developed a new methodology to measure the in situ lengths and forces within the ACL during both passive knee flexion and extension, and with application of external loads. Furthermore, the stabilizing effects of the quadricep and hamstring muscles on knee kinematics were examined. These studies were followed by evaluating the effects of ACL deficiency on knee kinematics. And finally, methods of ACL reconstruction were studied with particular emphasis on examining the role of graft placement and initial graft tension on knee function and graft force.

Baseline Data on the Tensile Behavior of the ACL

Although the human ACL has been studied for many decades, only a few published works exist in the literature that characterize its stiffness and strength properties. This is in part due to the obstacles encountered in specimen procurement as well as difficulties in methods of testing such a ligament that has a significant anatomic and geometric complexity. Some investigators have chosen to isolate the ACL into individual portions [5, 6] to allow for a uniform distribution of tensile stresses within the ligament, thus resulting in more credible stress-strain curves to represent its mechanical properties. However, in these studies the structural properties of the entire FATC, as represented by the load-elongation curve, could not be obtained without additional mathematical modeling and analyses.

Information about the tensile properties of the FATC as a structure can be used to serve as a basis of selection and design of biologic replacements or synthetic materials. In our laboratory, a new experimental device was recently developed [7] whereby the FATC could be tested at any angle of knee flexion and in different specimen orientations with respect to the applied load. We chose to orient the whole ligament to align with the applied tensile load while maintaining the normal angles of ACL insertion to the bones (called the "anatomical orientation"). Contralateral knees were tested with the tensile load aligned vertically with the tibia and ACL without maintaining the normal angle of ACL insertion to the bones (called the "tibial orientation"). The effects of loading direction on the structural properties of the paired human FATC were then evaluated.

Twenty-seven pairs of cadaveric human knees were equally distributed into three age groups: 22–35 years (mean of 29) representing the younger group, 40–50 years (mean of 45) as middle-aged group, and 60–97 years (mean of 75) constituting the older group. All soft tissues around the knee except the ACL were dissected away, and the femur and tibia were cut to appropriate lengths and secured in clamps for tensile testing. One knee from each pair was tested in the "anatomical" orientation while the contralateral knee was tested in the "tibial" orientation. While the knee angle was maintained at 30° of flexion, the FATC specimen was loaded to failure at a rate of 200 mm/min. The resulting load-elongation curves, representing the structural properties of the FATC, were recorded and the parameters of linear stiffness, ultimate load, and energy absorbed to failure were obtained.

Typical load-elongation curves for paired knees are shown in Fig. 1. These two curves demonstrate that the structural properties of the FATC were affected by specimen orientation. Those tested in the anatomical orientation resulted in higher linear stiffness, ultimate load, and energy absorbed to failure than those tested in the tibial orientation. Statistically significant differences

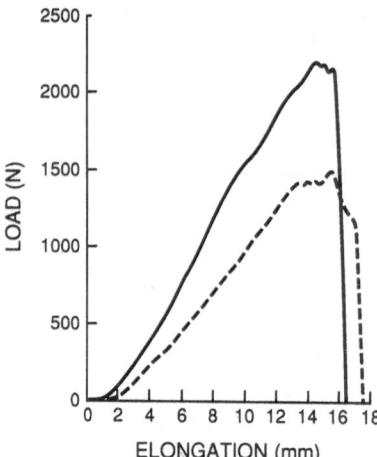

Fig. 1. Effect of specimen orientation on the structural properties of the FATC. *Solid line*, anatomical; *dashed line*, tibial

between specimen orientation for all three age groups were noted, but the differences diminished with increasing age. On the other hand, age had a profound effect on the structural properties, linear stiffness, and energy absorbed to failure of FATCs. For the younger specimens tested in the anatomical orientation, the ultimate load was 44% higher than that of the middle aged group, and 228% higher than that of the older group (Fig. 2).

The large changes in the structural properties of FATCs with increasing age are hypothesized to be, in part, a result of the decreases in types and levels of physical activity as well as changes in joint geometry and knee kinematics with aging. The differences observed with respect to specimen orientation during tensile loading reflect that while testing the FATC along the anatomical axis of the ligament, a greater proportion of its fiber bundles may be loaded simultaneously, resulting in higher stiffness and ultimate load.

Distance Between ACL Insertion Sites and Forces Experienced by the ACL in the Intact Knee

Determining the forces experienced by the ACL during normal function is a necessary prerequisite for understanding the role of the ligament in kinematics of the knee. Various methods have been implemented to ascertain ACL length changes during knee flexion-extension, including the use of liquid mercury strain gauges [8–10] and Hall effect strain transducers [11, 12]. Others chose to directly measure the forces in the ligament using buckle type force transducers [13, 14]. Recently, an implantable transducer has been used to measure ligament and tendon forces [15]. However, these techniques all required the attachment of mechanical devices to this short and broad ligament, possibly altering its geometry and thus affecting the results. Hefzy and Grood [16] used a spatial linkage to determine the length patterns of fiber bundles in the ACL.

Fig. 2. Effect of age on the linear stiffness and ultimate load of the human FATC tested in the anatomical orientation

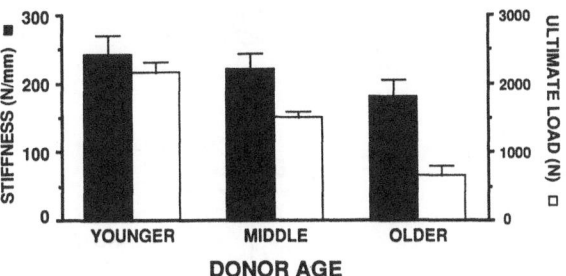

Variable changes in length of different areas of the ACL were examined by Butler et al. [17] by digitizing the ACL insertion sites and points along the ligament.

In our laboratory, Hollis et al. [18] used a kinematic linkage device to determine the distances between the ACL insertion sites during passive knee flexion and extension as well as when the knee is subjected to anterior-posterior or varus-valgus loads. The six df of the tibial motion with respect to the femur were measured by three translation (anterior-posterior, proximal-distal, and medial-lateral) and three rotational (varus-valgus, axial-tibial, and flexion-extension) transducers. The methods used to determine the elongation of the ACL consisted of the following steps. First, the human knee specimen without muscular soft tissues but with intact capsule and ligaments was fitted with the kinematic linkage. The knee was flexed and extended while the six df of knee motion were measured and recorded by the kinematic linkage. Second, A-P drawer force of $\pm 100\,\text{N}$ and a varus-valgus (V-V) bending moment of $14\,\text{N-m}$ were applied at knee flexion angles of $0°$, $30°$, $45°$, and $90°$ and similar six df knee motion were recorded. Then, the joint capsule and knee ligaments, except for the ACL, were dissected away. The ACL was considered to consist of the anteromedial (AM), posterolateral (PL), and intermediate (IM) portions, and the femoral and tibial insertion sites of these three portions of the ACL were located and digitized. Finally, through a series of coordinate transformations, the distances between the femoral and tibial insertion sites (representing the lengths of each portion) were calculated.

Passive Knee Flexion-Extension

Ten human knee specimens from donors with an average age of 30 years (range of 22–44 years) were used. The knees were dissected of muscle and other surrounding soft tissues except for the joint capsule and ligaments, and then the kinematic linkage was attached to the knee. The knee was cycled between $0°$ and $90°$ of flexion, and after 10 cycles of preconditioning the kinematic data were collected for one cycle and stored in the computer.

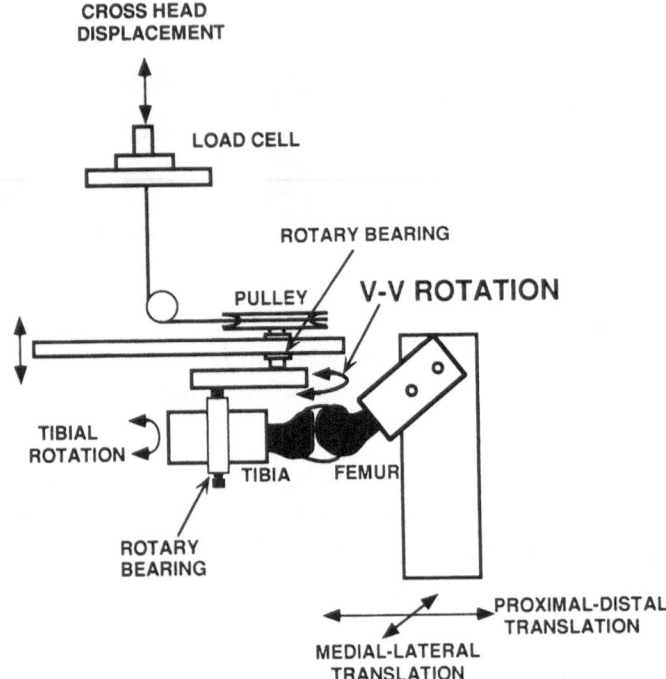

Fig. 3. Schematic diagram of apparatus used for applying a varus-valgus moment to the knee joint

Knee Flexion-Extension with Varus-Valgus or Anterior-Posterior Loading

To apply external loads, the knee with the kinematic linkage attached was mounted on an Instron materials testing machine, and a mechanism was used to apply either a V-V moment or an anterior-posterior (A-P) load. The V-V moment was applied to the tibia through a rotary bearing, pulley, and cable system with the knee flexion angle fixed and the remaining five df free. The cable was connected to the load cell of the Instron which measured the applied load (Fig. 3). The system was modified to apply an A-P load perpendicular to the tibia in the A-P direction.

The kinematic linkage was attached to the knee which was then placed in the loading device at 45° of flexion. V-V moments of ±14 N-m were applied for five cycles. Subsequently, the V-V moment versus knee motion information was recorded and stored in the computer. The tests were repeated with the knee at 30°, 90°, and finally at 0° of flexion. Similarly, the device was adjusted so that A-P loads of ±100 N could be applied perpendicular to the tibia. Cyclic tests were again performed at the same flexion angles and the data recorded.

Fig. 4. Location of markers on the femoral and tibial insertions of the ACL

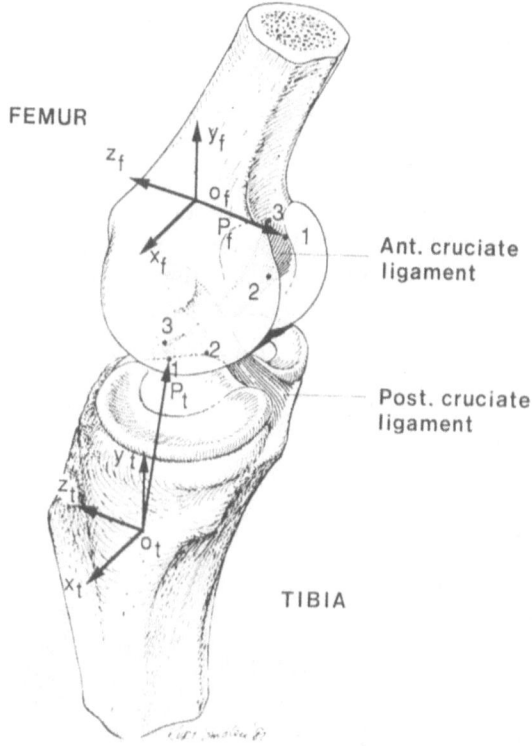

Lengths and Length Changes of Portions of the ACL

To calculate the distances between the insertion sites for the measured knee motion, all soft tissues except for the ACL were dissected away from the knee. Small Kirschner wire markers, 0.7 mm in diameter, were placed in the tibial insertion sites of each portion. The fibers of each portion were then traced up to their femoral insertions, and corresponding markers placed at these locations (Fig. 4). Thus, marker 1 corresponded to the AM portion, 2 to the PL portion, and 3 to the IM portion. The position of the tibia with respect to the femur was recorded using a coordinate system on both the tibia (O_t) and the femur (O_f), and the three-dimensional coordinates of each marker on both the femur and tibia were obtained with a pointer attached to the kinematic linkage. The distance (representing the length of a portion of the ACL) between the femoral and tibial markers was then determined [18].

At 30° of the knee flexion, the lengths of the AM, PL, and IM portions were 35.9 ± 0.9 mm, 19.3 ± 1.1 mm, and 28.1 ± 1.1 mm, respectively, indicating that the AM portion is approximately 17 mm longer than the PL portion at 30° knee flexion. A one-way analysis of variance (ANOVA) showed a significant difference among the lengths of the three portions. These portions also experienced different patterns of length change during passive knee flexion from 0° to 90°. The AM portion increased in length from 34.4 ± 1.0 mm to 38.0

± 0.8 mm, while the PL portion decreased in length from 22.5 ± 1.2 mm to 15.4 ± 1.2 mm. Both of these changes were significantly affected by the flexion angle. The IM portion, on the other hand, showed little change in length with passive knee flexion-extension. The length for the AM portion was the shortest at 0°, indicating that this portion of the ACL may be lax or buckled at full extension, and the stability of the knee may be maintained by other portions. Indeed, the PL portion was shown to be longest at full extension.

Application of V-V moments to the knee produced appreciable V-V rotation as well as coupled axial tibial rotation, but insignificant medial-lateral, proximal-distal, or anterior-posterior translations. The V-V rotation between

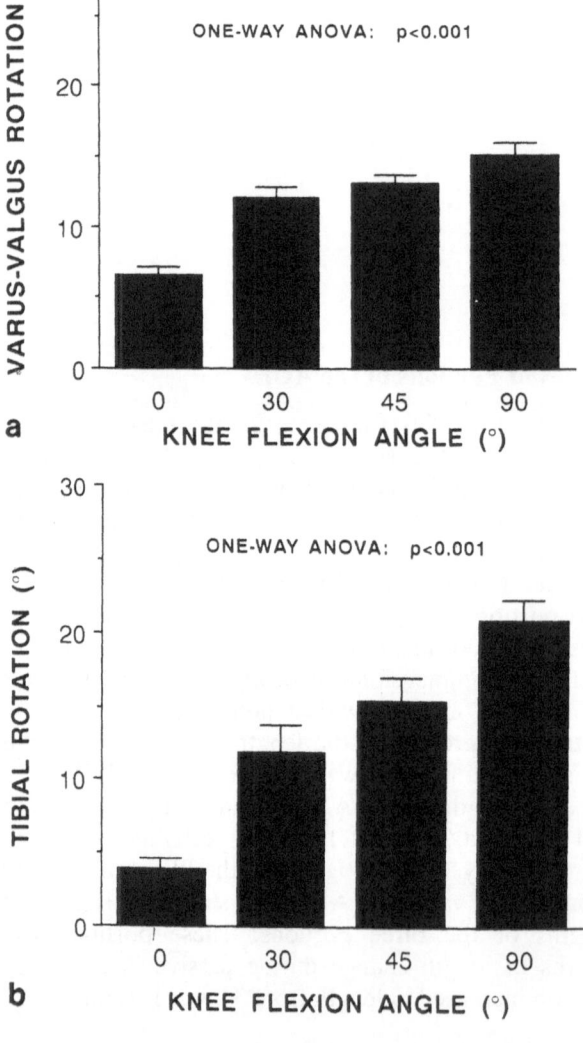

Fig. 5. a Varus-valgus rotation and **b** coupled axial tibial rotation as a function of knee flexion when a ±14 N-m varus-valgus moment is applied to the tibia

±14 N-m moment was found to increase with the knee flexion angle (p < 0.001; ANOVA) (Fig. 5), and the tibia was found to rotate internally with valgus loading and externally with varus loading. Application of A-P loads to the knee produced appreciable A-P tibial translation with respect to the femur as well as axial tibial rotation, but again there was little or no medial-lateral translation, proximal-distal translation, or varus-valgus rotation. The A-P displacement between ±100 N of tibial load was also found to change with the knee flexion angle but did not follow a monotonic change with flexion angle (Fig. 6). The tibia always rotated internally with anterior loading and externally with posterior loading, but the degree of rotation did not always increase with increasing knee flexion.

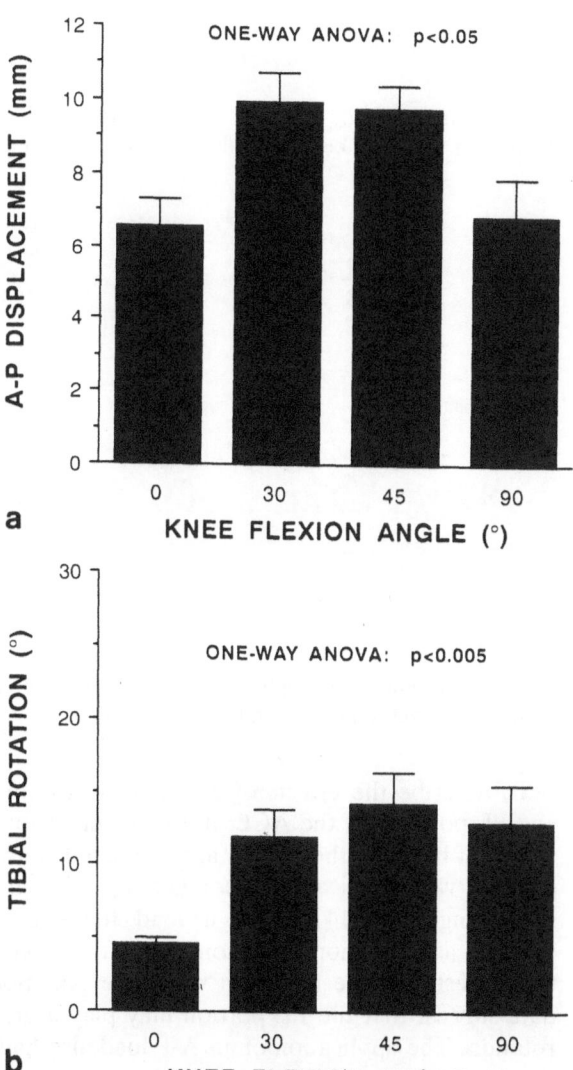

Fig. 6. a Anterior-posterior displacement and **b** coupled axial tibial rotation as a function of knee flexion when a ±100 N force is applied to the tibia

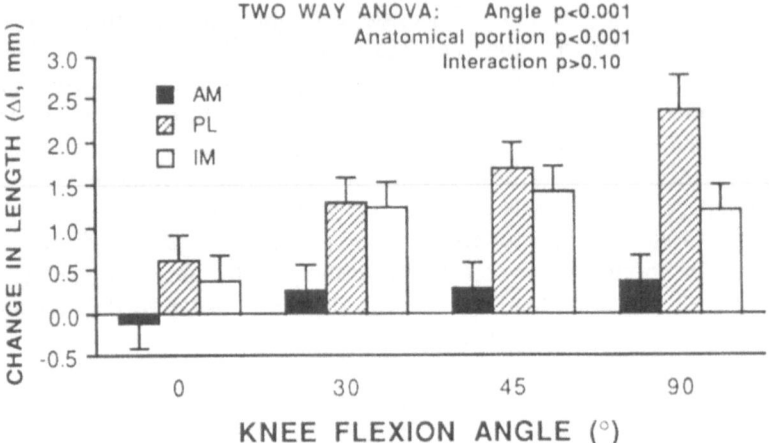

Fig. 7. Differences in lengths of the anteromedial, posterolateral, and intermediate portions of the ACL between unloaded and valgus loaded knees

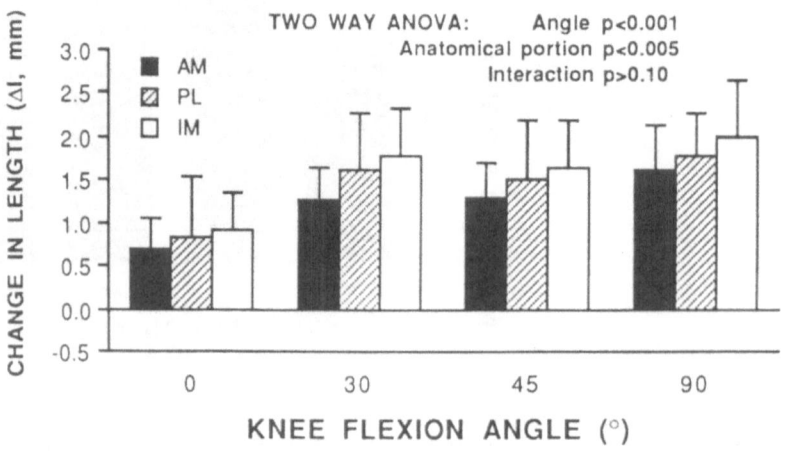

Fig. 8. Differences in lengths of the anteromedial, posterolateral, and intermediate portions between unloaded and anterior loaded knees

To describe the effects of external loading on the lengths of the AM, IM, and PL portions of the ACL, it is convenient to express the data as the change in length between the loaded and unloaded states at the same flexion angle. A positive value indicates a lengthening, whereas a negative value denotes a shortening. For a 14 N-m valgus load, the length changes for each of the three portions as a function of flexion angle are shown in Fig. 7. The PL portion had the largest increase in length while the AM portion had the smallest. These data suggest that the PL portion may play a greater role in restraining valgus rotation. The application of an A-P load also had an effect on the length of the three portions, with increases again noted for increasing flexion angle. With

100 N of anterior load, the lengths for the AM, IM, and PL portions again increased with the flexion angle (Fig. 8). Changes in the length of the ACL depended greatly on both the degree of knee flexion and the anatomic portion.

Forces Experienced by the ACL

With the information on the lengths of the portions of the ACL during various knee motions, the forces experienced by the ACL can also be determined. After the kinematic linkage had recorded the motion between the femur and tibia and the insertion sites had been digitized, the three portions of the ACL were dissected with their bony attachments from the knee for tensile testing [19]. From this test, a relationship between length and tensile force was determined for each portion of the ACL. These length-tension relationships were then used to calculate the forces in the three portions based on the lengths obtained. Preliminary results indicate that there is virtually no force applied to the ACL during knee flexion and extension in the absence of external loads and muscle forces. With the application of anterior loads to an intact knee, the AM portion sustained a majority of the applied force. The PL portion also sustained a fraction of the applied force at knee extension, but this value decreased with knee flexion. This demonstrates the important distinction between ligament length changes and ligament forces, as some portions of the ACL may undergo relatively large length changes during knee motion even though no load is applied.

Effects of Quadriceps and Hamstrings Loading

The role of the ACL in the joint is dependent not only on the knee flexion angle or external loading, but also on the activity of the musculature surrounding the knee. Further studies in collaboration with Drs. Shoemaker and Daniel have been directed towards analysis of the quadriceps muscle-ACL interaction [20]. Using eleven human cadaver knees (mean age of 79 years; range 66 to 88 years), the knee kinematics were measured with an experimental device called the Oxford Rig (originally designed by J.J. O'Connor and associates in Oxford, England) which allowed knee specimens to be tested throughout a range of flexion-extension in a simulated vertical stance and with quadriceps stabilization [21]. Anterior-posterior tibial translation was measured directly with a linear potentiometer mounted rigidly onto the tibia. Data was collected as the knee was extended from 100° flexion both in the presence and absence of a quadriceps force. In the quadriceps absent state, the knee was extended manually by raising the crosshead assembly of the Oxford Rig. In the quadriceps stabilized state, the tendon was connected to a servo-electric motor which generated an extension moment sufficient to overcome a flexion moment created by a 45 N weight applied at the level of the hip.

Knee flexion angle had a significant effect on the measured anterior-posterior tibial translation. As the intact knee was extended from 100° in the absence of a quadriceps force, the tibia translated posteriorly 24.8 ± 1.8 mm

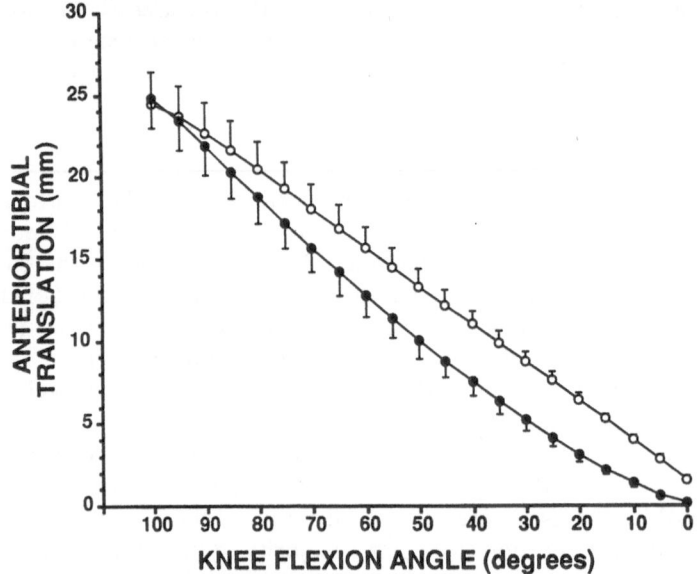

Fig. 9. Anterior tibial translation versus knee flexion for the intact knee in the quadriceps absent and quadriceps stabilized states. *Solid circle*, quadriceps absent (hip load: 0N); *open circle*, quadriceps stabilized (hip load: 45N) (modified from Shoemaker et al. [20])

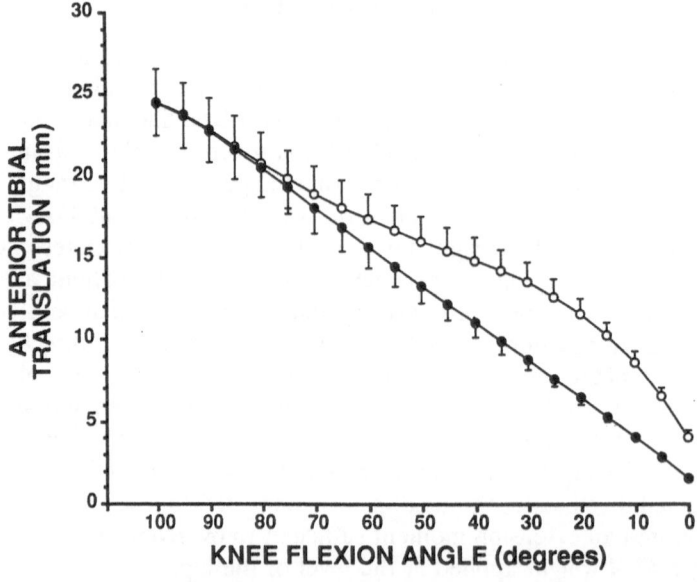

Fig. 10. Anterior tibial translation versus knee flexion in the quadriceps stabilized knee before and after ACL sectioning. *Solid circle*, intact; *open circle*, ACL cut (modified from Shoemaker et al. [20])

relative to the femur (Fig. 9). In the stabilized state, the quadriceps force necessary to extend the intact knee against a flexion moment created by a 45 N load decreased from 943 ± 71 N at 100° knee flexion to 93 ± 13 N at full extension. With quadriceps stabilization, the anterior tibial translation increased and was significantly different from that in the quadriceps absent state but was highly dependent upon knee flexion angle (Fig. 9). The largest difference (3.5 mm) was observed between 40° and 25° of flexion.

Additional studies were performed to examine the role of the hamstrings in stabilizing the knee with Drs. More and Daniel [22]. Ten human cadaver knees (mean age of 70 years) were tested. Hamstring forces of 45 N and 90 N were applied evenly between the medial and lateral sides. On the medial side, the force was again evenly divided to represent the contribution of the semimembranosus and semitendinosus muscles. On the lateral side, the force was applied through the biceps tendon. The quadriceps force did not change significantly with the addition of hamstrings force (p > 0.05), whereas anterior tibial translation decreased with the addition of hamstrings force (p < 0.05). Relative to full extension, the tibia was internally rotated throughout knee flexion, and the addition of 90 N hamstrings force reduced the maximum rotation in the intact knee by 3.3 ± 0.6°, or 23% (p < 0.01).

Kinematics of the ACL Deficient Knee

Sectioning the ACL and repeating the knee motion tests as described in the previous section led to increased anterior tibial translation at all flexion angles in the intact state, independent of quadriceps force [20]. In the presence of quadriceps force, the change in translation was small as the knee was extended from 100° to 70° of flexion, and gradually increased to a maximum of 5 mm at 20° of flexion, and then decreased to 2.5 mm at full extension (Fig. 10). These results demonstrate that abnormal motion in the ACL deficient knee differs with loading conditions and flexion position. The antagonistic functions of the quadriceps muscle and the ACL were also highlighted.

With the addition of hamstrings force, anterior tibial translation decreased during knee flexion and extension (p < 0.05) but remained higher than in the intact knee; the increase observed in anterior translation due to quadriceps force was only partially balanced by the addition of hamstrings force. Relative to full extension, the tibia was internally rotated throughout knee flexion. The addition of hamstrings force reduced the internal rotation throughout knee flexion, although not restoring it to the intact state. This demonstrates that the increase in internal rotation of the tibia was not controlled by the addition of quadriceps and hamstrings muscle forces.

ACL Reconstruction

During reconstruction of the ACL, the distal end of the graft is usually routed through a bone tunnel originating from the anterior insertion of the ACL to the tibia and extending anteriorly and medially through the tibia. The proximal end of the graft is commonly routed either "over the top" (OTT) of the posterior aspect of the lateral femoral condyle or through a second femoral bone tunnel ("double tunnel" or DT). While the DT technique places the graft in an anatomical configuration, it is more difficult to perform surgically than the OTT technique and results in more abrasion to the ACL replacement at the edge of the femoral bone tunnel.

In addition to graft placement, the initial tension applied to the graft is critical to the immediate post-operative stability and kinematic behavior of the ACL reconstructed knee. The application of 10–40 N initial graft tension has been advocated to restore knee stability. Yet, the initial tension applied surgically can vary considerably and often is not controlled quantitatively during the procedure.

OTT vs DT Graft Placement

The effects of graft placement were studied using patellar tendon grafts [23]. A 10 mm wide patellar tendon allograft with bone blocks at each end was used with its distal end routed through a bone tunnel in the tibia and joined to a device designed to measure graft force. The proximal end was routed either through a 9 mm diameter femoral bone tunnel (DT) or "over the top" of the posterior aspect of the lateral condyle. The allograft was initially tensioned to

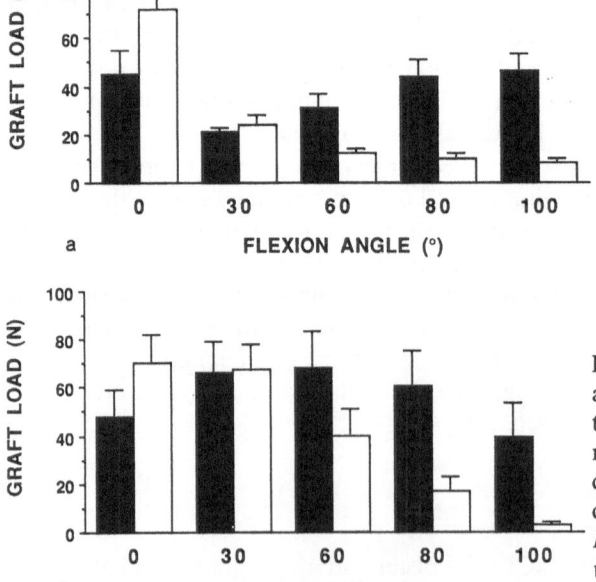

Fig. 11. Force experienced by an ACL graft during "double tunnel" and "over the top" reconstruction in the **a** quadriceps absent state and **b** quadriceps stabilized state. *Black area*, double tunnel; *white area*, over the top (modified from Adams et al. [23])

Table 1. Change in graft load with the addition of quadriceps force (quadriceps stabilized/quadricepsabsent). *$p < 0.05$ **$p = 0.08$ (modified from Adams et al. [23])

Initial tension	Flexion angle			
	0°	30°	60°	80°
10 N	1.4	8.6*	4.6**	0.9
40 N	1.0	4.7	2.4	1.5

20 N at 20° knee flexion, and graft force was recorded as the knee was extended from 100° to 0° of flexion, both passively and then with quadriceps force.

Significant differences in graft force between the DT and OTT graft placement were demonstrated during knee flexion ($p < 0.05$). The force was less for OTT reconstruction than for DT at 60°, 80°, and 100° of knee flexion (Fig. 11). The highest graft force for OTT occurred at full extension even with the addition of quadriceps force. However, graft force was shown to increase with the addition of quadriceps force only at 30° and 60°.

Initial Graft Tension

Using a synthetic graft that has a load-elongation behavior similar to that of the human ACL, the effects of initial graft tension on graft forces were studied [23]. The ACL was reconstructed in the "double tunnel" configuration and connected to a device used to apply initial graft tension and measure graft forces. Initial graft tension of either 10 N or 40 N was applied at 20° of flexion in the absence of quadriceps force, and knees were extended passively from 80° to 0° both with and without quadriceps stabilization. Increasing initial graft tension from 10 N to 40 N resulted in a significant increase in measured graft force at all flexion angles, both in the absence and presence of quadriceps force (see Table 1). In the absence of quadriceps force, the graft forces were significantly higher at full extension than at flexion angles for both levels of initial tension ($p < 0.05$). The variation in graft force demonstrates that true isometry is not attainable within the range of motion studied. The substantial change in graft forces due to quadriceps force is important to understanding the interaction of muscle forces. Graft forces, therefore, should be taken into consideration during knee rehabilitation following ACL reconstruction.

O'Connor et al. [24] also estimated the forces in the ACL during a quadriceps loaded leglift. A simple optical method was used to record the track of one bone relative to the other from full extension through 135° of knee flexion. Subsequently, the ACL was then sectioned and replaced with a wire, and the flexion motion repeated. The tension needed to restore the track of the bones to that of the intact joint was then assumed to be the force in the ACL. Although estimation of these forces demonstrated considerable scatter, the results did indicate that the ACL force increased as full extension was approached and decreased for higher angles of flexion.

Discussion

Through many recent investigations, we are beginning to comprehend the role the ACL plays by gathering data on its forces and length changes during complex knee motion. Experimental testing has provided some baseline data on the tensile characteristics of the ligament as a function of specimen orientation and age. Various methods have been implemented to measure the length changes in different portions of the ACL when the knee is subjected to passive motion or external loading. Estimation of the forces in the ACL during knee motions have also been attempted. However, the muscle forces around the knee remain unknown and therefore more experimental investigation is necessary to elucidate the functional forces in the ACL. Future approaches should include an inverse dynamics approach to calculate the resultant joint forces and joint moments [25] followed by mathematical modeling to determine the forces in the ligaments.

It should be recognized that data on forces in the ACL during specific activities are needed to aid in the design of ligament replacements and to establish proper reconstructive techniques. Further, as the different parts of the ACL are known to be loaded during knee flexion and extension, the current methods of replacing the anatomically twisted ACL with a simple graft may be inadequate. Only through investigations in ACL biomechanics can crucial information on restoring normal function in an ACL deficient knee be obtained.

Acknowledgements. This work was supported by NIH Grant AR-39683 and done in collaboration with Drs. S.C. Shoemaker, R.C. More and D.M. Daniel, of the SKC Joint Mechanics Laboratories, San Diego, California, USA.

References

1. Newton PO, Horibe S, Woo SLY (1990) Experimental studies on anterior cruciate ligament autografts and allografts. In: Daniel DM, Akeson WH, O'Connor JJ (eds) Knee Ligaments: Structure, Function, Injury, and Repair. Raven Press, New York, pp 389–399
2. Noyes FR, Butler DL, Grood ES, Zernicke RF, Hefzy MS (1984) Biomechanical analysis of human ligament grafts used in knee-ligament repairs and reconstructions. J Bone Joint Surg [Am] 66A:344–352
3. Arnoczky SP, Warren RF, Ashlock MA (1986) Replacement of the anterior cruciate ligament using a patellar tendon allograft. J Bone Joint Surg 68A:376–385
4. Ballock RT, Woo SLY, Lyon RM, Hollis JM, Akeson WH (1989) Use of patellar tendon autograft for anterior cruciate ligament reconstruction in the rabbit: A long term histological and biomechanical study. J Orthop Res 7:474–475
5. Butler DL, Kay MD, Stouffer DC (1986) Comparison of material properties in fascicle-bone units from human patellar tendon and knee ligaments. J Biomech 19:425–432

6. Newton PO, MacKenna DA, Lyon RM, Akeson WH, Woo SLY (1989) Comparison of the mechanical properties of the medial collateral and anterior cruciate ligaments of the rabbit knee. ASME/AMD 98:53–56.

7. Lyon RM, Woo SLY, Hollis JM, Marcin JP, Lee EB (1989) A new device to measure the structural properties of the femur-anterior cruciate ligament-tibia complex. J Biomech Eng 114:350–354

8. Kennedy JC, Hawkins RJ, Willis RB (1977) Strain gauge analysis of knee ligaments. Clin Ortho Rel Res 129:225–229

9. Meglan D, Zuelzer W, Buck W, Berme N (1986) The effects of quadriceps force upon strain in the anterior cruciate ligament. Trans Orthop Res Soc 11:55

10. Monahan JJ, Grigg P, Pappas AM, Leclair WJ, Marks T, Fowler DP, Sullivan TJ (1984) In vivo strain patterns in the four major canine knee ligaments. J Orthop Res 2(4):408–418

11. Arms SW, Pope MJ, Johnson RJ, Fischer RA, Arvidsson I, Eriksson E (1984) The biomechanics of anterior cruciate ligament rehabilitation and reconstruction. Am J Sports Med 12(1):8–18

12. Arms SW, Renstrom P, Stanwyck TS, Hogan M, Johnson RJ, Pope MH (1985) Strain within the anterior cruciate during hamstrings and quadriceps activity. Trans Orthop Res Soc 13:139

13. Lewis JL, Lew WD, Hill JA, Hanley P, Ohland KJ, Kirstukas S, Hunter RE (1989) Knee joint motion and ligamental forces before and after ACL reconstruction. J Biomech Eng 111:97–106

14. Lew WD, Lewis JL (1982) The effect of knee prosthesis geometry on cruciate ligament mechanics during flexion. J Bone Joint Surg 64A:734–739

15. Xu WS, Glos DL, Butler DL, Stouffer DC, Grood ES (1990) Analytic sensitivity studies of implantable force transducer in goat patellar tendon. First World Congress of Biomechanics, La Jolla, CA, 2:321

16. Hefzy MS, Grood ES (1986) Sensitivity of insertion locations on length patterns of anterior cruciate ligament fibers. J Biomech Eng 108:73–82

17. Butler DL, Martin ET, Kaiser AD, Grood ES, Chun KJ, Sodd AN (1988) The effects of flexion and tibial rotation on the 3-D orientations and lengths of human anterior cruciate ligament bundles. Trans Orthop Res Soc 13:59

18. Hollis JM, Horibe S, Adams DJ, Marcin JP, Woo SLY (1988) Force distribution in the anterior cruciate ligament as a function of flexion angle. ASME/BED 13:41–44

19. Woo SLY, Gomez MA, Seguchi Y, Endo CM, Akeson WH (1983) Measurement of mechanical properties of ligament substance from a bone-ligament-bone preparation. J Orthop Res 1:22–29

20. Shoemaker SC, Adams DJ, Daniel DM, Woo SLY (1992) Quadriceps/anterior cruciate ligament interaction: An in vitro study of joint kinematics and ACL graft tension. Clin Ortho Rel Res in press

21. Biden EN, O'Connor JJ (1990) Experimental methods used to evaluate knee ligament function. In: Daniel DM, Akeson WH, O'Connor JJ (eds) Knee Ligaments: Structure, Function, Injury, and Repair. Raven Press, New York, pp 135–151

22. More RC, Karras BT, Neiman R, Fritschy D, Woo SLY, Daniel DM (1991) Hamstrings — An anterior cruciate ligament protagonist: An in vitro study. Am J Sports Med in press

23. Adams DJ, Shoemaker SC, Harris SL, Daniel DM, Woo SLY (1990) Effect of initial graft tension and quadriceps force on anterior cruciate ligament (ACL) graft load. Trans Orthop Res Soc 15:508

24. O'Connor J, Biden E, Bradley J, Fitzpatrick D, Young S, Kershaw C, Daniel D, Goodfellow J (1990) The muscle stabilized knee. In: Daniel DM, Akeson WH, O'Connor JJ (eds) Knee Ligaments: Structure, Function, Injury, and Repair, Raven Press, New York pp 239–277
25. Herzog W (1987) Individual muscle force estimations using a non-linear optimal design. J Neuro Meth 21:167–179

Spine

Clinical Biomechanics of the Spine

JOHN H. EVANS[1]

Introduction

The spine lends support to the body and forms the fulcrum on which muscles act to cause motion and to resist inertial and external forces. In clinical terms it is most significant that it also provides flexible armor to the spinal cord and cauda equina. Owning in part to its unique, dual roles of support and protection and to the number of pain and other neurological problems arising in the spine, it has received widespread attention from scientists as well as clinicians. In addition to the obvious concern of physicians and surgeons in disorders of the spine, the allied professions also seek a better understanding of the nature of pain-related and otherwise disabling abnormalities. Others with a professional interest in the spine include those seeking a more basic understanding of its structure and function, both normal and abnormal, and of its tolerance to adverse environments and its susceptibility to damage.

It is hardly surprising then that a wealth of information is being amassed but it is disappointing that our knowledge is still very meager. The links between theoretical and experimental findings and clinical practice are often tenuous.

A comprehensive review of spinal biomechanics would be daunting and, possibly, inappropriate and unhelpful. However we are at a stage when more rational explanations for clinical observations are helping to confirm diagnosis, dispel myths and, in general, offer improved prospects for management. In this context relevant biomechanical issues will be addressed and in the process illustrations will be drawn from a few of the fields in which there have been significant developments in recent years.

---

[1] Rehabilitation Engineering Centre, Hong Kong Polytechnic, Hong Kong

Clinical Problems

Surgical intervention is indicated in many acute and chronic spinal disorders where there is an underlying mechanical dysfunction or frank derangement. The causes and nature of such disorders vary greatly but the clinical aberrations which must be addressed relate principally to pain and other neurologic sequelae and to deformity and structural inadequacy. These may often present together as they are almost inextricably linked. Often structural failures are frank and readily attributable but some induced or naturally acquired derangements are less easily discernable.

Spinal fractures, with or without neurological involvement, such as those resulting from impact loading from falls or vehicular accidents, will impart a mechanical deficit and may lead to local or columnar instability. Similarly erosion of the bony spine through disease processes such as TB and neoplasm will impart instability but also loss of structural tissues. Biomechanically these conditions present a common challenge, that of restoring strength while preserving or reconstituting the essential anatomy. More commonly mechanical derangements leading to clinically significant neurological sequelae may not be associated so much with frank mechanical dysfunction as with more subtle impairments which test existing diagnostic practice to its limits.

Microscopic and macroscopic structural damage may be caused to any of the connective tissues, hard or soft, thus contributing to mechanical dysfunction. While such damage can be induced by overtly mechanical agencies such as shock, vibration [1], and other overload conditions, tissue "degeneration" can also be initiated and mediated by a whole host of different agencies.

Pain may result from abnormal range or nature of motion at the segmental level and although it is still often not possible to identify the precise anatomical source of such pain, spinal load or motion may elicit this response [2]. Arguably the local causes of radiating pain and claudication are more readiy identified as in the case of nerve root entrapment. Abnormal motions and gross deformations can usually be detected if not characterized and quantified.

Although not commonly recognized as mechanical failures, scoliosis, and similar, gross structural deformities, which may lead to catastrophic collapse or impairments such as cardiopulmonary dysfunction, are also managed by surgical interventions aimed at restoring regional anatomy and stabilizing the spinal column.

The Normal Spine

Much research, has been directed to acquiring a better understanding of the host of mechanisms of spinal failure and collapse in the hope of being able to improve differential diagnosis and subsequent clinical management. Considerable attention has been given to the normal spine as an essential step in acquiring a proper perspective on abnormal behaviour. In this quest, which is by no means yet won, it has been common practice to make the problem more tractable by considering the spinal column as comprising a contiguous series of repeat units, the motion segments. Within a specific region of the spine (i.e., thoracic or lumbar) these segments will have similar structural characteristics in that the vertebra may be of similar shape and their soft connective tissue linkages follow a specific pattern. The limitations of analyzing the behaviour of a single motion segment in isolation are generally appreciated and the interaction with neighboring segments, the influence of soft tissues which span over two or more segments, and the influence of musculature are gradually being incorporated in analysis of the spine as more precise data are gathered on regional anatomy [3–5] and greater consideration given to the natural muscle control strategies. Consideration of the entire spine, in comprehensive terms, is a daunting task which has yet to be attempted. Existing studies are based on specific inquiry and models are limited by bold assumptions, simplifications, and approximations [6, 7].

The traditional concept of the spinal motion segment is that of a discrete, self-contained, osseoligamentous structure comprising neighboring rigid, bony vertebrae and the intervening soft connective tissues. Motion of the entire spine is seen as that of the cumulative motions of each of these segments. These motions are induced by the forces applied to the spine, internally by muscles and inertia, and externally by interaction with the environment, frequently as forces transmitted though the arms and pelvis. The relationship between such forces and the movement of the spine is governed by the mechanical characteristics of the spine and ultimately the soft tissues which constitute the motion segment [8, 9].

The mechanical characteristics of the individual segment depends on the bony geometry and the distribution, dimensions, and nature of the soft tissues. The relative contribution of each tissue structure to resisting motion is dependent on the magnitude and direction of the applied forces [5, 8–12] and, to a lesser extent, on the duration or temporal nature of these forces — some will be relatively persistent or static, as those due to body weight, whereas others may be intermittent or of short duration, such as the inertia forces resulting from vibration and shock.

A great deal has been written about the mechanical characteristics of motion segments as deduced from testing of excised segments and their components in vitro [8–10, 13, 14]. Systems of forces have been applied to specifically produced motions, in the sagittal, frontal, and transverse planes [8, 14]. In recent years much attention has been directed to considering the more complex motions [9] some of which may be intrinsically coupled (e.g., lateral flexion and rotation). Within the motion segment the structures of particular interest are the intervertebral disk [10, 15–19], the zygapophyseal joints and the posterior ligamentotendo-fascial complex [5, 16, 38]. The disk and zygapophyseal joints are the principal load-carrying components which are known to degenerate. These have also been identified as the source of much pain. However the soft tissue complex attaching to and having influence on the spinous processes is now being recognised as playing a more significant mechanical role than had previously been assumed [4, 5, 14].

Intervertebral Disk

The disk comprises the nucleus pulposus, a highly hydrated, structured gel bounded and constrained by the cartilaginous end-plates of the vertebral bodies and by the collagenous annulus fibrosus. The biochemistry and biophysics of the nucleus has been extensively researched, in particular its metabolic demands and nutrient supply [10, 15–19]. Relatively little information is available on the annulus fibrosus although specific functions have been somewhat intuitively ascribed and inferred from clinical observations. Perhaps it had been tacitly assumed that this avascular structure was functionally simpler and clinically less significant than is now proving to be the case.

Let two examples serve to make this point, both of which relate to the role of the annulus fibrosus. Recent histomorphological studies conducted by myself and a coworker indicate that the fibers of the outer lamellae are the first to be significantly stretched when the intact motion segment is flexed and also when it is subjected to axial torsion (J.H. Evans, J. Foster, 1987, unpublished work). Thus, in addition to constraining the pressurized nucleus, the annulus serves to restrain the movement of the segment and provide natural limits to its range of motion. Similarly in axial torsion the outer fibers of the anterolateral quadrant appear to work in concert with the zygapophyseal joint structures in resisting motion. It is worth recalling that the outer annulus is richly innervated as is the anterior longitudinal ligament [2] with which it is intimately linked. In these studies of intact motion segments there was no indication of high stretch in the fibers of the posterior guardant and it is thus unlikely that physiologic loads could cause the annular failures in this vicinity which are observed clinically.

The behaviour of the nuclear material under sustained segmental loading is essentially that of an incompressible fluid, developing hydrostatic pressure which acts equally on the inner annulus and on the vertebral end-plates, thus resisting their apposition. Under load the nucleus loses volume as some fluid is expelled, primarily through the cartilaginous end-plates. Low-molecular-weight, water-soluble, radio-opaque dyes injected into the nucleus have revealed this preferential pathway for fluid expulsion (J.H. Evans, J. Foster, 1987, unpublished work). The resulting increased osmotic gradient likely helps replenish the nucleus on reduction of joint loading and this 'pumping' action probably plays a significant role in the nutrition of the disk [15, 18, 19]. Gradually increasing deflection under loads, whether compressive or bending [18, 20], is partly attributable to these fluid movements. This creep response is recoverable. The anterior column, comprising relatively rigid vertebra connected by these specialised disk joints, forms a flexible, segmented compression structure.

Supports and Constraints

It is understood that the posterior column of the spine acts in tandem with the anterior column to constrain spinal motion at the extremes of range both in sagittal plane bending and in axial rotation. The spinous processes provide leverage for the action of the spinal muscles which cause motion within this physiologic range. However a number of connective tissue structures may also play a significant role in balancing the external forces acting on the spine within this range of motion. This apparent anomaly arises partly because the posterior spine is commonly considered to be truly segmented — the ligamentum flavum being discontinuous at each lamina, the capsular structures being associated with a single zygapophyseal joint, and the interspinous ligament being confined to the interspace. Only the supraspinous ligament has been attributed with a multisegmental role. However, recent thorough anatomical dissections have revealed structural relationships and connections which provide evidence of considerable mechanical interactions among these soft tissues [3–5].

In the mid lumbar region where the soft connective tissue structures of the thoracolumbar fascia (TLF) merge and intertwine with the ligamentous system in close association with the tendoaponeuroses of the paraspinal muscle, there is evidence of significant mechanical interaction. Indeed it has be argued that the paraspinal muscles [5] both in their active and passive states, and the abdominals [4, 5, 14, 21] may help to control the spine especially when flexed against on external force.

Paraspinal Muscles

In the fully flexed position the osseoligamentous spine is commonly considered to be the sole supporter of the bending moments transmitted by the trunk [8, 9]. Thus it is often stated that "we hang on our ligaments" when adopting or working in a stooped position as the paraspinal muscles have been observed to be "electromyographically quiet." However the passive resistance to stretch which is exhibited by most muscles, increases dramatically near the extreme of elongation. The actual length of contractile muscle tissue is sensibly constant in the paraspinals, and as this represents a relatively small proportion of the length of the total muscle-tendon structure, it has been estimated [3, 22, 23] that force resisting stretch at full flexion is not only significant but may exceed that produced by active contraction of the muscle. Thus it could be said that, in this extreme posture, the spine hangs on its ligaments and (passive) muscles.

Abdominal Muscles

It has been observed that an increased intra-abdominal pressure often accompanies load carriage in the stooped position. This increase has been postulated as being due to a reflex mechanism, involving contraction of the appropriate abdominal muscles, which helps support the spine and augments erector activity. This mechanism is commonly described as involving the effect of pressure acting on the diaphragm and on the pelvic floor causing an axial force which generates extensional moment about the spine [21, 24]. However it has been recognized that as the involved abdominal muscles take origin from the TLF, this connective tissue structure must also be subject to tension when the pressure in the abdominal cavity is raised in this manner [5, 14]. The tension thus developed will be transmitted in turn to the bony spine, either directly via the tips of the transverse processes for example, or indirectly by virtue of the intricate mechanical interaction of components of the TLF and the mid-line ligamentous system [5], the intraspinous and supraspinous ligaments.

The forces acting on the spine as a consequence of such abdominal muscle contraction have been calculated and measured in cadaveric experiments as being substantial and their contribution to extending and stabilizing the spine have been demonstrated [5].

The true significance of these supportive effects has yet to be determined but it would be prudent to consider that the integrity of these soft connective tissue structures be preserved or reconstituted. Procedures such as posterolateral fusions can destroy the continuity of these soft tissue structures and connections and are thus likely to weaken or otherwise disturb the mechanical characteristics of neighboring segments. Even interbody fusions, involving surgical procedures which reduce the need to damage or destroy the ligamentous systems, are thought to weaken neighboring segments. However recent clinical and cadaveric studies [25] suggest that the apparent hypermobility of joints neighboring a fused segment is less significant than had been feared hitherto, there being only a relative increase in mobility and not an absolute increase.

Failure of the Spine

In clinical terms the spine may be deemed to have failed if there is frank structural disruption or distortion resulting in the inability to provide mechanical strength or columnar stability and, more controversially, if pain is generated as a consequence of abnormal behaviour of the motion segment. The former type of failure includes, in addition to spinal fracture, conditions such as scoliosis, Schuermann's disease, and spondylosis. Most if not all, of these conditions can result in pain or neurological deficit. However the origins of low back pain are often less well identified although it is widely accepted that they are frequently linked with mechanical dysfunction at the segmental level [2, 26, 27].

Most of the structures of the motion segment could be a source of pain. However much attention has been paid to the intervertebral disk and the facet joints as, although the inner disk and articulating surfaces of the facet joints are insensate, the encapsulating soft tissues are rich in nociceptive nerve ending [2]. The avascular, insensate, load-bearing structures appear to be the first to fail as a consequence of cumulative microdamage and the ravages of time. Failure of these biomechanically critical structures leads to failure, or at least serious overload, of the sensitive surrounding tissues due to altered anatomy, redistribution of stress, and a mechanochemical degenerative cascade. The causes of the degeneration and subsequent failure of articular cartilage have been extensively researched but less is known of the mechanisms involved in degeneration of the nucleus [10, 16].

It seems likely that the nucleus forms the focus of this internal disk disruption and that, when no frank causes are evident, the cascade is triggered by impaired nutrition. The high anaerobic metabolism of the disk leads to a high lactic acid content and it has been observed that the most oxygen-deficient region of normal disks is at the junction between the nucleus and inner annulus. As the principal route for nutrient flow is through the end-plates, it appears that although encouraged by normal disk motion, vibrational loads or movements impair the flow of some critical metabolites [15]. Thus disk degeneration can be initiated and accelerated through both underuse [16] and overuse, a common pattern now emerging in the vital balance of such tissues.

The sensitivity of the lumbar spine to vibrational and shock loadings, as are experienced in riding various forms of vehicle [1], has long been recognized. Indeed the established whole body tolerance thresholds are highly dependent on the spine, it being the "weakest link" in seated subjects, forming the only true structural link between pelvis and thorax. The detrimental effects of prolonged inactivity are probably of greater clinical significance [2, 16, 28]. Extended bed rest and bracing should be avoided and surgically immobilized disks will also deteriorate.

The effect of subcatastrophic mechanical insults appears to be in part cumulative and the clinical signs are often delayed, making cause-effect correlation more difficult to establish. One significant causal link, which is gaining wide acceptance, relates to the microscopic and usually self-limiting failure of the cancellous bone in the subchondral region of the vertebral body. Failure of individual trabeculae can be detected at relatively low loads [29] which may not uncommonly be experienced by vertebrae, and more extensive regions of cancellous bone can be microscopically disrupted under severe compressive loading [29], even if transient in nature. These local failures of cancellous bone which are not readily detected radiographically could weaken the vertebra and thus render it more liable to catastrophic damage in the period before healing has occurred. Perhaps more significant is the fact that the regions of bone in which microfractures have subsequently healed contain numerous microcallus and trabecular overgrowths [30, 31] and can become sclerotic. This in turn will both stiffen the end-plates, reducing their ability to attenuate sudden compressive loads, and impede the flow of nutrients into the nucleus.

There are probably other mechanisms, equally difficult to identify in regular examination, which underlie the biomechanical susceptibility of the spine to everyday loading and which perhaps explain why these effects appear to be cumulative in nature. This area of work is bound to receive greater attention as more powerful diagnostic tools come into regular use, and preventive measures may then be developed.

Abnormal movement can occur in the motion segment as a consequence of changes in material properties and/or structural integrity of soft tissues and bone alike. These abnormal motions may be considered as resulting in "clinical instability" [12, 27, 32–26] when they are of such a nature or magnitude that they compromise the sensitive tissues that the spine normally protects. Only infrequently are these abnormal movements sufficiently gross as to constitute mechanical instability where structural collapse is imminent.

Surgical Intervention — Fusion

The technique of fusing one or more segments of the spine is widely used and usually involves the placement of bone grafts to speed the process of bony union. Whatever the primary intention the procedure will inevitably involve destruction or removal of tissue, its nerves, and blood supply. The common aim of fusion is to restore structural integrity to the spinal column [2, 37–40], but very often the primary intention is to prevent pain-eliciting segmental motion.

Indications and Aims

The efficacy of spinal fusion in general has been clouded by the issue of suitable indications: it may be a sound procedure employed for the wrong reasons, or in the wrong cases. In the literature, there is poor correlation between the rate of fusion and the relief of low-back pain. While this brings up the question of adequate and consistent clinical assessment of bony fusion, it also recognizes the fact that fusion per se is often not clearly indicated. Three factors which have great bearing on the eventual outcome of spinal surgery are selection of the patient, establishment of a structural diagnosis, and choice of operative procedure [2, 39, 41]. Of particular significance is the link between the identified problem and the surgical solution, there being no universally applicable technique. Until a structural diagnosis is made on the basis of clinical and laboratory testing there is no indication for such surgery.

As an aside it should be recalled that pseudarthroses resulting from ununited fusions and degenerate, fibrosed disks often provide stable and relatively flexible joints which are pain free. It can be argued that from a biomechanical perspective such joints may be superior to a true fusion. The clinical problem lies in knowing when such a joint will effectively relieve symptoms and also in how to reliably produce this type of joint. The quest for a synthetic intervertebral disk is, in part, promoted by such observations but the necessary and desirable characteristics of the resulting joint are not yet adequately defined. Despite the scorn previously poured on these ideas there has been a resurgence of interest in this form of potential solution.

The syndrome of spinal nerve entrapment is considered an important surgical indication by many. However knowledge is too meager to establish the proper place for surgical intervention, and methods of avoiding postoperative scarring are not well established.

Fusion is frequently employed to stabilize the spine when structural integrity has been severely compromised, but often more rigorous demands are made on the procedure. For example, it is maintained that in many cases if motion is effectively eliminated, then the associated pain will be alleviated. Not so well recognized are the advantages of restoring the segmental anatomy or of deliberately decompressing sensitive structures. These advantages may be most apparent after resection of the disk when segmental collapse would otherwise occur.

Thus the aims of fusion may be multifaceted, involving restoration of load-bearing capacity and anatomy and also reduction of aggravating mechanical stress in sensitive tissues and it must not be forgotten that, in the process of surgical intervention, such tissues may be removed or, at least desensitized.

Grafting

A great variety of techniques involving arthrodeses of numerous anatomic sites has been proposed and used to good effect [2, 26, 39]. While biomechanical considerations of obtaining a strong, rigid fusion are usually paramount, there is a multifactorial overlay which makes for a wide choice of candidate constructs. The technical demands of a particular procedure, the nuances of each particular case, and, of course, personal preference and experience will influence the final choice of technique. The need to harvest graft material of differing quality, quantity, and shape will determine whether further surgery is required at sites remote from the site of fusion or whether allograft or even bone substitutes should be considered. Additionally the choice of the different techniques of internal fixation and postoperative immobilization draws as much from personal preference and particular clinical demands as it does from purely biomechanical factors. As a consequence, there exists a plethora of techniques that may differ fundamentally or in detail only.

There are five anatomic sites within a motion segment at which bony fusion is commonly attempted: the posterior spinous processes, the transverse processes, the laminae, the facet joints, and the vertebral body [2]. A number of the most commonly used techniques are based on obtaining fusion at more than one of these sites, for example the posterior-lateral fusion technique that can avoid the midline structures is now widely employed.

It is not appropriate to review the full range of techniques nor to analyze the different constructs in biomechanical terms. Rather, the cardinal factors will be outlined to provide the basis on which rational comparisons may be drawn. It is hoped that this approach will outline the biomechanical rationale so that, when techniques are adapted to accommodate those "out of the ordinary" cases that occasionally present, the most biomechanically advantageous grafts can be constructed.

Paradoxically, the aims of grafting are not necessarily identical to the aims of fusion. In biomechanical terms the critical phase in the progress towards satisfactory fusion is the early postoperative period, before any bony union has been established or adequate fibrous connections developed which can impart segmental stability. The fused segment will eventually have to transmit forces comparable to those experienced by a normal spinal segment. Prior to union, the spine may be protected by internal or external fixation and by other suitable patient management schemes. Adequately proportioned grafts of appropriate tissue, when correctly sited, can impart considerable strength and stiffness to the freshly operated segment [42]. The strength of a well grafted segment can immediately tolerate body weight, and the stiffness imparted can prevent irritable motion and encourage bony union.

The conditions which lead to graft incorporation or substitution appear to include the following biomechanical factors: the stress, or intensity of force acting at the graft/host bony interface and the relative motion between them. The stress should not exceed the failure value of graft or host material but, over a period of time, compressive stress will be beneficial so long as no excessive relative motion is induced at the interface.

These desirable biomechanical conditions imply that a good graft will bear load without moving in the immediate postoperative period. In this context, it must be remembered that the graft will normally be subject to compression, although it may be required to resist some shear force that tries to slide it across the surface of the host bone. Not until union is begun can tensile forces be transmitted, as the graft will simply lift free of the host.

Based on these essential criteria and "desirable conditions" it is possible to explore the biomechanical potential of various surgical constructs. Recall that the immediate postsurgical biomechanical aims are to carry reasonable loads that might correspond to moderate body movement in the standing position and to be stiff and unmoving under these loading conditions. It is important to note that if these conditions are satisfied, the united segment will also provide excellent strength and stability once fusion is achieved. Discussion will therefore be centered on the biomechanical characteristics of the construct in this critical phase, immediately after operation.

Biomechanically Sound Constructs

It is obvious that many clinically successful fusion techniques do not meet the stringent requirements laid down in the previous section. Clearly, these surgical constructs do not afford inherent stability in the initial phase and rigorous adherence to post-surgical immobilization routines will be demanded. By adopting the inherently stable graft constructs, a far less arduous and demanding convalescence can be anticipated and mobility can be maintained in adjacent tissue structures. The benefits to patients and care providers are very real and immediately apparent.

Structure

The various techniques, which have been specifically developed to address differing presenting conditions or have evolved with increasing awareness of desirable outcomes and improving technologies, are myriad, yet a biomechanical analysis reveals the existence of certain basic constructions. Two such basic choices can be characterised by analogy with a "tripod" and a "flagpole" supported by tie ropes [24]. Both provide inherent postoperative stability. Flexural stiffness and graft compression can be provided by many other means but similar structural analogs may prove useful in planning and optimizing the total construct. Let us consider these mechanical analogs as they will clarify some of the biomechanical factors involved in grafting for stability.

The tripod, like a three-legged stool, is stable when placed on a rigid surface. The three legs are spread out so as to reduce the chance that one leg will lift (the condition when there is no compressive force between leg and ground), with ensuring catastrophic instability. Should it not be possible to spread the legs sufficiently to provide stability under the maximum toppling (bending) moment that is anticipated, then two potential solutions exist: fastening the legs to the ground so that both compressive and tensile forces can be transmitted across the interface, or attaching between the tripod and the ground a central wire that is under tension — the higher the tension, the greater the resistance to toppling.

Interestingly, an analogous structure or construct is the flagpole, supported by guy wires. The minimal, stable structure comprises a central compressive element, which can simply rest on the ground, and three tension wires that spread out equally around the pole. Again, the higher the tension in the wires, the higher the compression in the pole, and the greater the resistance to toppling under the influences of "wind loads."

In these illustrations, it is obvious that there will be a catastrophic collapse if the applied forces cause any one element, either in tension or compression, to fail. Equally, the flexibility (the inverse of rigidity) will depend on both the stiffness of the structural elements and their spatial separation.

In translating these concepts into practical fusion constructs, several constraints are imposed. The base on which to sit the tripod, as it were, is clearly limited to the bony elements of the motion segment, unless the construct incorporates additional internal supports or fixators. Similarly, as far as the flagpole technique is concerned, the pole itself will inevitably be based on the vertebral body, and thus the construct will be centered around an interbody fusion.

In both of the above graft arrangements, existing soft connective tissues will normally serve as the tension members. However, the stabilizing action of these tissues may be augmented by wiring or other internal fixators [2, 24, 38].

It is important to recall that adequate stability of the construct is dependent on the elements being preloaded and this can be achieved by using the graft to distract the segment, thus stretching the appropriate soft-tissue interconnections. Wiring and adjustable fixators may also be used to this end.

Material

The consequences of inadequate fixation can be gross graft "settling" or even total collapse, and possibly pseudarthrosis in place of true bony union. The potentially catastrophic failure involving graft extrusion can also be the result of inadequate fixation. However, host site and graft preparation are crucial, not only to ensure optimal conditions for fusion but also to guarantee a strong, stable construct.

Adequate compressive strength can be attained with an interbody graft if the force is uniformly distributed across a large area of bone graft. In order to carry body weight without detectable crushing of the bone tissue, the graft must cover approximately two-thirds of the area of the end-plate [24, 42]. Rarely can a true "unipour" graft be incorporated, so that the individual components of the graft should be congruent and thus bear load equally [2, 24, 38].

Preparation of the host site is vital for the success of an interbody fusion; decortication is recommended, or at least perforation of the end-plate to ensure graft incorporation. Biomechanical factors are also of considerable importance. The crush resistance of the cancellous bone of the intervertebral bodies is markedly site-depended — the more central the site, the weaker the bone. Thus, grafts should cover the face of the body and extend to as near the periphery as possible. Additionally, the cancellous bone is strongest immediately beneath the end-plates and weakest at mid-height. Thus, in preparing the vertebra for grafting, minimum bone should be removed commensurate with providing a flat, bleeding surface.

The choice of graft material is governed primarily by biological and mechanical factors although the availability of the material and the ease with which it may be harvested obviously have a bearing on the final choice. In determining the desirable mechanical characteristics, it is important to consider both the ideal graft and the concept of damage limitation, should the grafted segment be caused to fail due to overloading or inadequate immobilization. Ideally the graft should closely match the strength and stiffness of the bone at the host site as this should lead to uniform interface stress and minimum interfacial movement resulting from tissue deformation. Should collapse occur due to overload of the grafted segment, subsequent management, which may include revision surgery, is facilitated if the graft rather than the bone fails at the host site. Clearly extrusion of the graft fragments must be contained so as not to cause impingement on the cord or other sensitive or vulnerable tissues.

What Does the Future Offer?

In respect of the particular problems addressed in this chapter there are few certainties in prospect. Considering the importance of stability, it is necessary to improve diagnostic methods to delineate the normal and demonstrate the pathologic motion segments in the spine. There must also be improvement in diagnostic techniques and methods of fixation. Present surgical procedures too often leave the motion segment unstable and lead to fusions which fail to unite in a satisfactorily high proportion of patients. Perhaps the clearest evidence points to the conclusion that there is likely to be far less major surgery as our knowledge and imaging abilities improve.

A better understanding of the causes and consequences of pain and deformity will not only spawn more effective preventative measures but will capitalize on the potential of the rapidly developing field of medical imaging, both structural and functional, to provide vastly improved diagnosis. When surgery is indicated there should be a clearer aim and an improved range of tools available to ensure the desired outcome. Many of the bulky and rigid devices used for attaining correction or fixation may be replaced by less invasive and compliant systems which gradually allow the body tissues to accept in full the structural demands placed on the spine. This could be achieved, in part, by more mechanically sophisticated, biodegradable materials and devices. Similarly, substitutes for both hard and soft connective tissues will proliferate and some will find temporary or even permanent roles.

The major investment in developing hip and knee joint prostheses, for example, will likely benefit similar endeavours relating to the spinal joint. For, while the structural differences may be great, the tissues, their substitutes, and interactions are virtually identical. Probably technical solutions will be easier to attain than a clearer definition of the problem. In the event it is not unreasonable to anticipate that the techniques of intervention and the development of candidate materials will proceed abreast of research into the mechanisms involved in the onset and progression of spinal disorders.

References

1. Boshuizen HC, Bongers PM, Hulshof CTJ (1992) Self-reported back pain in fork-lift truck and freight-container tractor drivers exposed to whole-body vibration. Spine 17(1):59–65
2. O'Brien JP (1983) The role of fusion for chronic low back pain. Orthop Clin North Am 14(3):639
3. Bogduk N, Macintosh JE, Pearcy MJ (to be published) A universal model of the lumbar back muscles in the upright position. Spine
4. Bogduk N, Twomey LT (1987) Clinical anatomy of the lumbar spine. Churchill Livingstone, New York
5. Tesh K, Shaw-Dunn J, Evans JH (1987) The abdominal muscles and spinal stability. Spine 12(5):501–508
6. Siereg A, Arvikar RJ (1975) A comprehensive musculoskeletal model of the human vertebral column. Advances in Bioengineering. In: Proceedings of the Winter Annual Meeting of Bioengineering Divsion of ASME, Houston, Texas. American Society of Mechanical Engineers, pp 74–75 (Advances in Bioengineering Series)
7. Yettram AL, Jackman MJ (1980) Equilibrium analysis for the forces in the human spinal column and its musculature. Spine 5:402–411
8. Adams MA, Hutton WC, Stott JRR (1980) The resistance to flexion of the lumbar intervertebral joint. Spine 5(3):245–253
9. Lin HS, Liu YK, Adams KH (1978) Mechanical response of the lumbar intervertebral joint under physiological (complex) loading, J Bone Joint Surg [Am] 60-A(1):41–55
10. Johnstone B, Urban JPG, Roberts S, Menage J (1992) The fluid content of the intervertebral disc. Spine 17(4):412–416
11. Crispo JJ, Panjabi MM (1992) Euler stability of the human ligamentous lumbar spine. Part 1: Theory. Clin Biomech 7:19–26
12. Crispo JJ, Panjabi MM, Yamamoto I, Oxland TR (1992) Euler stability of the human ligamentous lumbar spine. Part 2: Experiment. Clin Biomech 7:27–32
13. Adams MA, Hutton WC (1983) The mechanical function of the lumbar apophyseal joints. Spine 8(3):327–330

14. Gracovetsky S, Farfan HF, Lamy C (1981) The mechanisms of the lumbar spine. Spine 6:249–262
15. Holm S, Maroudas A, Urban JPG, Selstam G, Nachemson AL (1981) Nutrition of the intervertebral disc: Solute transport and metabolism. Connect Tissue Res 8:101–119
16. Holm S, Nachemson AL (1981) Nutritional changes in the canine intervertebral disc after spinal fusion. Clin Orthop 169:243–258
17. Holm S, Nachemson AL (1983) Variations in the nutrition of the canine intervertebral disc induced by motion. Spine 8:866–874
18. Panagiotacopulos ND, Pope MH, Bloch R, Krag MH (1987) Water content in human intervertebral discs: Part 2. Viscoelastic Behaviour. Spine 12(9):918–924
19. Urban JPG, Holm S, Maroudas A, Nachemson AL (1982) Nutrition of the intevertebral disc. Effect of fluid flow on solute transport. Clin Orthop 170:296–302
20. McGill SM, Brown S (1992) Creep response of the lumbar spine to prolonged full flexion. Clin Biomech 7:43–46
21. Fast A (1988) Low back disorders: Conservative management. Arch Phys Med Rehab 69:880–891
22. Macintosh JE, Bogduk N (1991) Attachments of the lumbar erector spinae. Spine 16(7):787–792
23. Macintosh JE, Bodgduk N (1987) The morphology of the lumbar erector spinae. Spine 12:658–668
24. Evans JH (1985) Biomechanics of lumbar fusion. Clin Orthop Rel Res 193:38–46
25. Chow DHK, Luk KDK, Leong JCY, Evans JH (1992) Segmental mobility of the lumbar spine after fusion — a radiological and a biomechanical study. In: Chan FHY, Chan KL, Mak AFT, Schindler F (eds) Proceedings of the Biomedical Engineering Symposium. Hong Kong Institution of Engineers, Hong Kong, pp 21–24
26. Farfan HF, Kirkaldy-Willis WH (1981) The present status of spinal fusion in the treatment of lumbar intervertebral joint disorders. Clin Orthop 158:198
27. Frymoyer JW, Selby DK (1985) Segmental instability: Rationale for treatment. Spine 10:280–286
28. Pope MH (1987) The biomechanical basis for early care programmes. Ergonomics 30:351–358
29. Thomas I, Evans JH (1988) Acoustic emission from vertebral bodies. J Mat Sci Letters 7:267
30. Hansson T, Roos B (1981) The relation between bone mineral content, experimental compression fractures and disc degeneration in lumbar vertebrae. Spine 6:147
31. Hansson T, Roos B (1981) Microcalluses of the trabeculae in lumbar vertebrae and their relation to the bone mineral content. Spine 6:375
32. Bergmark A (1987) Mechanical stability of the human lumbar spine. Doctoral dissertation, Lund Institute of Technology, Department of Solid Mechanics, Lund, Sweden
33. Crispo JJ (1989) The biomechanical stability of the human lumbar spine: Experimental and theoretical investigations. Doctoral dissertation, Yale University, New Haven, Connecticut
34. Kirkaldy-Willis WH, Farfan HF (1982) Instability of the lumbar spine. Clin Orthop 165:110–123
35. Nachemson A (1985) Lumbar spine instability: A critical update and symposium summary. Spine 10:290–291

36. Posner I, White AA, Edwards WT, Hayes WC (1982) A biomechanical analysis of the clinical stability of the lumbar and lumbosacral spine. Spine 7:374

37. Bohlman HH (1985) Treatment of fractures and dislocations of the thoracic and lumbar spine. Current concept review. J Bone Joint Surg [Am] 67(1):165–169

38. Lin PM (1982) Introduction of PLIF, biomechanical principles, and indications. In: Lin PM (ed) Posterior lumbar interbody fusion. Charles C Thomas, Springfield, IL, pp 3–57

39. Selby DK (1983) When to operate and what to operate upon. Orthop Clin N Am 14(3):577–587

40. Wetzel FT, La Rocca H (1991) The failed posterior lumbar interbody fusion. Spine 16(7):839–845

41. Spengler DM, Freeman C, Westbrook R, Miller J (1980) Low back pain following lumbar spine procedures. Failure of initial selection? Spine 5:356

42. Pearcy MJ, Evans JH, O'Brien JP (1983) The load bearing capacity of vertebral cancellous bone in interbody fusion of the lumbar spine. Eng Med 183–184

Hip Joint

Biomechanical Advances in Total Hip Replacement

PHILIP C. NOBLE[1]

Summary. Cemented hip arthroplasty is one of the most successful procedures of modern surgery. However, attempts to extend this procedure to younger and more active individuals using cemented or cementless methods of fixation have met with mixed success. Cementless procedures still have an unacceptable incidence of pain and limp at follow-up. These symptoms appear to be related to the presence of localized areas of instability of the stem/bone interface secondary to inadequate proximal or distal fixation. In addition, an alarming complication of some cementless prostheses is focal osteolysis which appears to arise from a foreign body reaction to particulate debris generated by motion between the implant and bone. Continuing advances in cementless arthroplasty will arise from a systematic approach to implant design leading to more anatomic stems, greater integration of implants and instruments, and increased attention to the impact of surgical technique on implant/bone fit and the stability of cementless fixation.

The results of cemented arthroplasty also continue to improve. Retrospective studies of retrieved implants have shown that the fundamental cause of aseptic loosening and focal osteolysis is fragmentation of the cement mantle. This complication can be minimized through use of implants specifically designed for cemented fixation, centralized within the medullary canal with proximal and distal centralizing devices which guarantee a minimum thickness of bone cement. In terms of the biomechanics of the hip joint itself, hip replacement can be enhanced yet further through greater attention to restorating the anatomic relationship between the femur and the acetabulum. This necessitates the use of femoral components which increased head offset and coordination of the position of the femoral head with the shape of the femoral stem.

In the acetabulum, cementless fixation has been particularly successful in the short term. However, dependable fixation of acetabular cups sometimes

---

[1] Orthopedic Research Laboratory, The Methodist Hospital, Houston, Texas, US

necessitates adjunctive the use of bone screws which has led to a small but serious incidence of neurovascular complications. The morbidity of screw fixation within the pelvis may be minimized if bicortical screw purchase is avoided and if screw placement is restricted to areas of thickest available bone.

Solutions are still awaited to many unsolved challenges in total hip replacement. These include increased wear-resistance of the articulating surfaces, a truly biocompatible implant/bone interface and greater retention of cortical bone in the presence of rigid component fixation.

Key words. Hip replacement — Biomechanics — Femur — Acetabulum — Anatomy

Introduction

The last decade has seen many developments in total hip replacement; some have enhanced the outcome of this procedure, while others have demonstrated the limitations of some new approaches. An abiding legacy of this experience has been expansion of the options available to many patients, particularly those who are young and active, or who require revision of a previous hip prosthesis. While many surgeons have been concerned with the relative merits of cemented and cementless modes of implant fixation, the fundamental aim of every arthroplasty remains restoration of normal, pain-free function of the hip joint. This necessitates that any prosthetic device recreate the normal anatomic relationships between the femur and the acetabulum. However, no artificial joint will perform successfully in the long term unless the method of fixation of its articulating components to host bone is biologically compatible.

At this stage in the development of total hip arthroplasty, many challenges exist to the continuing advancement of this procedure. These include:

1. Improvement of cementless fixation in the femur to ensure a more reliable clinical outcome
2. Prevention of focal osteolysis in cemented and cementless prostheses
3. Restoration of normal joint biomechanics
4. Refinement of the design and fixation of cementless acetabular cups to prevent osteolysis, wear and long term loosening

Although many advances have already been made in each of these areas, developments are still awaited to allow the indications for successful hip replacement to be extended still further.

Advances in Cemented Fixation in the Femur

Acrylic cement provides the surgeon with enormous flexibility in that it allows the variability of individual femora to be accommodated without compromising the fixation of the prosthesis. In cementless hip replacement, the shape of the implant dictates the geometry of the implantation site regardless of the distribution of suitable, supportive bone in each individual femur. In contrast, when cement is available as the fixation medium, the surgeon's preparation of the femur is not constrained by the shape or fit of the ultimate prosthesis. Thus, areas of weak or cystic bone may be removed intraoperatively, leaving the best available foundation to perform the new load-bearing function. In this way, cement can provide a perfect fitting implant for every patient, regardless of the variability of femoral anatomy.

Aseptic Loosening and Osteolysis

Despite the numerous advantages of cemented fixation, aseptic loosening and osteolysis remain significant long-term complications of cemented procedures, particularly in the young and active patient (Fig. 1). However, the incidence of these complications varies greatly between surgeons, primarily as a function of patient selection and cementing technique. This suggests that many of the failures of cemented hip replacement may be avoided through intraoperative measures.

The occurrence of focal osteolysis in association with cement ("cement disease") has been the driving force behind much of the interest in cementless

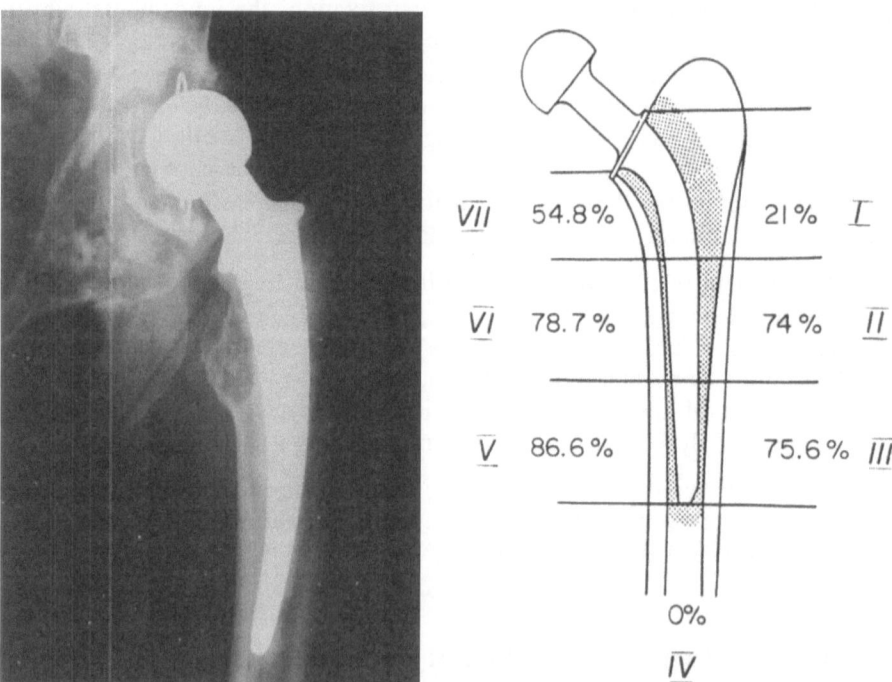

Fig. 1

Fig. 2

Fig. 1. A cemented hip replacement showing gross osteolytic destruction of both the femur and acetabulum

Fig. 2. The distribution of cement-deficient zones in cases of femoral osteolysis

hip replacement. Nonetheless, cement remains the method of choice for implant fixation in the majority of patients. In these cases, it is essential that we understand the causes of loosening and osteolysis if these complications are to be prevented in the future. In this regard, the work of Huddleston is extremely instructive. He systematically examined factors associated with femoral osteolysis in cemented hip replacements [1], and found that osteolytic lesions were often associated with defects and thin areas within the cement mantle (Fig. 2). This association was most pronounced over the distal third of the femoral component where almost 90% of cases with osteolysis had a deficiency of the cement mantle.

Similar observations have been reported by Anthony et al. [2] and Maloney et al. [3] (Fig. 3) and Vives et al. [4] These observations suggest that the osteolysis is not a reaction to bone cement per se, but to particulate debris caused by wear, erosion or fragmentation of artificial materials within the body. This concept is supported by recent reports of focal osteolytic lesions around cementless femoral components, often in areas of high relative motion between the bone and the prosthesis [5].

In the case of cemented hips, particles can arise from several sources. One is the wear of polyethylene, which appears to be a factor in the late loosening of acetabular cups and may explain the relationship between the incidence of loosening and the depth of wear of these components. However, large volumes of particles can be generated also by fragmentation of the cement mantle, which appears to be the key factor in the etiology of osteolysis within the femur. Based on this information, prevention of "cement disease" necessarily

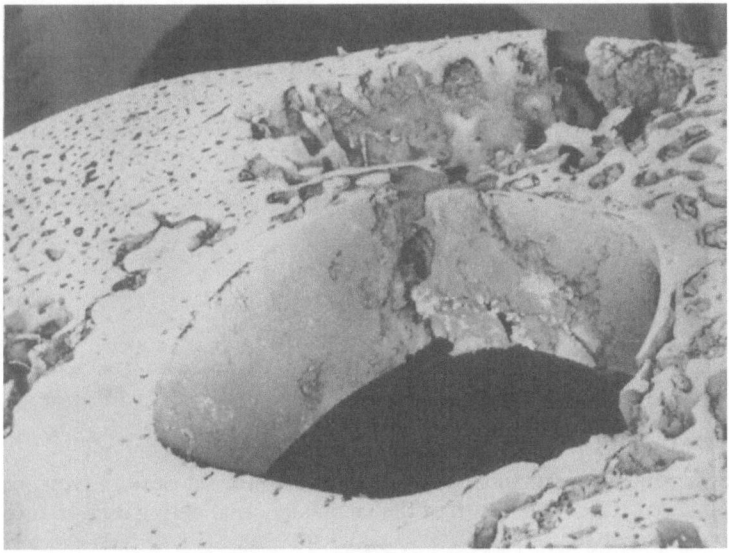

Fig. 3. Scanning electron micrograph cross-section through the bone and cement mantle of a specimen obtained at necropsy. Cortical resorption is evident due to osteolysis adjacent to a defect within the cement mantle

involves prevention of fragmentation and protection of the cement mantle from excessive stress [6].

The "Weak Links" of Cemented Fixation

The interfaces formed between the cement and the bone and the cement and the prosthesis are of profound importance in protecting the cement from excessive loading because of the relative weakness of the cement mantle [7–11]. In fact, the interfaces themselves, rather than the cement mantle, form the truly weak links in the chain of cemented fixation. Thus, any attempt to increase the durability of cemented fixation must address the strength of the interfaces and not simply methods of enhancement of the bone cement itself [12].

In many cases of prosthetic loosening, failure is initiated at the cement-bone interface. This may occur if weak, inadequate bone is left prior to cementing or if the interdigitation of cement into porous bone is inadquate due to poor pressurization or bone preparation [13, 14]. Intraoperatively, much can be done to provide intimate contact between cement and bone. The key steps are preparation of a strong, porous bony bed with open trabecular spaces and sustained pressurization of the cement so that mechanical interlock is achieved for several millimeters into the surrounding bone [15].

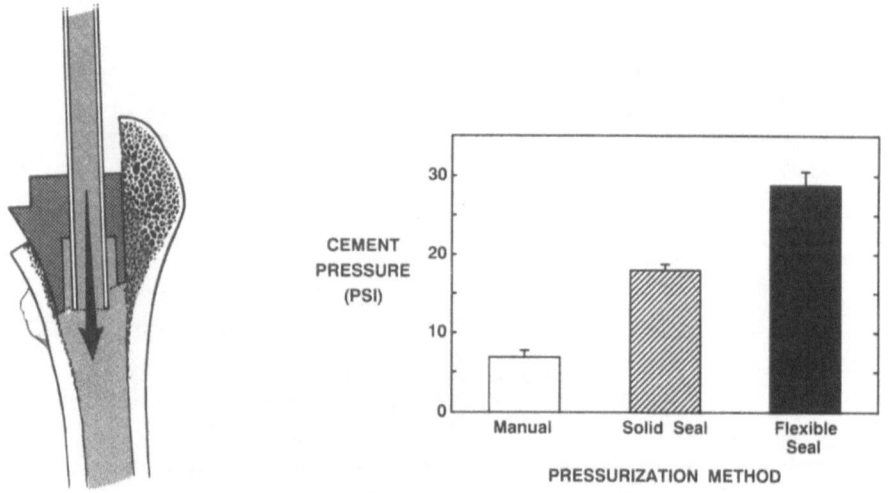

Fig. 4   **Fig.**

Fig. 4. Schematic representation of anatomically-contoured flexible pressurizing seal which is used to occlude the mouth of the medullary canal during introduction of bone cement via a cement gun

Fig. 5. Comparison of the average cement pressure generated during cement pressurization of the proximal femur using three alternate methods, manual ("two-thumbs") cementing, use of a solid seal with a cement gun, and use of the flexible seal illustrated in Fig. 4

While cement pressurization within the femur and acetabulum is easy to advocate, it is often difficult to achieve, particularly in the proximal femur, where the irregular cavity can be hard to effectively seal. Sealing devices are now available which have been specifically designed for the shape of the proximal femur (Fig. 4) [16]. These devices are relatively easy to use and are very effective in pressurizing the femur as the egress of cement is restricted under pressure (Fig. 5). This allows pressure to be maintained for longer periods as the cement gains viscosity. Advances had also been made to improve the cement-bone interface within the acetabulum. Angled nozzles are available for use with a cement gun, enabling cement to be introduced directly into fixation holes.

The durability of the cement mantle may be enhanced through two separate approaches. The first addresses the inherent strength of the cement itself and hence its resistance to fatigue failure [17, 18]. The second involves optimization of the design of the cement mantle to reduce the mechanical stresses developed during weight-bearing. While measures to improve the cement increase its average strength by 20–30%, attention to the optimum geometry of the mantle can lead to stress reductions in the 50–90% range [19].

Stresses within the cement mantle are concentrated in two areas: proximally, immediately below the femoral neck osteotomy, especially over the medial surface of the prothesis, and distally, around the tip of the stem where there is a large discontinuity in bending stiffness.

The precise geometry of the cement has a very profound influence upon the stresses developed during loading [9]. At the proximal end of the femur, a non-uniform distribution of cement provides the optimum level of cement stresses (Fig. 6). Finite element models predict that if the size of the femoral stem is increased, causing the cement to become progressively thinner, the stresses within the mantle increase alarmingly. Conversely, if the stem is

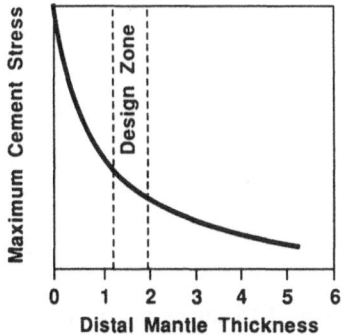 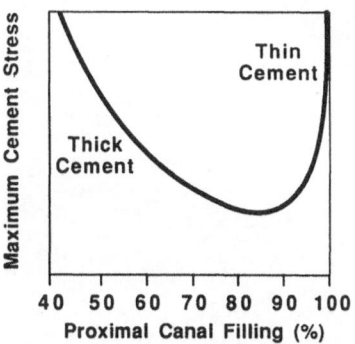

Fig. 6. Variation of the maximum stress within the cement mantle, distally (left), adjacent to the distal tip of the prosthesis, and proximally (right) near the level of the femoral neck osteotomy. Stresses are plotted as a function of the mantle thickness and the percentage of the proximal canal occupied by the femoral stem

undersized, leading to an extremely thick mantle, cement stress rises once again. Thus there is an optimum zone, corresponding to a medial cement thickness in the range of 3–7 mm where the cement stresses are at a minimum.

Distally, stresses within the cement are also dramatically reduced as the diameter of the stem becomes smaller leading to a greater proportion of cement to metal within the canal (Fig. 6). However, a critical point is reached where the stress generated within the stem itself becomes dangerously high. Thus, a compromise must be reached where the cement and the stem are protected simultaneously. This can be achieved by having a cement layer of reasonable thickness, generally of around 2 mm, at which point approximately three quarters of the stress concentration caused by the tip of the prosthesis is abated. Thus, on the basis of engineering analysis, the optimum cement mantle is of non-uniform thickness, ranging from a minimum of 2 mm distally to 3–7 mm proximally.

Some clinical support for these predictions is found in long term reviews of cemented hip replacements in which the effects of the geometry of the cement mantle have been examined explicitly [20, 21]. Sarmiento and Gruen looked at the effect of the thickness of the proximal medial cement on the incidence of significant calcar resorption, a phenomenon which appears to be a response to the generation of particulate debris. These investigators found that significant resorption became alarmingly prevalent once the thickness of the proximal cement was less than 2 mm.

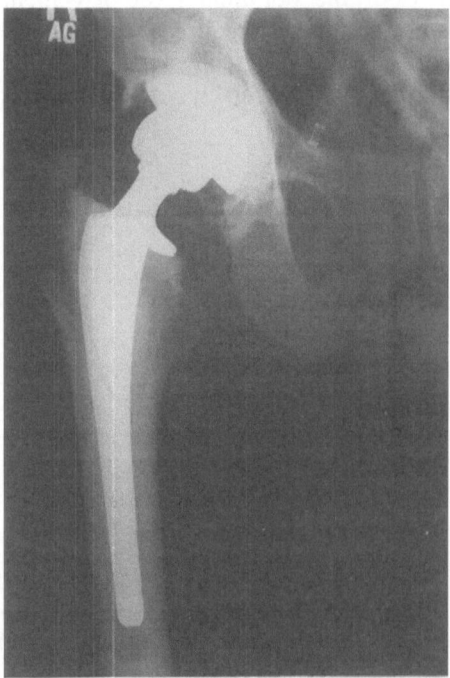

Fig. 7. Radiograph of a modern cemented hip replacement showing development of an ideal cement mantle through the use of an implant specifically designed for cement fixation and proximal and distal centralizing devices

Intraoperative Prevention of Cement Fragmentation

Several practical factors can be used at surgery to prevent the formation of thin cement and fragmentation of the mantle. These include centralization of the stem within the cement, and the use of measures to prevent impingement of the stem during its insertion into the femur. The cement mantle is also preserved by stems which are specifically designed to preserve an adequate thickness of cement, especially in the critical proximal and distal regions of the femur (Fig. 7).

In reviewing the radiographic results of conventional cemented hips performed by experienced surgeons, we have observed an alarming incidence

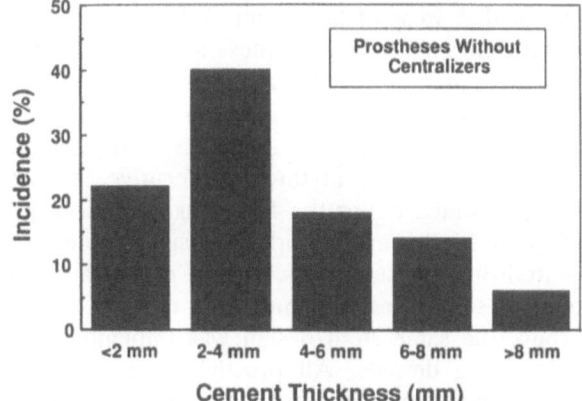

Fig. 8. Distribution of proximal medial cement thickness in cemented hips of conventional designs implanted without centralizing devices

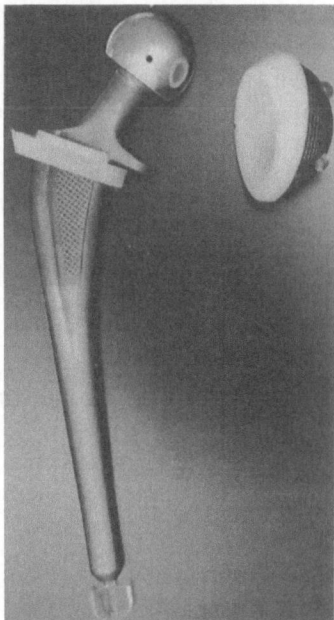

Fig. 9. Femoral and acetabular components with cement spacing and centralizing devices

of deficient cement mantles. In our study, less than 2 mm of cement remained between the medial cortex and the implant in 22% of the cases (Fig. 8). In addition, it was found that when proportional stems were implanted with bone cement, impingement between the anterior cortex and the distal tip of the prosthesis was extremely common and was evident on many lateral radiographs. In many cases, it was observed at the time of surgery that the cement mantle was structurally compromised proximally and distally where cement stresses were highest.

In order to correct these deficiencies, femoral prostheses have been designed on the basis of somatype analysis of the femur with modifications for the use of bone cement [27]. These implants are straighter than their cementless counterparts to allow for the development of thicker cement mantle in the proximal and medial areas of the femur. When used in conjunction with proximal and distal centralizing devices, these stems ensure that the cement mantle is of at least the minimum allowable thickness without any area of direct impingement between implant and bone (Fig. 9).

To evaluate the technical success of these innovations, a study was undertaken which looked at three consecutive series of patients in which surgery was performed under the following conditions [22]. In the first group, stems of proportional and standard designs were implanted without the use of centralizing devices. In the second group, a new design of stem derived from femoral somatypes was implanted using a distal centralizer, and in the third group, the same stem design was implanted using both proximal and distal centralizing deivces. All prostheses in the third group were implanted by orthopedic trainees under the supervision of an experienced implant surgeon.

The distal centralizer greatly reduced the frequency of stem/bone impingement from 37% in group 1 to 3% in group 2 and 0% in group 3. Proximally, the incidence of adequate cement mantles increased from only 35% in group 1, when conventional components and techniques were used, to 81% in group 2 using the somatype-derived stem with a distal centralizer. In group 3, acceptable cement mantles of 3–7 mm in thickness were observed in every case. This study demonstrates that proximal and distal centralizers can eliminate misalignment of the implant, even in the hands of the inexperienced surgeons. This further suggests that the use of a new stem design with centralizing devices may prevent the formation of inadequate cement mantles and greatly diminish the incidence of osteolysis in the years to come.

Optimization of Cementless Femoral Components

Great progress has been made in the development of cementless hip replacements. The impact of these advances has been most significant in revision arthroplasty and in primary replacements performed in the younger patient. Unfortunately, the clinical results of these procedures are not always comparable to those of cemented hip replacement. A significant incidence of pain and limp has been reported following cementless procedures, although these

Fig. 10. The relative displacement of a prosthesis with respect to the femur during repetitive loading

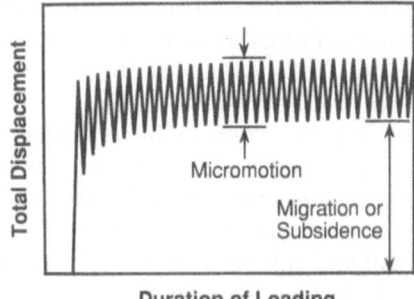

Fig. 11. Major components of interface motion between the femur and a cementless prosthesis

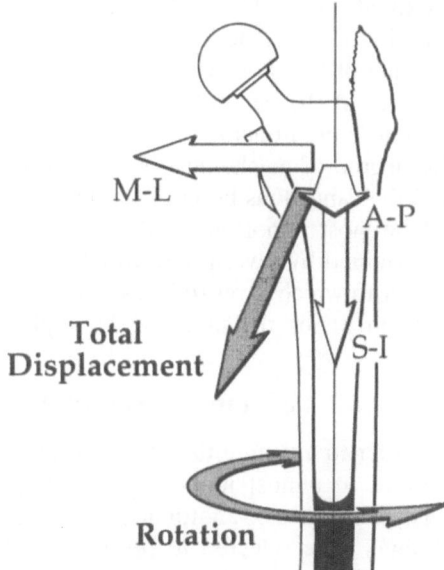

symptoms are generally not of such severity as to limit activity. These complications may be attributed to a variety of factors, principally excessive micromotion at the interface between implant and bone. Thus, the fundamental problem associated with the cementless technique is how to rigidly stabilize the femoral component in the femur without the use of acrylic cement.

All femoral prostheses move slightly within the femur during loading, leading to some permanent displacement. This is commonly referred to as migration or subsidence [23]. Typically, the subsidence of the prosthesis is greatest during the immediate post-operative period and decreases progressively with increasing weight-bearing (Fig. 10). A small amount of the total motion of the interface, which occurs with each loading cycle, is recovered following unloading of the hip joint and is referred to as "micromotion".

Significant components of micromotion and migration occur in the medial-lateral, anterior-posterior, and superior-inferior directions (Fig. 11). Rotational micromotion and displacement also occur about the femoral axis and appear to be responsible for much of the symptomatology of inadequately fixed components. In fact, rotational migration can become so extreme that, at revision, the prosthesis is found to be in frank retroversion.

To adequately characterize the motion taking place at the interface, it is necessary to measure micromotion and migration in each of the anatomic directions under physiologic loading conditions. Typically, this consists of cyclic compression and torsion. In the biomechanics laboratory, the normal loading of the hip joint may be simulated using an experimental technique in which a prosthesis is loaded in a test fixture in a mechanical testing machine. During loading, the relative motion of the prosthesis and the femur are followed using an array of extremely sensitive noncontacting transducers mounted on the proximal femur. Using this test procedure, each of the variables affecting the stability of the cementless hip replacement may be examined systematically.

Some of the most significant factors which have been found to influence the stability of the cementless implants are: The accuracy of proximal and distal implant/bone fit achieved at surgery, the surface texture of the prosthesis, and the degree of mechanical interference obtained between the proximal implant and the cancellous bone of the femoral metaphysis.

The most important of these factors, the fit of the implant to the femur, is determined by several key variables including the design of the prosthesis, the extent to which it reproduces femoral anatomy, and the effect of bone-shaping instruments on the shape of the implantation site.

Implant/Bone Fit in Cememtless Hip Replacement

In order to examine the fit of contemporary cementless stems to the femur, 15 femora of almost identical size, shape and geometry were obtained from a large collection of anatomic specimens [24]. Cementless stems of three implant designs were selected to represent the different designs most commonly employed in joint replacement, namely, a straight, "Austin-Moore" design (the AML prosthesis, DePuy Warsaw, In), a straight, "second-generation" design (the Harris-Galante prosthesis, Zimmer Warsaw, In), and an "anatomic", curved stem design (the Profile prosthesis, DePuy Warsaw, In). Components of each design were implanted into the femora using the instruments and surgical procedures recommended by the manufacturers. Each bone was sectioned into transverse slices at 5 mm intervals. Each slice was then examined to determine the nature of the interface achieved between the implant and surrounding bone.

It was found that over the majority of the available surface of the prosthesis, gaps larger than 1–2 mm were present between the implant and bone (Fig. 12). Though the distribution of cortical and cancellous contact varied from bone to bone, separation of stem and bone was a universal observation, regardless of the design of the prosthesis. These findings indicate that prostheses in which proximal porous coating is restricted to the anterior and posterior aspects of the femoral stem, function in a pressfit mode, as contact between the bone and implant is present over less than half of the ingrowth surface. These conclusions are consistent with some reports of cementless femoral stems retrieved

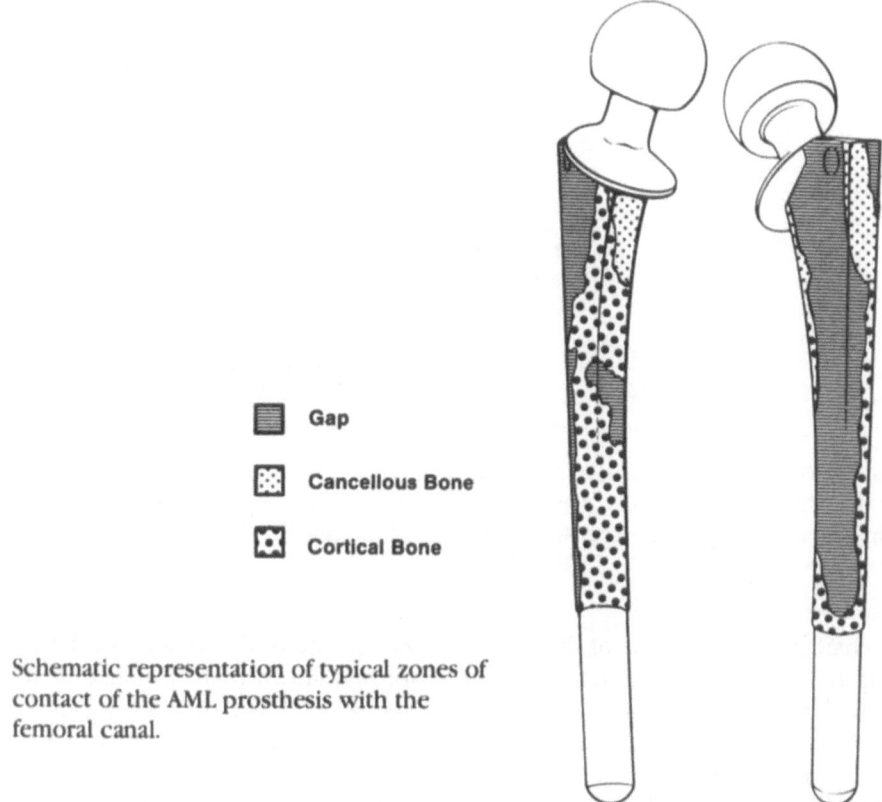

Gap

Cancellous Bone

Cortical Bone

Schematic representation of typical zones of contact of the AML prosthesis with the femoral canal.

Fig. 12. Schematic representation of typical zones of contact of an AML prosthesis with the femoral canal

at autopsy which show surprisingly little bony ingrowth into porous surfaces [25].

An overwhelming conclusion from this study is that cementless implants do not achieve a "hand-in-glove" fit within the femur. Nonetheless, for durable fixation, every implant must be supported by strong endosteal bone, despite the anatomic variations between femora. Moreover, the distribution of these areas of contact must be such that the multiaxial forces acting on the prosthesis are resisted without excessive micromotion. Research in this area is now focusing on the detailed geometry of the femur as stable fixation depends very critically upon our ability to obtain contact between the prosthesis and weight-bearing bone despite the anatomic variability of the medullary canal.

Femoral Anatomy and the Myth of Proportionality

In an attempt to systematically study the morphology of the femur, we have now examined over 800 specimens obtained from body donors. The intramedullary and extramedullary dimensions of the femur have been studied

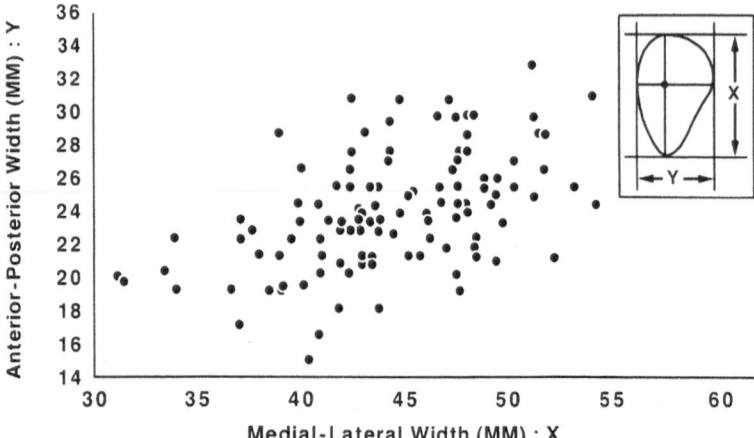

Fig. 13. The anterior-posterior and medial-lateral widths of the femoral metaphysis at the level of the femoral neck osteotomy. Notice the variability of the relative dimensions and hence the cross-sectional shape of the canal

to characterize the shape of the endosteal surface and to determine its relationship to overall bone size [26, 27]. This work has shown that the shape of the medullary canal is highly variable due to the adaptability of the form of the skeleton to its ultimate function. The dimensions of the femoral canal do not vary proportionally with the size of the bone. Individuals of the same stature have femora of similar lengths, but widely varying endosteal geometries. This is contrary to the previously held assumption that all femora were of one universal shape which could be scaled using standard anatomic dimensions to derive the canal geometries of larger and smaller bones [28]. In this sense, all femora were previously considered to be shaped proportionally.

In essence, the proportionality theory teaches that, as "normal" femora have one universal shape, implants of only one geometry, reduced or magnified in scale, will be needed to fit all bones in the normal anatomic range. This hypothesis can be tested by examining a cross-section through the femoral metaphysis at about the level of the proximal neck osteotomy (Fig. 13). It will be seen that a very large variation is present in the anterior-posterior and medial-lateral dimensions of the femur, leading to a weak correlation between the two [27].

The same considerations apply to the overall shape of the medullary canal in the AP and lateral views. For this purpose, we have developed a simple parameter called the canal flare index which expresses the overall shape of the medullary canal in the coronal plane (Fig. 14). This index is calculated by dividing the width of the femoral canal at 20 mm proximal to the center of the lesser trochanter by its width at the canal isthmus. The index can be used to identify three distinctly different shapes of medullary canals. Very straight or "stove-pipe" canals are observed when the canal flare index is less than 3.0,

Fig. 14. Diagrammatic representation of the canal flare index as determined from the standard AP radiograph

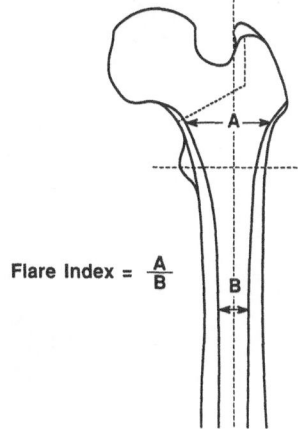

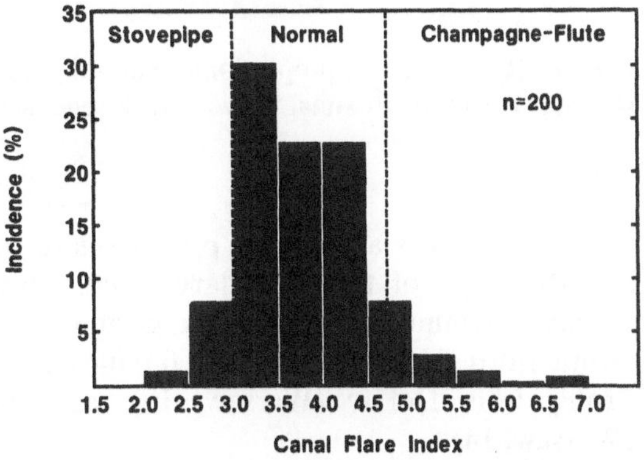

Flare Index = $\dfrac{A}{B}$

Fig. 15. Distribution of canal flare index across the normal population of femora. Zones corresponding to stove-pipe, normal and champagne fluted canals are shown

and are present in about 10% of the population. At the other extreme, highly flared or "champagne-fluted" bones have canal flare indices greater than about 4.7 and are also present in approximately 10% of cases. In between these two extremes lies the range of "normal" femora of typical radiographic appearance, with canal flare indices between 3.0 and 4.7 (Fig. 15).

Within the population, a continuous spectrum of femoral shapes is present, extending from the stove-piped to the champagne-fluted extremes. Although it is not possible to designate any one shape as "normal", some shapes are encountered more frequently than others within the population. This also confirms the conclusion that is is not possible to predict the dimensions of the medullary canal accurately by proportionally scaling one average shape to fit different sizes of femora. If this were the case, almost all bones would have the same canal flare index with only a few deviant cases with higher or lower values.

Despite these findings, many designers of femoral prostheses have adopted the proportionality theory to simplify the development of femoral stems as this

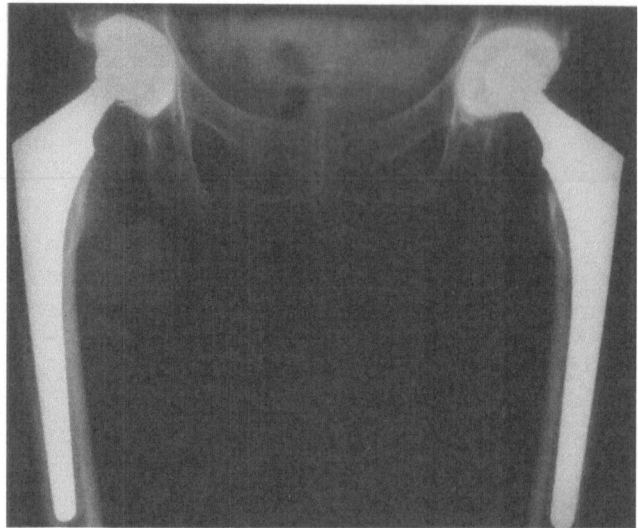

Fig. 16. Radiograph of proportional femoral prosthesis implanted in stove-pipe femora showing localized proximal fit with little endosteal contact distal to the level of the femoral neck osteotomy

approach enables all sizes of prostheses to be designed from one standard or average shape of the medullary canal. This concept has been utilized in a variety of commercially available designs including the ATS from Howmedica (Rutherford, NJ), the Omnifit/Omniflex from Osteonics (Allendale, NJ), the Profile from DePuy (Warsaw, In), and the Anatomic Hip from Zimmer (Warsaw, In).

Prostheses designed on the basis of proportionality are more flared than many femora encountered in joint replacement, particularly those with larger canals. This occurs because the predicted shape of the medial cortex is based upon the pooled average of all femora. Thus, on average, half the femora examined will be straighter and half will be more flared than a stem of average shape. The net result is some degree of mismatch in shape between the implant and the femur with limited contact between the prosthesis and the bony cavity (Fig. 16). In general, the medial surface of the prosthesis will be straighter or more curved than the medial cortex of any particular femur. Of particular concern are those cases in which the prosthesis is more curved that the femur; implantation of such a component can lead to increased risk of femoral fracture and impaired fixation (Fig. 17).

Another complication is canal impingement. Proportional stems increase in length in direct relationship with canal size; however, the distance from the lesser trochanter to the canal isthmus and to the anterior bow does not change as dramatically from small to large bones [27]. As a result, in larger femora, proportionally designed implants tend to be so long that the implant impinges upon the anterior cortex during implantation. This can undermine the proximal

Fig. 17. AP radiograph of a proportional femoral stem at the acute subsidence in a femur. Localized point contacts between stem and bone has resulted in a longitudinal fracture extending from the proximal femur to the canal isthmus, and distal to the tip of the stem

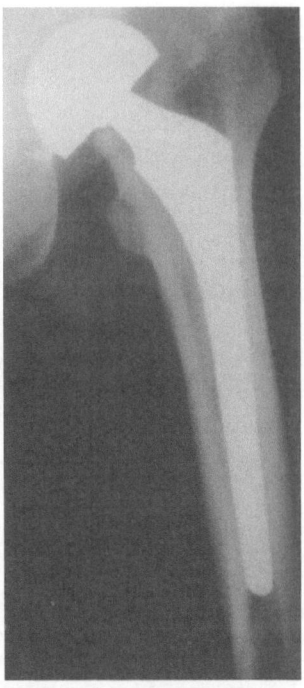

fit of the prosthesis and can cause fractures within the medullary canal (Fig. 18).

These observations have led to new methods for predicting femoral geometry and thus, in turn, to non-proportional designs of femoral stems. In these hip systems, each femoral component has been designed to best match the geometry of bones of a given size without any presumption that the shapes of implants of different sizes are geometrically related. This allows systems of eight or more stems to be created, each with its own unique shape.

Somatype Analysis of Canal Geometry

One of the most successful techniques for analyzing femoral geometry relies upon the concept of somatype analysis [27]. This is a mathematical technique in which the anatomic range of shapes and sizes of femora is expressed by a finite number of unique shapes called somatypes. It is possible to analyze a large number of femora and classify them into groups of similar shape, each approximating one somatype.

We have used this technique to classify a collection of 200 human femora. A set of 45 different somatypes was initially proposed to describe all possible shapes of femora in the normal population. However, after analysis of the fit of each somatype to the bones in our collection, only 17 of the possible 45 shapes were found to occur with a significant incidence. Starting from these 17 somatypes, it was then possible to reduce even further the number of

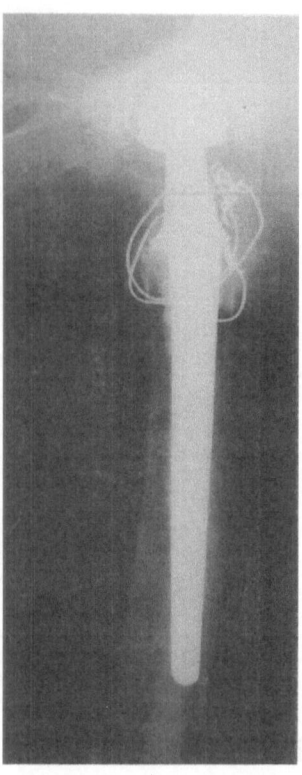

Fig. 18. Lateral radiograph of a proportional hip prosthesis cemented into the femur. The excessive length of the component has led to direct impingement between the stem and the anterior bone of the femur

somatypes required to describe the range of femora by varying the maximum error of fit permitted between any one bone and the closest somatype. For example, for an error of 2 mm, the number of different bone geometries was reduced to only 8, corresponding to a cementless hip system with 8 stems, each of a different geometry. Such a system would fit almost every bone encountered in clinical practice. If, however, the error of fit is increased to 3 mm, as in cemented hip replacement, the number of profiles can be reduced to the point where only 5 shapes can satisfactorily accommodate almost all human femora.

A further challenge to the fixation of cementless hip replacements is the variation in the proximal and distal dimensions of the femur. Several clinical studies have shown that simultaneous proximal and distal stabilization is important to minimize the incidence of symptoms following cementless hip replacement [29, 30]. In a recent study of the clinical performance of the AML cementless hip system, it was found that once the prosthesis filled more than 80% of the distal canal, clinical results that rated as only fair or poor were virtually eliminated [29]. However, simultaneous proximal and distal fill is extremely difficult to achieve with stems of fixed dimensions due to the variability of the proximal and distal dimensions of the medullary canal (Fig. 19). Thus, even if optimum stem fit is achieved in the proximal femur, the variation present within the isthmus prevents distal stabilization of many components. One novel approach to this problem is to use a modular stem with

Fig. 19. Distribution of measurements of the width of the medullary canal of the isthmus and the femoral neck osteotomy. For a given value of the proximal canal width, the width of the isthmus varies by ±4 mm. Conversely, the width of the canal isthmus is seen to provide almost no indication of the size of the proximal canal

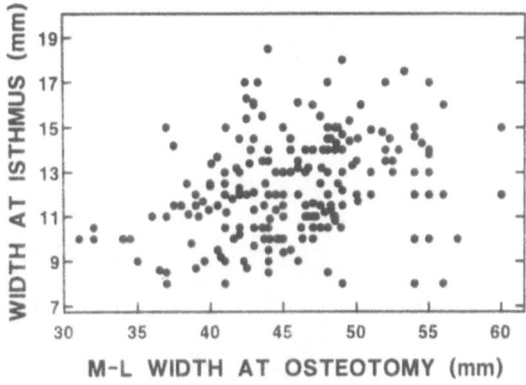

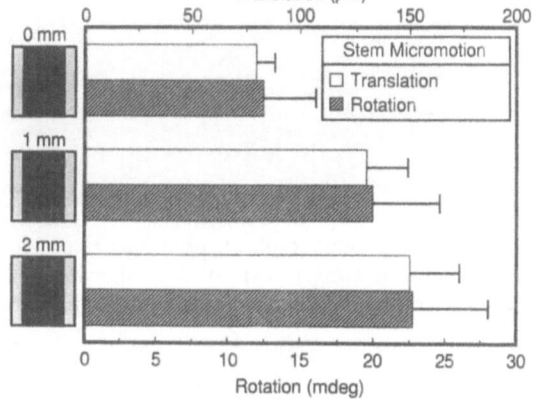

Fig. 20. The effect of distal canal fit on rotation and translational micromotion of the proximal stem/bone interface. The values for distal fit refer to the difference in size between the distal diameter of the prosthesis and the canal isthmus

a distal sleeve which allows independent fitting of the proximal and distal areas of the canal.

The importance of simultaneous proximal and distal fit is also reflected in the results of biomechanical measurements of micro-motion and migration of prostheses tested in the laboratory (Fig. 20) [23]. Experiments have been performed in which the size of the gap between the reamed femoral isthmus and the distal stem has been systematically varied through selection of centralizing sleeves of different diameters. These experiments showed that in the presence of constant proximal fit, a distal gap of only 1 mm between the bone and the implant led to an almost two-fold increase in the magnitude of the micro-motion of the proximal stem during cyclic loading. Similar results are also seen in rotational micro-motion which is thought to be the principal mechanism of aseptic loosening of cementless femoral prostheses.

One final consideration affecting the design and selection of hip prostheses is the profound influence of aging on the shape of the femoral canal [30–32]. As the body ages, the diameter of the isthmus of the femoral canal increases, primarily through reduction of cortical thickness, whereas the shape of the

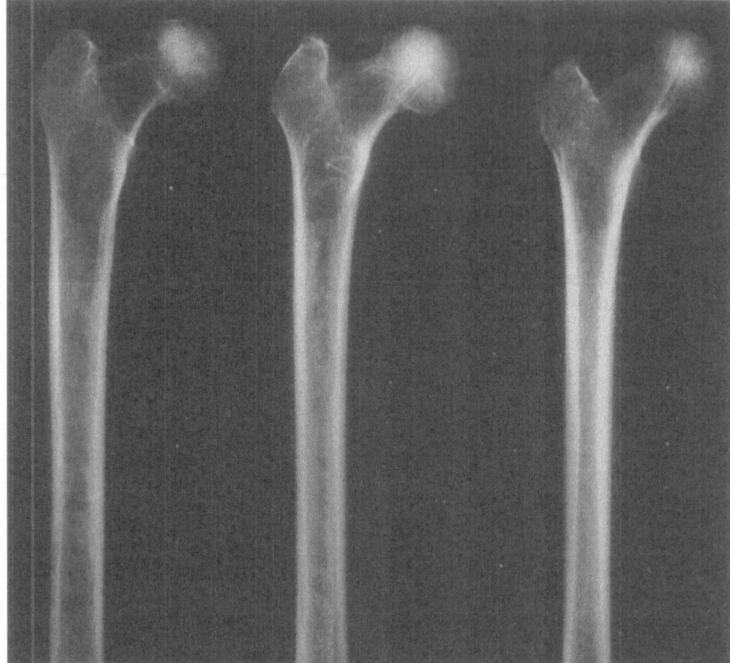

Fig. 21. An AP radiograph of three femora obtained of female donors of 80 (*left*), 65 (*center*), and 55 (*right*) years of age, showing marked differences of endosteal canal geometry associated with cortical thinning and diaphyseal expansion

metaphysis remains relatively constant. For this reason, aging causes the femur to become increasingly stove-piped, particularly in women of advancing years (Fig. 21) [32]. This variation of femoral shape has two significant implications for the selection of joint replacements. First, it is apparent that one single system of fixed stem geometry cannot meet the needs of both young and old patients. In other words, cementless hips designed for young people and cemented hips for old people cannot be accommodated by a single stem of one universal geometry. Second, implants which fit the proximal and distal femur in younger patients will only fit the proximal femur in older individuals, leaving the distal stem unsupported. Thus, some means of independent adjustment of the distal dimensions of the prosthesis is necessary to provide secure cement-less fixation.

Stem/Bone Interference

Another factor which profoundly affects the stability of cementless femoral stems is the presence of an interference fit between the cancellous bone of the metaphysis and the cementless prosthesis (Fig. 22) [23, 33]. This necessitates use of a prosthesis which is oversized compared with the corresponding

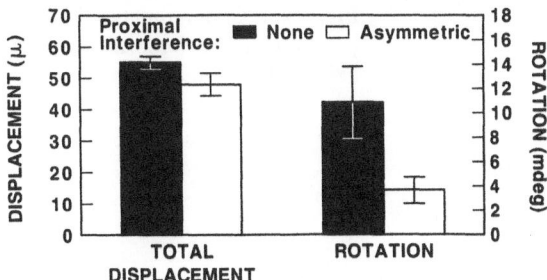

Fig. 22. The effect of oversizing the proximal section of the femoral component with respect to the size of the implantation cavity in terms of rotational and translational micromotion of the proximal interface. Two patterns of stem/bone interference have been studied: "line-to-line", where the broach and prosthesis are of identical shape, and "asymmetric", where the lateral, anterior and posterior aspects of the proximal implant are larger than the corresponding broach

broaching instrument which forms the cavity in the proximal femur. Although, in theory, the metaphyseal cavity formed by broaching is of exactly the same dimensions as the instrument itself, in practice, passage of the instrument enlarges the implantation site. Thus, when a prosthesis of exactly the same shape as the broaching instrument is implanted within the bone, gaps will inevitably be present at the interface. This prevents key areas of bone from providing any fixation regardless of the design of either the broach or the prosthesis. This problem may be minimized through use of a system in which the prosthesis is slightly larger than the corresponding broaching instrument. The prosthesis may be designed so that its interference with the implantation site varies with the elasticity of the bone at the interface. During implantation, the oversized component compresses the cancellous bone and fills many of the gaps left by broaching.

There are many factors which can undermine the fit and fixation achieved between the cementless femoral stem and each individual femur. Because of this inherent variability, selection of the cementless mode of implant fixation must be confirmed intraoperatively on the basis of the acute stability of the cementless prosthesis within the femur. Though few would claim that cementless fixation provides clinical results equal to cemented fixation, especially in the short term, new developments in cementless design and implantation techniques are reducing the incidence of many complications. It is hoped that with new innovations, cementless procedures will provide a long term solution to implant fixation with greater biological compatibility than has been possible in the past.

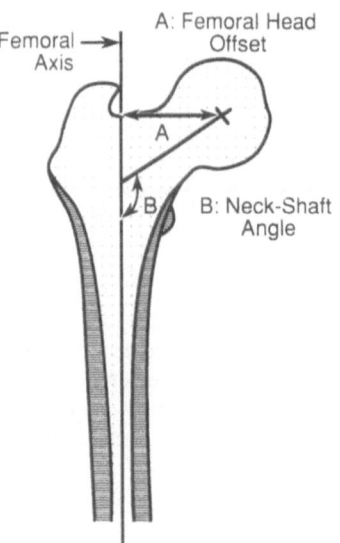

Femoral → Axis

A: Femoral Head Offset

A

B   B: Neck-Shaft Angle

Fig. 23. Diagrammatic representation of the position of the femoral head with respect to the axis of the medullary canal

Prosthetic Restoration of Femoral Head Position

In comparing pre-operative and post-operative radiographs of patients undergoing hip arthroplasty, a frequent observation is that the normal medial offset of the femoral head is not completely restored after surgery. The implications of this shift in head position are significant. Medialization of the femur reduces the mechanical efficiency of the abductor mechanism which, in turn, increases the abductor force necessary to provide balance to the hip joint. This leads to an increased joint reaction force and thus greater loading of the acetabular component. For this reason, there is increased interest in accurate restoration of the normal position of the femoral head after hip replacement.

In the coronal plane, the position of the femoral head is defined in terms of its medial offset from the femoral axis and its superior-inferior placement with respect to the lesser trochanter (Fig. 23). Enormous variation in head position in all of these directions has been observed in our studies of femoral morphometry. In the sagittal view, the variation in the anterior position of the femoral head is also considerable, corresponding to a range of anteversion of ±18 degrees. In the design of artificial joints, restoration of the normal joint is a sizable challenge because of the range of natural head positions. In the medial-lateral direction alone, a range of 35 mm is needed for adjustment of the position of prosthetic head for true anatomic restoration in virtually every femur (Fig. 24).

One way that this variability can be reduced is by varying the head offset with the size of the femur (Fig. 25). This reduces the range of anatomic variation to approximately 20 mm which, unfortunately, is still beyond the capacity of virtually every hip system currently available. A new approach to this problem is to have head offsets which vary with both the size of the

Fig. 24. Distribution of values of medial offset the femoral head

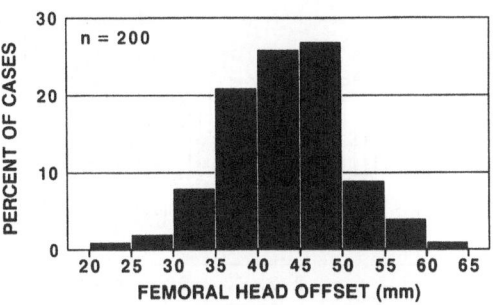

Fig. 25. The variation of medial head offset with the medial-lateral width of the medullary canal measured 20 mm distal to the center of the lesser trochanter

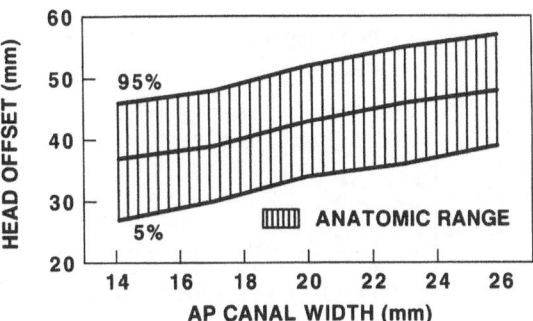

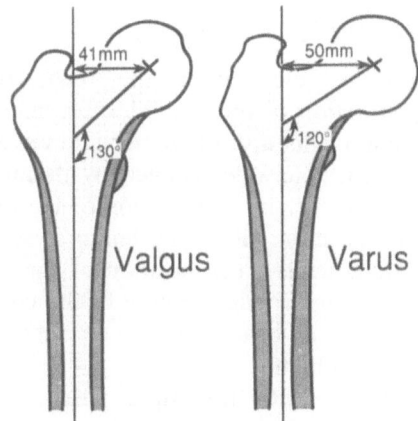

Fig. 26. The shape of the medial cortex is observed to be related to the neck/shaft angle of the femur and the medial offset of the femoral head

prosthesis and its medial curvature. This approach is based on the observation that femoral canals of a varus geometry tend to have large medial head offsets whereas in those with a valgus configuration the head offset is significantly smaller (Fig. 26). This strategy results in a set of implants consisting of valgus and varus somatypes with different neck/shaft angles and medial offsets. This enables much of the anatomic variation present within the normal bone population to be accommodated with a single implant system (Fig. 27).

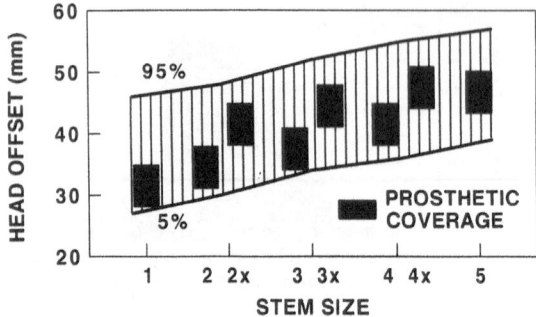

Fig. 27. Restoration of the normal medial offset of the femoral head is enhanced through the use of implants with varus and valgus somatypes with different ranges of head offset

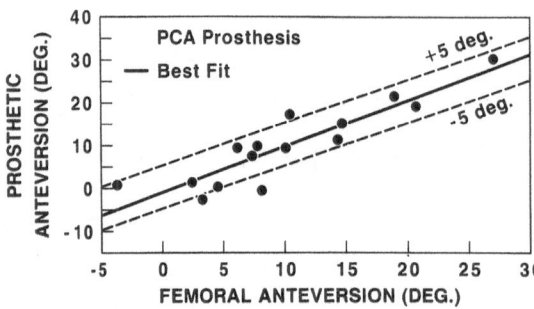

Fig. 28. The relationship between the anteversion of the PCA prosthesis and the original femur for 15 bones of widely varying orientation. In most cases, the natural anteversion is restored to within 5 degrees

Prosthetic Anteversion

Another element in the reconstruction of normal hip biomechanics is restoration of the natural anteversion of the proximal femur. As few femoral components are inherently anteverted, the anterior position of the prosthetic head is solely determined by its medial offset and the rotational position of the prosthesis in the femur. However, each prosthesis generally has only one position within the femur in which it achieves stable cementless fixation. Consequently, the position of the prosthetic head is primarily a function of the geometric distribution of the contact areas between the prosthesis and the femur and bears no direct relationship to the position of the head in the original bone.

We recently investigated this subject using a range of contemporary hip designs, two curved and one straight, in an attempt to determine how much normal anteversion remains following hip replacement. The results demonstrated that with each system, there was only a small difference between the average anteversion of the natural and prosthetic heads (Fig. 28). However, there was considerable variability of head position from femur to femur regardless of the design of the implant. In some instances, the femoral head was displaced by almost 10 mm posterior to its original position (Fig. 29). Surprisingly, there was no statistically significant difference ($P > 0.05$) in restoration of femoral head position between the conventional straight stem and an anatomically shaped stem with an anteverted neck.

Fig. 29. Average values of femoral anteversion, defined with reference to radiographic markers in the lateral trochanter, for a range of cementless prostheses with respect to the original femur

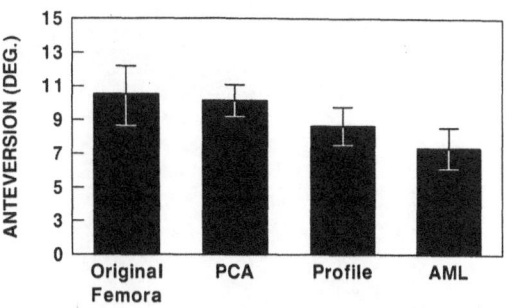

At this time, the only means available to overcome these deficiencies are modular and custom hip systems which allow the anteversion of the prosthetic head to be varied independently of the rotational position of the prosthesis within the femur. These approaches hold great promise for future use, especially in the context of revision hip replacement.

Anatomical Considerations in Acetabular Cup Design

In many parts of the world, the disturbing incidence of radiographic loosening of cemented acetabular cups has lead to the widespread adoption of cementless fixation in the pelvis. In some series this has led to some very disappointing results, particularly in the case of threaded components. Moreover, as cemented cups generally loosen late and generally fail by mechanisms not related to acrylic cement, it is conceivable that cementless components will also suffer failure of fixation in the long-term, principally through osteolysis generated by polymeric wear debris. Nonetheless, short-term results obtained using a variety of hemispherical cup designs, with and without transfixion screws, have been extremely encouraging, even though bony ingrowth into these components has not been reliable.

One of the key questions in the selection of cementless acetabular components is how to achieve fixation at surgery to maximize the chances of stable biological fixation. Currently, two methods are commonly used. One is mechanical interference which is achieved by implanting a shell 1–2 mm larger in diameter than the reamed acetabulum. The second is to achieve compression of the shell/bone interface with lag screws which are passed through the shell and into the reamed bony bed.

Although the use of screws may seem to be an obvious solution to guarantee the acute stability of cementless acetabular components, the use of these devices has resulted in a small, but tragic, incidence of neurovascular complications. These have occurred secondary to performation of vital structures within the pelvis by screws which have been passed through the inner pelvic wall in an attempt to obtain bicortical fixation. As a result, we have examined in some detail the intercortical thickness of bone available

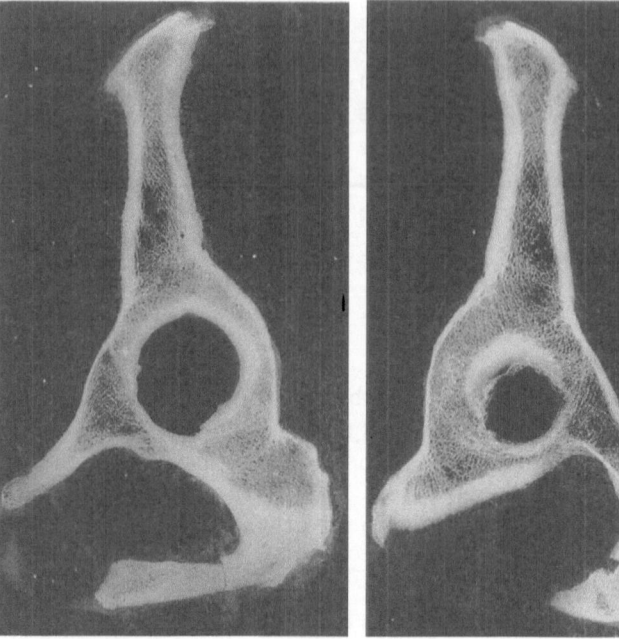

Fig. 30. Contact radiograph of a section taken through the pelvis, parallel to the acetabular rim at approximately 1 cm medial to the mouth of the acetabulum. The bone of the ischium visible at the bottom of the image, is significantly stronger than the weak cancellous structure of the pubis and ilium

Fig. 31. Contact radiograph of a slice taken near the floor of the acetabulum. Strong cancellous bone is visible posterior to the subchondral plate. Bone within the pubis and ilium is seen to be weak and of low density

around the acetabulum, the distribution of cancellous bone strength and the anatomic orientation of the neurovascular structures within the pelvis.

The Distribution of Bone Density

The distribution of pelvic bone density can be assessed through inspection of a series of sections taken through the pelvis parallel with the mouth of the acetabulum (Figs. 30, 31). Through use of this method, contact radiographs can be generated which can be subsequently digitized and computer enhanced to indicate areas of varying bone density. Examination of bone slices from the mouth of the acetabulum through to the sacroiliac joint reveals that the strongest bone stock is present beneath the acetabular rim and within the body of the ischium. Conversely, the cancellous bone within the ilium, immediately superior to the acetabulum, is of relatively low density and is less suitable for screw fixation. The bone within the public ramus is generally of low density, corresponding to inferior pull-out strengths of cancellous screws in mechanical testing.

By taking slices which penetrate deepter into the pelvis, it has been observed that very strong cancellous bone is also present within acetabular dome supermedial to the acetabulum due to the coalescence of the principal bony arcades of the pelvis. This is an ideal place for screw placement, given its orientation with respect to the joint reaction force. However, this zone is relatively small in comparison with the diameter of the acetabulum itself and so can be difficult to find on a reliable basis. It also requires longer screws which become more susceptible to large bending loads and the risk of mechanical failure. Our studies of the cancellous morphology of the pelvis indicate that the suitable positions for screw placement within the acetabulum occupy a relatively small area of the available surface of the shell. This is in contrast to the fact that many acetabular cups in common use come with a large assortment of hole positions which, in some cases, may compromise the remaining area available for porous ingrowth. Another complication of these designs may be long-term distortion of the bearing liner due to incomplete support of its concave surface.

The Location of Neurovascular Structures

A subject of great significance to the practicing orthopedic surgeon is the precise location of vital neurovascular structures within the pelvis beneath the acetabulum. This issue has been addressed in several studies within the laboratory. The first involved detailed examination of transverse sections of cadavers taken at the level of the acetabulum and the femoral head. These sections demonstrate the close proximity of the medial wall of the acetabulum to the bladder and the presence of the external iliac artery and vein behind the anterior edge of the acetabular surface. The obturator artery, vein and nerve are also found almost directly medial to the center of the femoral head as they pass down behind the quadrilateral surface through the obturator foramen.

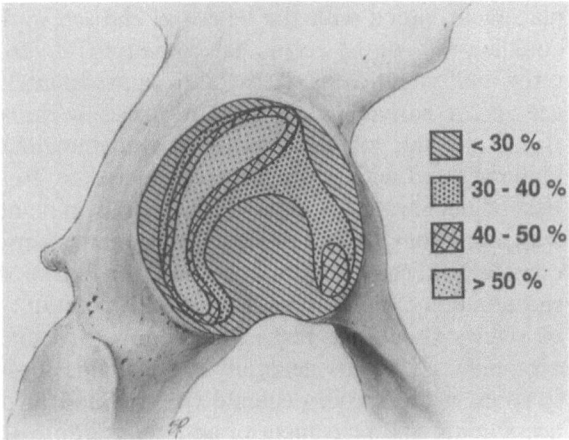

Fig. 32. Distribution of intracortical thickness of the pelvis for screws passing perpendicular to the acetabular surface. All thicknesses are normalized with respect to the diameter of the accetabulum

< 30 %

30 - 40 %

40 - 50 %

> 50 %

A second approach to visualization of these structures has been to over-ream the acetabulum within a cadaver and then to remove, by careful dissection, all of the periacetabular bone, leaving only the columns of the ilium superiorly, the pubis anteriorly, and the ischium inferiorly. This procedure exposes the neurovascular structures which lie immediately beneath the acetabulum and which would be encountered by the penetration of a bone screw. In such a preparation, the femoral artery and vein are found to be prominent and pass medial to the pubis. The femoral nerve and the obturator artery, vein and nerve pass laterally. A large proportion of the field of view is occupied by these vital structures, suggesting that the margin of error in screw placement is extremely small. This is particularly apparent for screws which attempt to gain purchase within the pubic ramus which becomes very narrow beneath the acetabular surface.

Acetabular Morphology

To guide the placement of screws within the acetabulum, we have also undertaken a morphological study of the distribution of thickness between the cortices of the pelvis as a function of the position of screw placement on the acetabular surface. Eight specimens spanning the anatomic range of the acetabular diameter were embedded in foam and divided into 16 pie-shaped slices. The outline of each transected slice was incorporated into a geometric model which simulated the path of bone screws passing through the inner and outer cortices at right angles to the surface of the acetabular shell. By analyzing each slice, we determined the maximum length of a screw which would engage cancellous bone without penetration of the inner wall of the pelvis (Fig. 32). By repeating the process for all bone slices, a three-dimensional map of the surface of the acetabulum was developed to guide the length and placement of screws used for acetabular fixation.

For the case of screws oriented at right angles to the shell (i.e. passing through the geometric center of the acetabular surface), the zones of safe screw placement varied with the length of the screw. When the length of the screw equalled 40% of the acetabular diameter (i.e., 16–24 mm), positions for "safe" screw placement were restricted to approximately one quarter of the available acetabular surface, almost entirely within the posterior half of the socket. Based on this observation, it is recommended that acetabular shells be manufactured with screw holes restricted to areas known to provide minimal risk of neurovascular injury. In existing implants which provide the surgeon with numerous holes for screw placement, screws should be positioned with considerable care, avoiding the anterior half of the acetabular surface and the rim and dome on the posterior side. In addition, whenever possible, the length of screws should be restricted to only 40% of the diameter of the reamed acetabulum. As this generally corresponds to short screws providing limited fixation, acute stability should be obtained through accurate reaming of the acetabulum and placement of an oversized prosthetic component.

Conclusions

Hip arthroplasty continues to advance through due recognition of the lessons of the past and continuing expansion of the visions of the future. There is no doubt that hip replacement, when performed in the older patient, has been one of the most successful procedures of modern surgery. However, we need to extend this procedure to meet the needs of the young, the active and the patient undergoing revision of a prior hip replacement. We now know many of the weaknesses and limitations of this procedure but still need to perfect solutions and improvements. Advances are ongoing in fixation, biomechanics, and in the restoration of human anatomy. However, developments are still awaited in the biocompatibility of implant materials, improvements in wear resistance of articulating surfaces and methods to preserve bone stock in the face of rigid implant fixation. We have now passed the era when an excellent outcome was the sole province of the extraordinary surgeon and his fortunate patient. Today, the challenge to the surgeon, the designer and the engineer is to bring enduring restoration of normal hip function within the grasp of every surgeon and the expectations of every patient. Then we will have truly succeeded.

References

1. Huddleston HD (1988) Femoral lysis after cemented hip arthroplasty. J Arthroplasty 3:285–297
2. Anthony PP, Gie GA, Howie CR, Ling RSM, Perkins RD (1990) Localised endosteal bone lysis in relation to soundly fixed femoral components of cemented tatal hip replacements: A possible mechanism. J Bone and Joint Surg 72B:532
3. Maloney WJ, Jasty M, Burke DW, O'Connor DO, Zalenski EB, Bragdon C, Harris, WH (1989) Biomechanical and histologic investigation of cemented total hip arthroplasties: A study of autopsy-retrieved femurs after in vivo cycling. CORR 249:129–140
4. Vives P, de Lestang M, Jarde O, Decoopman M (1987) Interest du contact direct entre la tige femorale et I'os diaphysaire dans les prostheses totales cimentees. Rev Chir Orthop [Suppl] 2:218–220
5. Maloney WJ, Jasty M, Harris WH, Galante JO, Callaghan JJ (1990) Endosteal erosion in association with stable uncemented femoral components. JBJS 72A: 1025–1034
6. Noble PC, Tullos HS, Landon GC (1991) The optimum cement mantle for total hip replacement. American Academy of Orthopaedic Surgeons, Park Ridge, Illinois (Instructional Course Lectures, vol XL)
7. McNiece GM, Amstutz HC (1976) Finite element studies in hip reconstruction. In: Komi PV (ed) Proceedings of the 5th International Congress on Biomechanics, University Park Press, Baltimore, pp 339–405
8. Ahmed AM, Raab S, Miller JE (1984) Metal cement interface strength in cemented stem fixation. J Orthop Res 2:105–118
9. Huiskes R (1980) Some fundamental aspects of human joint replacements: Analysis of stresses and heat conduction in bone-prosthesis structures. Acta Orthop Scand 185:109–200

10. Krause WR, Krug WH, Miller JE (1982) Strength of cement-bone interface. Clin Orthop 163:290–299
11. Crowninshield RD, Tolbert JR (1983) Cement strain measurement surrounding well-fixed femoral component stems. J Biomed Mater Res 17(5):819–828
12. Noble PC, Scheller AD, Tullos HS, Levy RN, Turner RH (1987) Applied design criteria for total hip prosthesis. In: Stillwell T (ed) The art of total hip arthroplasty. Grune and Straton, Orlando, pp 51–68
13. Beckenbaugh RD, Ilstrup DM (1978) Total hip arthroplasty: A review of three hundred and thirty-three cases with long-term follow-up. J Bone Joint Surg [Am] 60:306
14. Ling RSM (1980) Prevention of loosening of total hip components. In: Mosby CV (ed) The hip Proceedings of the Eighth Open Scientific Meeing of the Hip Society. St Louis, pp 292–307
15. Miller J, Johnson JA (1987) Advances in cementing techniques in total hip arthroplasty. In: Stillwell T (ed) The art of total hip arthroplasty. Grune and Straton, Orlando, pp 277–291
16. Noble PC, Hammerman SM, Alexander JW, Green RB, Marltry JA, Tullos HS (1987) Innovations in cementing techniques in total hip replacement. Scientific Exhibit at the American Academy of Orthopaedic Surgeons 54th Annual Meeting, San Francisco
17. Davies JP, Jasty M, O'Connor DO, Burke DW, Harrigan TP, Harris WH (1989) The effect of centrifuging bone cement. JBJS 71B:39–42
18. Burke DW, Gates EI, Harris WH (1984) Centrifugation as a method of improving tensile and fatigue properties of acrylic bone cement. JBJS, 66A:1265–1273
19. Levy RN, Noble PC, Scheller A, Tullos HS, Turner R (1988) Prolonged fixation of cemented total hip replacement. Surgical Rounds for Orthopaedics, pp 15–22
20. Sarmiento A, Gruen TA (1985) Radiographic analysis of a low modulus titanium-alloy femoral total hip component. JBJS 67A:48–56
21. Llinas A, Ebramzadeh E, McKellop H, Gogan W, Sarmiento A (1991) Femoral component size and position in cemented total hip prostheses. In: Proceedings of the American Academy of Orthopaedic Surgeons 58th Annual Meeting, Anabheim, California, March 7–12
22. Scheller A, Levy RN, Noble PC, Turner R (1989) A comparative analysis of total hip component position and cement technique. In: Proceedings of the American Academy of Orthopaedic Surgeons 56th Annual Meeting, Las Vegas, Nevada; February 9–14
23. Noble PC, Kamaric E, Alexander JW, Tullos HS (1989) What makes cementless implants work? Scientific Exhibit at the American Academy of Orthopedic Surgery 56th Annual Meeting, Las Vegas, Nevada
24. Noble PC, Alexander JW, Granbery ML, Granberry WM, Maltry JA, Tullos, HS (1988) The myth of "press-fit" in the proximal femur. Scientific Exhibit at the American Academy of Orthopaedic Surgeons 55th Annual Meeting, Atlanta, Georgia
25. Cook SD, Barrack RL, Thomas KA, Haddad RJ (1988) Quantitative analysis of tissue growth into human porous total hip components. J. Arthroplasty 3:249–262
26. Noble PC (1990) Proximal Femoral Geometry and the design of cementless hip replacements. Orthop Rel Sci 1:86–92
27. Noble PC, Alexander JW, Lindahl LJ, et al. (1988) The anatomic basis of femoral component design. Clin Orthop 235:148–165

28. Averill RG, Pachtman N, Jaffe WL (1980) A basic dimensional analysis of normal human proximal femora. In: Proceedings of the Eighth Annual Northeast Bioengineering Conference, Cambridge, Massachusetts, pp 352–356
29. Granberry ML, Noble PC, London GC, Tullos HS (1989) Factors influencing the clinical results of cementless femoral stems in total hip arthroplasty. In: Proceedings of the American Academy of Orthopaedic Surgeons 56th Annual Meeting, Las Vegas, Nevada; February 9–14
30. Ericksen MF (1979) Aging changes in the medullary cavity of the proximal femur in American blacks and whites. Am J Phys Anthropol 51:563
31. Ruff CB, Hayes WC (1982) Subperiosteal expansion and cortical remodeling of the human femur and tibia with aging. Science 217:945
32. Smith RW, Walker RR (1964) Femora expansion in aging woman: Implications for osteoporosis and fractures. Science 145:156
33. Manley MT, Capello WN, Averill RG, Cohen R (1990) Effect of stem design parameters, stem fit, and bone quality on the torsional stability of femoral stems. In: Proceedings of the American Academy of Orthopaedic Surgeons 57th Annual Meeting, New Orleans, Louisiana, February 8–13

Knee Joint

Biomechanics of Knee Replacement

JOHN J. O'CONNOR[1] and JOHN W. GOODFELLOW[2]

Introduction

The development of knee replacement over the past 30 years has been quite different to that of hip replacement. All hip prostheses use the ball and socket geometry of the natural joint and development has concentrated on the design of stems and on methods of fixation of components to the bone. In contrast, the optimum design of the articular surfaces of knee prostheses is still controversial and many fundamentally different designs are used in current practice. In this chapter, we review the range of designs of knee prostheses and classify them according to the extent that they use and rely on retained ligaments. The review is introduced by a description of the functions of the ligaments of the knee. The chapter concludes with a discussion of the design of surface-replacement knee prostheses which allow the retention of all the ligaments, and the need for meniscal bearings in such designs is explained.

Geometry and Mechanics of the Hip

The center of the hip joint lies at the center of the femoral head. This is the point about which the two bones move relative to each other. The geometric center of the joint is not only its center of rotation, but also its mechanical center. Figure 1b shows an external load W pointing medial to the joint, balanced by a muscle force T acting lateral to the joint. The level-arm of the load and the lever-arm of the muscle are measured from the geometric center of the joint. The moments of the two forces (lever-arm length times force magnitude) must balance; the reason why muscle forces are quite often larger than the external loads is because their lever-arms at the joints are shorter [1].

---

[1] Department of Engineering Science, Parks Road, Oxford, OX3 3PJ, UK
[2] Nuffield Orthopaedic Centre, Headington, Oxford, OX3 7LD, UK

Fig. 1a,b. Coronal view of the leg with a load W applied through the foot and passing **a** lateral or **b** medial to the hip. Muscle forces T resist **a** abduction and **b** adduction at the hip. The forces W and T are balanced by the contact force N acting on the articular surface of the femoral head. The lever-arms available at the hip to the load, b_w, and to the muscle force are the perpendicular distances from the centre of the femoral head to the line of action of each force. Muscle forces are often larger than the loads because their lever-arms are shorter; the corresponding contact forces are even larger

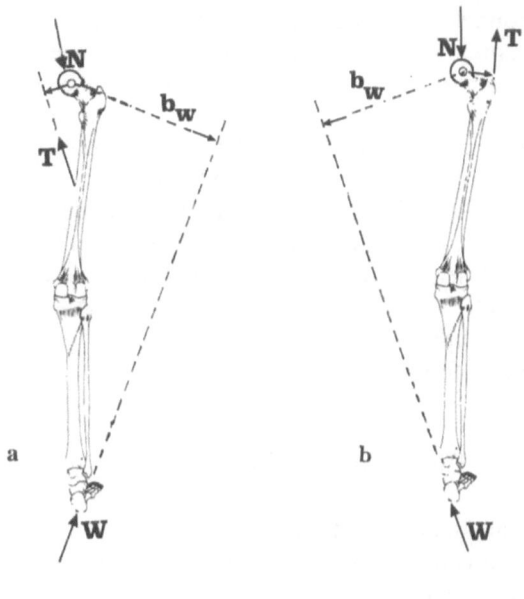

The geometric center is therefore the mechanical center, the fulcrum about which the external loads and the muscle forces exert their leverage.

Design of Hip Prostheses

Hip prostheses used in current practice differ from each other in a number of ways. They use different materials. The femoral stems have different shapes. The radius of the femoral head differs from design to design. The acetabular component may be metal-backed. Some of the components are porous-coated. However, they have one thing in common. They all use a spherical ball and socket, like the articular surfaces of the natural joint. They might then be expected to reestablish the natural movements of the bones on each other and the natural lever-arms of the muscles, giving a reasonably physiological functional result.

Design of Knee Prostheses

Knee replacements in common use today also differ from each other in materials, in the use of metal backing and porous coating, and in the design of fixation stems, but they differ more fundamentally in the shapes of their articular surfaces. There is no common agreement about the correct shapes of the articular surfaces, if indeed there is a single set of correct shapes.

The range of available designs probably reflects different interpretations of the complex geometry of the surfaces of the natural joint. The femoral condyles which make contact with the tibia are spherical [2] with definable centers of curvature. In the sagittal plane, the medial tibial plateau is slightly concave whereas the lateral plateau is slightly convex. There is therefore no single geometric joint center which can be located on an X-ray picture.

We will show that the instantaneous centre of the knee, the point about which the bones flex and extend relative to each other in the sagittal plane, is defined by the geometry of the ligaments and that it moves relative to both bones during flexion/extension movements. For this reason, the ligaments play an essential role in controlling the active as well as the passive stability of the joint. We will show that the fundamentally different types of knee prosthesis currently available differ from each other in the use they make of the retained ligaments.

The Ligaments of the Knee

The femur and tibia are connected by the cruciate ligaments in the intercondylar region and by the collateral ligaments, the lateral collateral and the medial collateral, along the sides of the bones. The articular surfaces hold the bones apart and resist interpenetration by transmitting compressive stresses across their interfaces. The ligaments hold the bones together and resist distraction by transmitting tensile stresses along the lines of their fibers. The ligaments limit the movements of the bones on each other and control bone movements within those limits. Within the limits of motion, they collaborate with the muscles and articular surfaces in transmitting load across the joint.

Ligaments Define Motion Limits of the Bones

It is usual to emphasize the role of the ligaments in limiting the movement of the bones upon each other, in defining the range of passive movement, and in controlling the passive stability of the joint. In the simplest terms, the anterior cruciate ligament (ACL) resists anterior subluxation of the tibia, the posterior cruciate ligament (PCL) resists posterior subluxation, the medial collateral ligament (MCL) resists abduction, the lateral collateral ligament (LCL) resists adduction, the posterior capsule resists hyperextension. However, this description is obviously far too simple. The ligaments act together in groups to limit these various motions. For instance, all the ligaments act together to limit distraction of the joint. The cruciate ligaments act with the collateral ligaments in limiting abducting or adducting movements (Fig. 2) and in limiting long axis rotation of the tibia. Daniel and Stone [3] have summarized the large body of experimental evidence defining the primary and secondary ligamentous restraints to various types of movement.

Movement of the bones beyond the motion limits is resisted by progressive tightening of the ligaments. The tension forces set up in the ligaments to limit movement are balanced by compression forces between the articular surfaces of the bones.

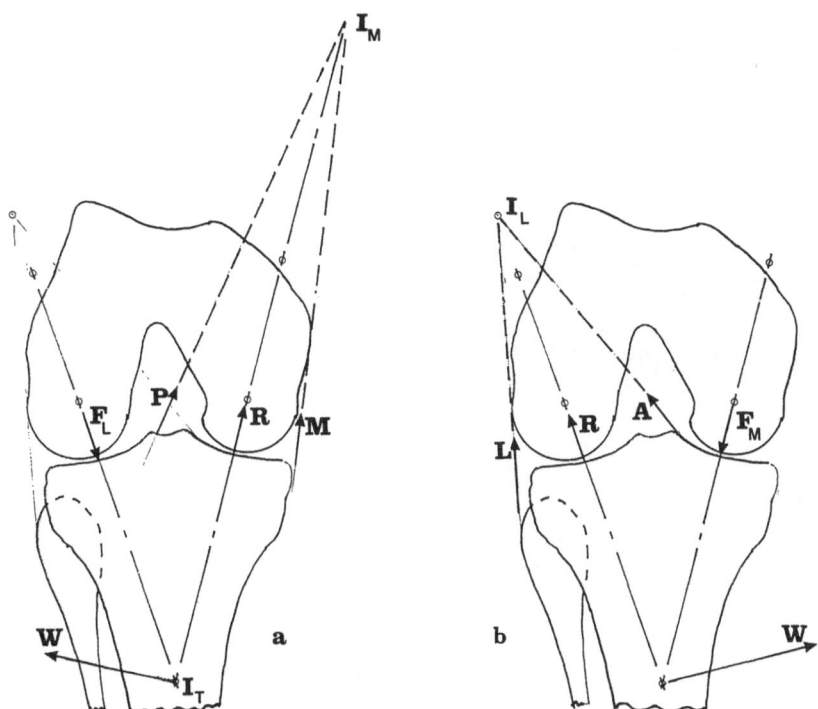

Fig. 2a,b. Coronal view of the knee with a load W pushing the tibia **a** into valgus and **b** into varus. Contact forces **a** F_L in the lateral compartment and **b** F_M in the medial compartment resist interpenetration of the bones. Lift-off of the opposite compartment is resisted by a force R, the resultant of forces **a** in the posterior cruciate (P) and medial collateral (M) ligaments and **b** in the anterior cruciate (A) and the lateral collateral ligaments (L). I_M, intersection of P and M; I_L, intersection of A and L; I_T, intersection of F_M and F_L

Ligaments Control Movement Within the Motion Limits

Whereas the ligaments tighten to offer increasing resistance to movement of the bones beyond their motion limits, they rotate isometrically about their points of origin and insertion on the bones to allow movement within those motion limits. These actions can be explained by reference to a computer model of the joint.

The ligaments and the bones form a mechanical linkage or mechanism which guides the movements of the bones upon each other. A simplified version of the linkage is the four-bar cruciate linkage ABCD shown in Fig. 3 [4, 5]. Two links represent the anterior cruciate ligament, AB, and the posterior cruciate ligament, CD, and the two links, BC and AD, join their points of attachment on the femur and on the tibia. Changes in the angle of flexion of the joint result in equal changes in the angle between the femoral link BC and the tibial

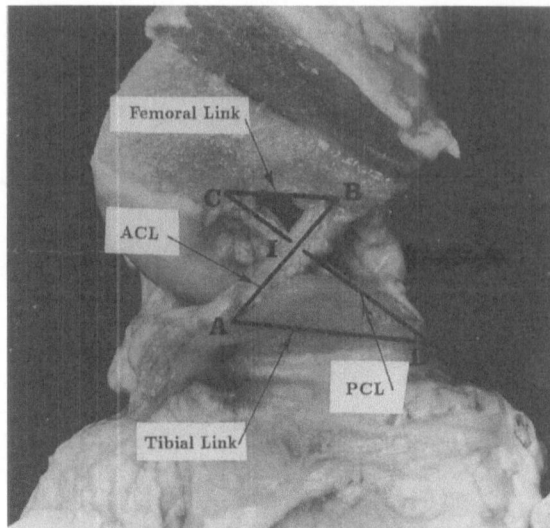

Fig. 3. Section through a human knee, exposing the cruciate ligaments with a diagram of the four-bar cruciate linkage *ABCD* superimposed. *ACL*, anterior cruciate ligament; *PCL*, posterior cruciate ligament

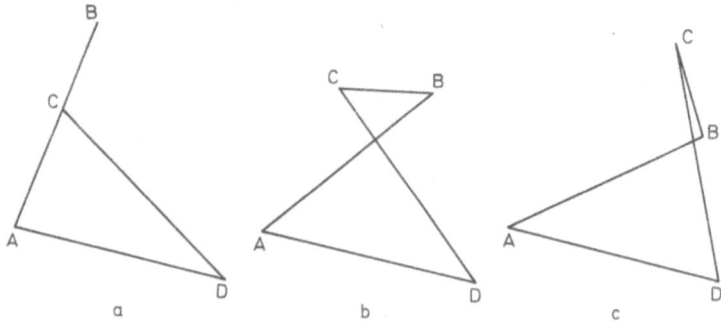

Fig. 4a–c. The cruciate linkage **a** at full extension, **b** at 70°, and **c** 140° of flexion

link AD. Figure 4 shows the linkage at extension, 70°, and 140° of flexion. The shape of the linkage changes as the ligaments rotate about their attachment points. The three diagrams of Fig. 4 are shown superimposed in Fig. 5. With the femur flexing and extending on a fixed tibia (Fig. 5a) the femoral attachments B and C of the cruciates rotate on circular arcs about their tibial attachments, A and D respectively, while the ligament links AB and CD remain isometric. Similarly, in Fig. 5b, with the tibia flexing and extending on a fixed femur, the tibial attachments, A and D rotate isometrically about their femoral attachments, B and C, respectively.

   The Instantaneous Center of the Linkage. The instantaneous center of the linkage lies at the point I (Figs. 3, 5) at which the ligament links cross. The flexion axis of the joint, about which the bones flex and extend, passes through I. Because the geometry of the linkage changes during flexion and extension, the instantaneous center moves backwards and forwards relative to the two

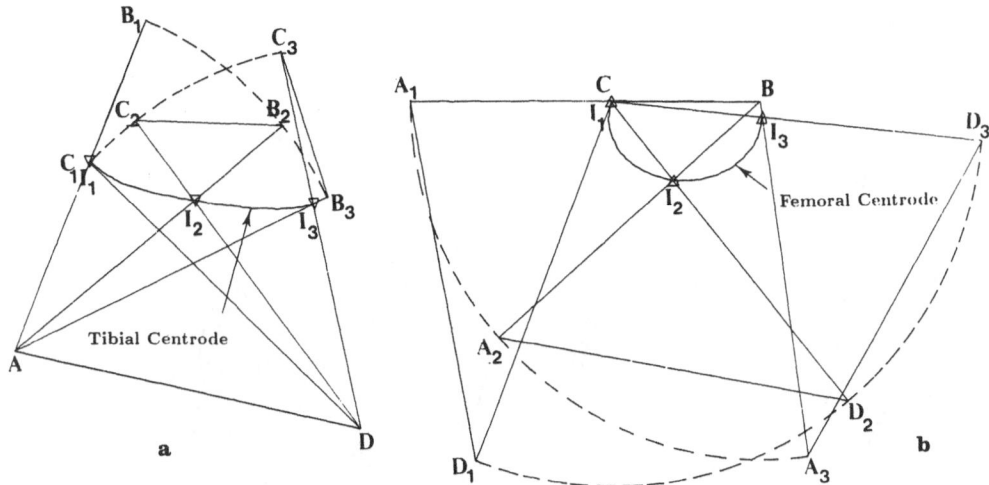

Fig. 5a,b. The three figures of Fig. 4 superimposed **a** with the tibial link AD fixed and **b** with the femoral link BC fixed. The instant center of the linkage I moves from position I_1 to I_3 along the tibial centrode in **a** and along the femoral centrode in **b**. The paths of the instant center relative to the two bones differ

bones. It moves along the curve marked "tibial centrode" (Fig. 5a) relative to the tibia and along the curve marked "femoral centrode" (Fig. 5b) relative to the femur. As a result, the femur rolls as well as slides on the tibial plateau. Similar rolling movements in the human knee were described by W.E. Weber and E.F.W. Weber in 1836 [6].

In contrast to the hip, the centre of the knee is determined not by the geometry of the articular surfaces but by the geometry of the ligaments. Unlike the ball and socket hip joint, the centre of the knee moves relative to both bones.

An elaboration of the computer model is shown in Fig. 6 demonstrating the rolling movements of the femur on the tibia. In addition to the cruciate linkage, the model now includes representations of the femur, the tibia, and of the quadriceps, hamstrings, and gastrocremius muscles. The tibial plateau of the model is flat. The shape of the corresponding femoral condyle is nearly circular. The contact point between the bones follows the instantaneous center and moves backwards during flexion and forwards during extension.

Relation between Ligament Geometry and Articular Surface Shape. The problem which confronts the designer of a surface replacement knee prosthesis is to choose surface shapes which are compatible with ligament geometry. In terms of the model in Figs. 3–6, compatible surface shapes allow flexion and extension while maintaining the links of the cruciate mechanism isometric. If one chooses the shape of one of the articular surfaces arbitrarily, what should be the compatible shape of the other surface?

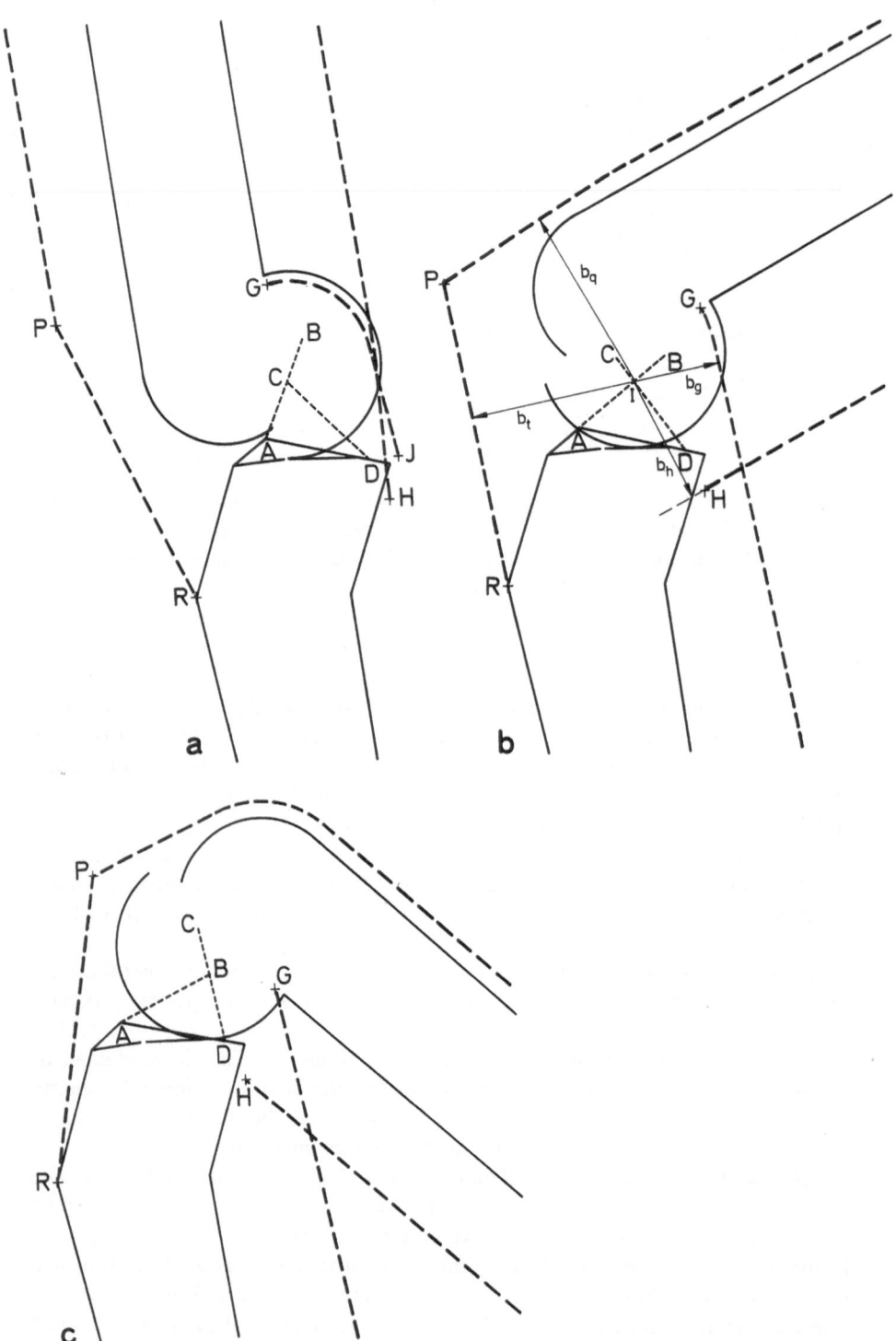

Fig. 7. Determination of the shape of the femoral condyle compatible with ligament geometry and a flat tibial plateau. Points on the femur move on circular arcs about the instant center I relative to the tibia so that, during flexion, a contact point F_1 on the femur would interpenetrate the tibia or lead to ligament stretch, whereas a contact point F_2 would lead to separation of the bones or slackening of the ligaments. For continuous contact, the bones must touch at F where the perpendicular to the tibial plateau passes through I. Calculation of the position of F relative to CB gives the shape of the femoral condyle

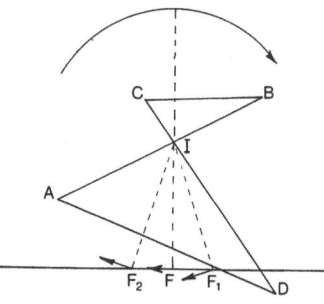

The theoretical solution to the problem is illustrated in Fig. 7 where the shape of the tibial surface has been chosen to be flat and the shape of the femoral surface is sought. An alternative statement of the problem is to ask where the compatible femoral surface should touch the chosen tibial surface in Fig. 7. If the position on the tibia of the contact point can be found, its position relative to the femoral link BC can be calculated. When this is done over the flexion range, the shape of the compatible femoral condyle is obtained. In Fig. 7, contact must occur at the point F where the perpendicular to the tibial plateau passes through the instant center. Since all points on the femur rotate in circles centered at the instant center, only the point F on the femur can move tangential to the tibial plateau, without separation or interpenetration of the surfaces and without slackening or tightening of the ligaments.

The shape of the femoral surface in Fig. 6 above was calculated on this basis. It is very similar to a circle [4]. Figure 8 shows sections through the medial and lateral femoral condyles of a human knee, with circles superimposed on the posterior aspects of the femoral condyles. The circles fit the shapes of the

---

Fig. 6a–c. Models of the femur and the tibia, connected by the cruciate linkage $ABCD$ and spanned by the quadriceps, hamstrings, and gastrocnemius muscles, **a** at extension, **b** at 70°, **c** at 140° of flexion. The patellar tendon is inserted into the tibia at R and meets the quadriceps tendon at P, the centre of force of the patella. The hamstrings tendon inserts into the tibia at H. Both quadriceps and hamstrings lie parallel to the femur but, at high flexion angles **c**, the quadriceps tendon wraps around the front of the femur to form the "tendo-femoral" joint. Gastrocnemius arises from G on the femur and inserts into the back of the foot (below the diagram). Near extension, **a**, **b** it wraps around the back of the femoral condyle. Near extension **a**, it wraps around the back of the tibial plateau at J. In **b**, the lever-arms of the muscles are the perpendicular distances from the instant center I to the muscle tendons. b_h, b_g, b_t, and b_q represent the lever-arms of the hamstrings, gastrocnemius, patellar, and quadriceps tendons, respectively

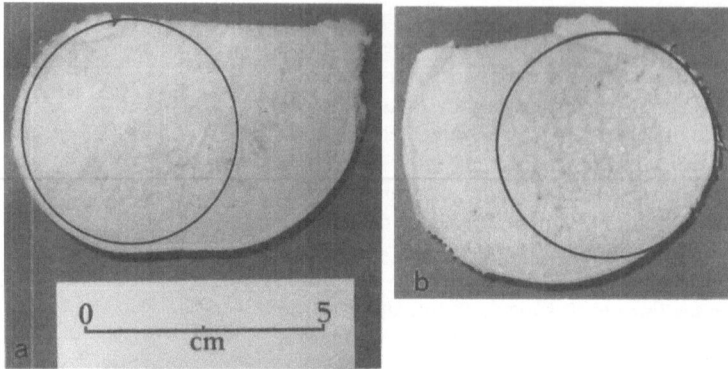

Fig. 8a,b. Sections of **a** a lateral and **b** a medial femoral condyle of a human knee, with *circles* superimposed on the posterior facets

condyles quite well, consistent with the conclusion of Kurosawa et al. [2] that the posterior aspects of the femoral condyles are spherical.

The Muscle Tendons. Lines representing the tendons of the quadriceps, hamstrings, and gastrocnemius muscles are included in the model shown in Fig. 6. The figures show how the lines of action of the muscle, ligament, and contact forces all vary systematically during flexion and extension as the tendons and ligaments rotate about their points of insertion and as the femur rolls backwards and forwards upon the tibia. Since the ligaments control the rolling movements of the bones upon each other, they also influence the directions of the muscle tendons and the moments which they can exert about the knee.

Lever-arms of the Muscles. If muscle power is to be restored to normal and a physiological result obtained after arthroplasty, it is necessary to restore the natural leverage of the muscles. This is best done by restoring ligament geometry. The lever-arms of the muscle tendons are their perpendicular distances from the instant centre, I (Fig. 6b). Their lengths have been calculated from the geometric model of the knee and are shown plotted against flexion angle in Fig. 9. The lever-arm of the patellar tendon is longer than that of the hamstrings which in turn is longer than that of the gastrocnemius muscles over the whole range of movement. Each varies with flexion because the directions of the tendons changes and because the flexion axis moves. The general pattern of this variation agrees with deductions made from EMG signals by Baratta et al. [7].

Ligament Forces in Activity

There have been many studies of ligament strain or ligament forces when they define the limits of passive motion at the knee and equally large numbers of published papers dealing with ligament strain distributions during passive

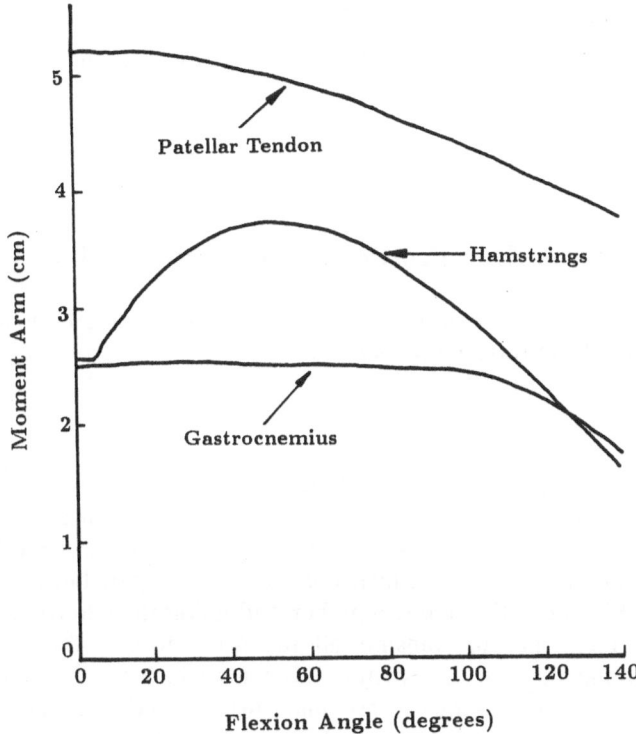

Fig. 9. Calculated lever-arm lengths of the muscle tendons plotted against flexion angle

flexion/extension. Their potentially more stressful roles during activity have largely been reglected. Morrison's work in calculating ligament forces in walking [8, 9], further developed by Harrington [10], has stood alone until the recent paper by Collins and O'Connor [11].

EMG measurements during normal walking have shown that only one or other of the three main muscle groups spanning the knee, quadriceps, hamstrings or gastrocnemius, is active over most of the normal walking cycle. In these circumstances, the value of the muscle force is determined by the condition that its moment should balance the combined moments of the external and inertial loads about the flexion axis of the joint. It is then very unlikely that the component of the muscle force parallel to the tibia plateau in the sagittal plane should also balance exactly the corresponding component of the loads. The only structures available to make up the difference are the ligaments. The studies of walking which have attempted to account for the ligaments have shown that they transmit substantial force during most of the gait cycle and, in particular, during stance phase, (see the chapter by Harrington). The ligaments therefore can play an essential role in controlling the active stability of the joint.

Summary

The ligaments are the key to the understanding of the kinematics and mechanics of the knee, reflecting their roles in limiting movements of the bones and in controlling movement and transmitting load within the limits of motion. Our review does not deal with their possible role in proprioception.

Relation Between Knee Prosthesis Design and Retained Ligaments

We can distinguish between different designs of knee prosthesis according to the use made of the retained ligaments.

The Hinge

There may be joints so badly damaged that the ligaments are completely destroyed and the surgeon has no option but to use a hinged prosthesis [12, 13]. The hinge, with a fixed axis of flexion/extension, replaces the function not only of the articular surfaces which hold the bones apart but also that of the ligaments which hold the bones together and guide their movements on each other. The hinge has to transmit all the shear forces, the abducting and adducting moments, and the twisting moments which the ligaments normally transmit in association with the articular surfaces and, for that reason, very long stems are needed to gain attachment to the bones. There has been a history of loosening of many of these devices although their proponents report quite satisfactory survival rates. Lettin et al. [14] reported a cumulative survival rate of 84% at 10 years for the Stanmore prosthesis while Heinert and Engelbrecht [15] reported 12-year survival rates of 87% for their first generation St. Georg hinge and 94% for their second generation. These rates are somewhat lower than those reported for some condylar forms of knee arthroplasty [16]. Hinges are very invasive and may make infection more likely. The consequences of failure can be very serious.

Much better clinical results are expected for hinges which allow not only flexion and extension but also rotation about the long axis of the tibia [17]. But such devices are still very invasive.

Condylar Prostheses

The condylar prosthesis [18], bicompartmental or tricompartmental, retains the collateral ligaments of the knee and seeks to restore their contribution to mediolateral and long axis rotational stability. However, it usually involves sacrifice at least of the anterior cruciate ligament and most often of both cruciate ligaments. It is one of the paradoxes of meetings on knee surgery that a lecture from one expert on the surgical repair of the torn anterior cruciate ligament is often followed by a lecture on knee replacement in which the sacrifice of the cruciate ligaments is routinely recommended.

When the cruciates are sacrificed, many designs seek to replace some of their functions by stabilizing the prosthesis in the anteroposterior direction. The simplest way to do that is to use a dished concave plateau on the tibia, deeper than the natural medial tibial plateau [18, 19]. Such a device transmits to its fixation not only the compressive forces of the natural joint, but also shear forces which tend to rock the tibial component on its fixation. Clinical success was achieved by using sufficiently long stems to gain fixation to the bones [16].

Other forms of stabilized prostheses have been developed with protrusions on one component interlocked with appropriate sockets on the other component. The Attenborough knee [20] used a rod hinged to the femoral component which slid up and down inside an appropriate hole in the tibial component. The Deane knee [21] had a spherical protrusion on the tibial component and a corresponding socket on the femoral component; the articulation was intended to reproduce the rolling movement of the femur on the tibia during flexion and extension. The Spherocentric design [22] also used a ball and socket arrangement. The Insall-Burstein Posterior Stabilised [23] knee brought this form of prosthesis to clinical success by appropriate design of fixation necessary to transmit the associated shear forces and rocking moments to the bone. Very low loosening rates have been reported [16].

PCL-retaining Condylar Prostheses

More recently, families of prostheses have been introduced which allow retention of the posterior cruciate as well as the collateral ligaments [24–28]. These involve sacrifice of the ACL. The articular surfaces are more or less anatomical, with a rounded component on the femur but a flat component on the tibia. In many of such devices, the tibial plateau has a lip at the front to prevent anterior subluxation of the femur but is relatively flat at the back with no obvious means of preventing anterior subluxation of the tibia.

Articular surfaces of dissimilar shapes can touch only over small areas, either at a point or along a line in the unloaded state. The contact pressure under load is equal to the value of the compressive contact force divided by the area of contact. The value of the contact force varies with the mechanics of the activities of daily living and is outside the control of the prosthesis designer. The value of the contact area is determined absolutely by the designer. If the contact area is small, the contact pressure is large and the classical theory of wear of sliding surfaces predicts a high wear rate [29, 30]. The nonconforming incongruous shapes of the articular surfaces of many modern prostheses might be expected to lead to high wear rates.

Wear in Knee Replacement

Analysis of components retrieved from clinical practice has established that the wear occurring in the polyethylene components of knee replacements is much greater than that found in wear studies of acetabular components [31, 32]. That is not to say that wear in hip replacements is negligible. Dowling et al. [33]

examined 21 acetabular cups retrieved at postmortem examination, having been clinically successful, and found extensive surface cracking of the polyethylene and penetration of the femoral head into the cup. Wroblewski [34] measured the penetration rate of the femoral head into the acetabular socket of 23 Charnley hip replacements retrieved at revision. He found an average penetration rate of 0.19 mm per annum (1 mm every 5 years); the range of penetration rates reported was from a minimum of 0.01 mm to a maximum of 0.5 mm per annum. Rose et al. [35–37] attributed differences in wear rates to differences in molecular weight and to differences in the frequency and distribution of fusion defects in the polyethylene component. Harris [38, 39] has associated bone resorption after hip replacement with the response to polyethylene wear debris. Nolan and Buckhill [40] have recently reported a similar effect in uncemented knee replacement.

The higher wear rate of knee components compared with hip components arises from the larger contact presures expected with nonconforming incongruous surfaces [31, 32, 41]. The high wear rate of nonconforming prostheses was not only unexpected, it was predicted in very clear terms. Black [42] warned that "problems of excessive wear and of creep encountered in TKR design are the direct result of failing to allow adequately in the design for the properties of the materials." Rostoker and Galante in 1979 [43] showed from laboratory studies that the wear rate of polyethylene increases *exponentially* with increasing contact pressure, possibly reflecting the change of mechanism from mild to severe wear observed with increasing load by Burwell and Strang [44]. Rose et al. [37] confirmed the exponential dependence of wear rate on contact pressure. These experiments fully explained very high wear rates reported from clinical practice with nonconforming components [17, 45–47] designed in the 1970s. It is becoming clear that these lessons were not learnt by some designers in the 1980s.

Single case reports [48–50] showed early catastrophic wear of the polyethylene tibial trays of nonconforming metal-backed prostheses [25]. This outcome has been confirmed by retrieval analysis of several more extensive series [51, 52]. Clinical failure due to wear has been accelerated by the adoption of metal backing of the tibial component (to facilitate porous coating), so that the associated very thin polyethylene layers wear through more quickly [49–51].

The Patello-Femoral Joint

Clinical results of bicompartmental condylar femoral components were often marred by problems with the retained patello-femoral joint [53]. Tricompartmental femoral components were introduced which covered the trochlear facet of the distal femur as well as the condyles and gave much better pain relief [19]. Resurfacing of the patella also became common practice [19]. Initially, patellar buttons were all-polyethylene. Later, metal backing was introduced but catastrophic early failures have been widely reported, involving wear of the polyethylene coating down to the metal backing and fracture of the metal backing [54]. These problems were in addition to those previously reported: fracture [55] or dislocation [56] of the patella. The wear problems at the patello-femoral joint largely arise from the incongruous nature of the articular surfaces, with surfaces in contact which have curvatures of opposite signs and very small contact areas under load. These curvatures are necessary to facilitate rotation of the patella about anteroposterior and mediolateral axes. Overloading problems may arise if the femur lies too far anterior on the tibia, so that the angle between the quadriceps and patellar tendon forces becomes more acute than normal with consequent increase in the patello-femoral contact force [57]. Backwards rolling of the femur on the tibia in flexion, as in Fig. 6, helps to protect the patella from excessive forces.

Leg Alignment

Great emphasis has recently been placed on the need to restore natural leg alignment and very elaborate sets of equipment have been devised for this purpose [25]. The stated object is to restore the mechanical axis to the leg, i.e., the Maquet line [58] joining the center of the hip to the center of the ankle should pass through the center of the replaced knee. It is asserted that the tibial component should be implanted with its plateau perpendicular to the mechanical axis when seen from in front; that the leg should be realigned so as to restore the mechanical axis; that small errors in alignment lead to early failure due to wear or loosening.

While perfect alignment is obviously desirable, clinical evidence relating malalignment to failure is hard to find. The paper by Lotke and Ecker [59] is often quoted in support of such assertions, but it merely describes the design of a prospective study of knee arthoplasties in which a correlation would be sought between an alignment score and a clinical score. The follow-up was too brief and the number of failures at that time too few for definite conclusions to be drawn.

The only papers which show a correlation between post-operative alignment and failure are those describing a small series of ICLH arthroplasties by Bargren et al. [60] and a recent paper by Denham et al. [58]. The latter paper showed that loosening rate was 3% in knees in which the postoperative Maquet line lay within the middle third of the tibial plateau and 24% in knees in which it lay outside that limit. However, their data also demonstrates that 73% of their failures were in the 25% of their knees which had been in varus preoperatively ($P < 0.05$).

Load Distribution Between Compartments of the Knee. The emphasis on correct alignment of the prosthesis and restoration of the mechanical axis of the leg is based on the assumption that these factors strongly influence the distribution of the compressive contact force between the medial and lateral compartments of the joint. Some evidence in support of that proposition may be derived from the work of Johnson et al. [61] and Hsu et al. [62].

We have recently measured load distribution in cadaver knees which were mounted in a rig simulating quadriceps-controlled flexed knee stance [63]. The articular surfaces were replaced by unconstrained prosthetic components which allowed retention of all the ligaments. Instrumented meniscal bearings were inserted between the articular surfaces and measured the compressive force in each compartment independently.

It was found that, when the quadriceps force was small and the resultant compressive force transmitted by the knee correspondingly small, the distribution of force between the two compartments depended sensitively on small changes in direction, ±5°, of the applied load relative to the limb. The medial compartment carried the bulk of the compressive force when the line of action of the load lay just medial to the joint, the lateral compartment was overloaded by a load passing through the lateral compartment. This behaviour is to be expected from analyses such as that of Fig. 2 above. However, when the quadriceps force was large relative to the external load and the resultant compressive force transmitted by the knee correspondingly large, the compressive forces in the two compartments were more nearly equal and independent of the line of action of the load.

The force applied by the quadriceps is transmitted to the tibia through the patellar tendon which lies in the centre of the joint in the coronal plane. The line of action of the patellar tendon force is therefore independent of the line of action of the external load and, in the coronal plane, is fixed relative to the two compartments of the knee. Our experiments demonstrate that, when the muscle and contact forces are large in relation to the external load, the fixed line of action of the patellar tendon force in the center of the joint dominates and equalizes the distribution of compressive force in the two compartments.

It follows that load distribution in the coronal plane is determined by those factors which give rise to large muscle forces, mainly the line of action of the external load in relation to the flexion axis of the joint in the sagittal plane. When the lever-arm of the load about the flexion axis is longer than that of the muscles (Fig. 7), the muscle forces are larger than the load and it is the lines of action of the muscle forces rather than that of the load which then determine the distribution of compressive force between the compartments. Sagittal plane mechanics therefore determines force distribution in the coronal plane.

Small variations in the implantation of a prosthesis in relation to the line of action of the load will have little effect on load distribution when the muscle forces are large, as confirmed by our experiments. In the same circumstances, small residual postoperative varus or valgus deformities would have similarly small effects on compartmental load distribution. The advice that the leg should be realigned within a very few degrees of the Maquet line may be unnecessarily demanding.

Soft Tissue Release

We showed in Fig. 2 above that the cruciate ligaments act with the collateral ligaments to control the alignment of the normal limb. When the joint is slightly flexed (Fig. 10), the lateral collateral ligament slackens while some fibers in the ACL, the PCL, and the MCL remain tight, maintaining leg

Fig. 10. a Coronal view of a slightly flexed knee with fibers of the cruciate and medial collateral ligaments tight whereas the lateral collateral ligament is slack. **b** When the cruciates are sacrificed during arthroplasty, the knee has to be pushed into varus to tighten the lateral collateral ligament so that a medial release is necessary to realign the limb

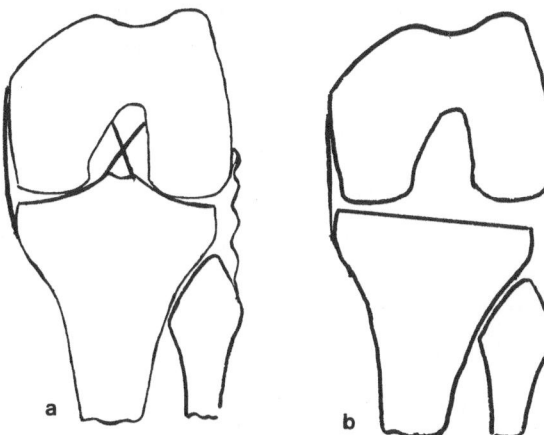

alignment [64–67]. If the cruciates are sacrified while the collateral ligaments are still of normal length, the LCL can be tightened only be adducting the tibia (Fig. 10b). In order to implant a prosthesis, a rectangular gap rather than the trapezoidal gap of Fig. 10b is required and release of the MCL is necessary. Sacrifice of one or both of the cruciates, therefore, also involves compromising the integrity of the MCL, even if it has not contracted in the disease process.

The Anterior Cruciate Ligament in Arthritis

In early arthritis, the articular surfaces are irreversibly damaged while all the ligaments are still intact. In a series of 449 arthritic knees operated in Oxford by JWG between 1976 and 1986, the state of the ligaments was recorded at surgery in 416 cases. The PCL and collateral ligaments were intact in all. The ACL was described as "normal" in 232 (56%), "damaged" or "absent" in 184 (44%). This experience is confirmed by the Swedish figures [68] for the period 1976–1983 in which, of the 4,446 knees operated for gonarthrosis, about 50% were unicompartmental, many of them presumably with intact ligaments. All the ligaments, including the anterior cruciate ligament, are therefore intact in many patients requiring knee replacement. It would seem obvious that they should be retained and restored to function whenever possible.

Design of Surface Replacement Prostheses

Consider now the problem of designing a surface replacement which will replace the worn articular surfaces of the bones with components which will transmit only compressive stresses from one bone to the other, while retaining all the ligaments to transmit tension stresses.

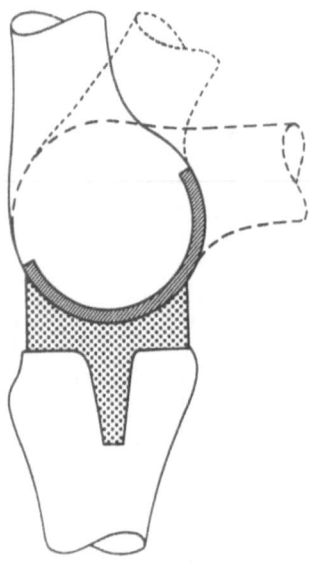

Fig. 11. Fully congruous contact gives large contact areas and low wear rates but allows only fixed axis flexion and extension, like a hinge. It is incompatible with the cruciate linkage

Following the arguments of Figs. 7 and 8 above, we should use surfaces that are very similar in shape to the natural surfaces; spherical surfaces on the femoral condyles, flat surfaces on the tibia (Figs. 7, 8). But such surfaces make contact over small areas. The contact stresses would be very high and, as we have shown above, high wear rates have been found in such designs.

Rapid wear can be avoided by using conforming surfaces (Fig. 11) but such surfaces absolutely fix the axis of flexion and extension and do not allow rotation. The structures which control these rolling movements of the femur, the cruciate ligaments, must therefore be sacrificed. Conforming surfaces transmit shear forces and torques and therefore require large stems to gain attachment to the bones [58].

The Dilemma of Surface Replacement Design

Components compatible with the ligaments have to be incongruous and will wear out. Components made congruous to minimize wear are incompatible with the geometry and mechanics of the ligaments and require ligament sacrifice. They require long stems for fixation and are inconsistent with the ideal of a minimally invasive surface replacement.

Solution of the Dilemma, the Meniscus

Nature interposes mobile flexible menisci between the femur and the tibia. They allow the movements required by the ligaments while bringing the surfaces into conformity. The menisci move to the front of the tibial plateau in extension and to the back in flexion to allow the rolling movements of the femur on the tibia [1, 67]. During long axis tibial rotation, they move

Fig. 12. Components of the Oxford Meniscal Knee assembled

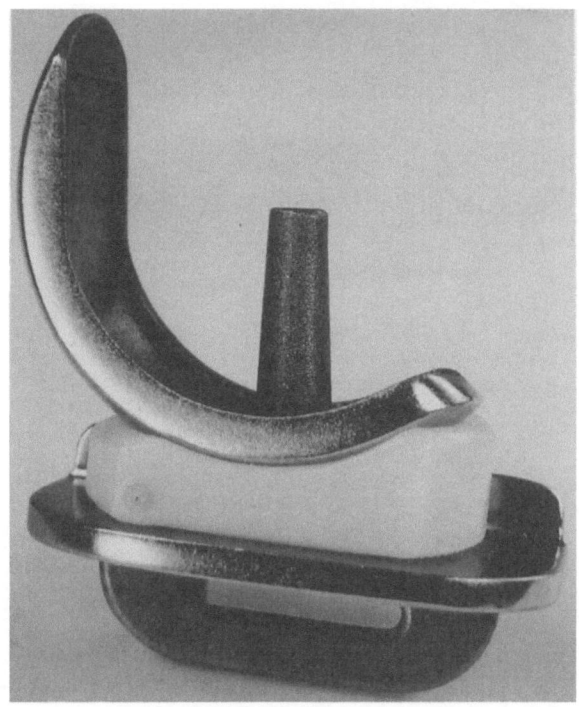

reciprocally, one backwards one forwards [67]. The menisci fill the gaps between the incongruous articular surfaces of the bones. We showed in 1973 [69, 70] that the menisci transmit all the compressive load between the femur and the tibia. Seedhom et al. [71, 72] and Walker and Erkmann [73] came to similar conclusions.

These functions of the natural menisci can be replicated in a knee prosthesis by using fully conforming but unconstrained meniscal bearings interposed between components fixed to the bones.

A Meniscal Surface Replacement Prosthesis

Figure 12 shows the components of a meniscal surface replacement prosthesis. We use spherical metal components on the femoral condyles, flat metal components on the tibial plateaux, and interposed polythylene meniscal bearings between them. The meniscal bearing has a spherical socket as its upper articular surface, of the same radius as that of the femoral component. It has a flat lower surface, matching the tibial component, so the surfaces can conform in all positions and give large contact areas.

The meniscal bearings are made in a series of nine thicknesses, varying in steps of 1 mm. After the metal components have been implanted, we use the gap gauges to measure the gaps which have been left between them. We found

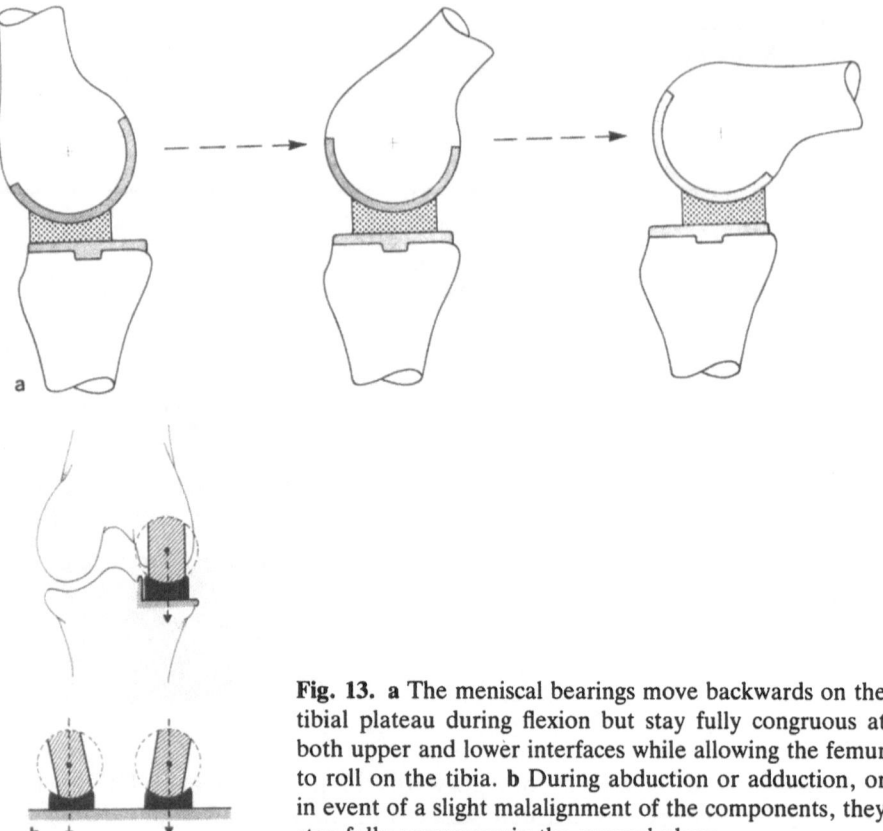

Fig. 13. a The meniscal bearings move backwards on the tibial plateau during flexion but stay fully congruous at both upper and lower interfaces while allowing the femur to roll on the tibia. **b** During abduction or adduction, or in event of a slight malalignment of the components, they stay fully congruous in the coronal plane

by experiment that we needed to choose the meniscal bearing thickness to an accuracy of 1 mm in order to restore ligament tension and to avoid instability. Having chosen the appropriate thickness, the meniscal bearing could then be pushed into position.

In the cadaver, the meniscal bearings moved to the front of the tibial plateau when the joint was brought into extension. When the joint was flexed, they moved to the back [41]. These movements were absolutely systematic and occurred every time. If one tried to prevent the movements of the meniscal bearings the joint would lock and would neither extend nor flex. In the cadaver specimens the bearings moved backwards about 1 cm in 100° of flexion [4]. These movements of the meniscal bearings accommodate the rolling motion of the femur of the tibia (Fig. 6) under the control of the cruciate linkage (Fig. 3). During tibial rotation, the bearings move reciprocally, one forwards one backwards [41]. The use of these mobile elements allows the bones to move upon each other in the way dictated by the ligaments. Similar movements of the meniscal bearings occur in living patients. We made measurements of the movements in 24 patients up to 5 years post-operatively [74] and found values roughly half those measured in the cadaver specimens.

The meniscal bearings allow complete congruity in all positions with complete mobility (Fig. 13a), not only when seen from the side, but also when seen from the front (Fig. 13b). The spherical surfaces at the upper interface allow conformity even during abduction or adduction. Indeed, conformity is still retained if the femoral components are implanted with a slight degree of varus or valgus malalignment and point contact which would be obtained with cylindrical femoral components is avoided.

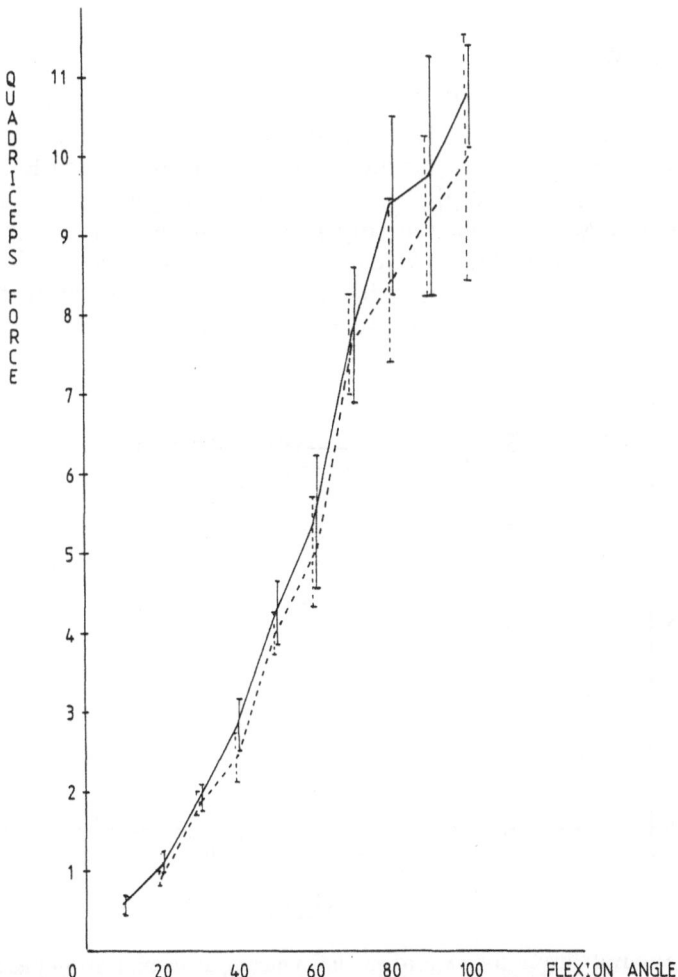

Fig. 14. Quadriceps tendon force divided by vertical load plotted against flexion angle for seven cadaver specimens before and after meniscal arthroplasty in a simulation of flexed knee stance. *Solid line*, intact joints; *dashed line*, replaced joints

Restoration of Muscle Power

Restoration of ligament function restores natural movements of the bones and the natural leverage available to the muscles. We tested seven specimens in a rig simulating flexed knee stance [75]. The specimens were held in various positions of flexion under vertical load by the tension force in a wire sewn to the quadriceps tendon. We tested the specimens before and after meniscal knee replacement. The measured quadriceps force is plotted against flexion angle in Fig. 14. The results obtained from the intact joint are indistinguishable from those obtained from the replaced joint. On this basis, restoration of ligament function should lead to restoration of full muscle power and should yield a functional result.

Clinical Results

The clinical results for the Oxford Meniscal Knee have been published in a number of papers since 1986 [76, 77]. It has been found to give satisfactory pain relief and restoration of function. Failure rates have been low (5% at 7 years) in knees in which all the ligaments were intact at surgery. In knees in which the ACL was described at surgery as "damaged" or "absent," failure rates have been higher (20% at 7 years) and more constrained devices are now preferred. In gonarthritis, in knees in which the ACL is still intact, the disease is limited to the medial compartment in more than 80% of cases. Medial

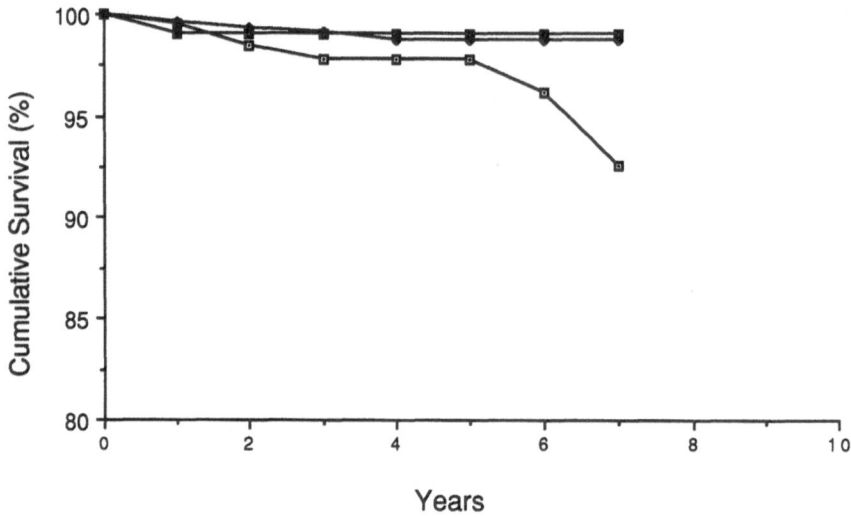

Fig. 15. Survival curves for Oxford medial unicompartmental arthroplasty compared with those of the Posterior Stabilised and Total Condylar prostheses [16]. Cumulative survival rate plotted against years followed-up. *Solid Squares*, Oxford Knee (medial unicompartmental); *diamonds*, Posterior Stabilized TCP; *open squares*, Total Condylar Prosthesis (TCP)

unicompartmental arthroplasty is now the main application of the Oxford Meniscal Knee.

We have recently reviewed 121 consecutive cases of Oxford medial unicompartmental meniscal arthroplasty [78], followed up from 24 to 100 months (mean 44 months). There has been one failure. The survival curve is shown in Fig. 15, where it compares well with recently published curves for the Total Condylar and Posterior Stabilised prostheses [16], devices said in the USA to represent the "gold standard" for knee replacement. It will be seen that all three devices give comparable and reliable results in the medium term. The Swedish data [68] suggest that 50% of knees requiring replacement for gonarthrosis are most appropriately treated by medial unicompartmental arthroplasty. The doubt about unicompartmental arthroplasty — that it is not as reliable as tricompartmental arthroplasty — has been overcome in significant numbers of carefully selected cases by an unconstrained meniscal design which is conservative, reliable, and physiological.

Wear in Meniscal Knee Replacement

We have measured the penetration rate of the metal components of Oxford knee replacements into 23 meniscal bearings retrieved at revision 1–9 years postoperatively [79]. It was found to be very small, about 1 mm every 40 years on average, about seven times slower than in Charnley hip replacements. This experience confirms the benefits of using fully conforming meniscal bearings. However, many of the retrieved bearings showed signs of pitting and abrasion wear; one or two of them were very badly damaged. This experience shows that polyethylene is far from an ideal material for joint replacement but its weaknesses can be minimized in knee replacement by the use of fully conforming meniscal bearings.

Conclusion

The surgeon needs an armamentarium of devices for knee replacement, depending on the state of the ligaments. A hinge may be appropriate when the ligaments are completely destroyed. When the cruciate ligaments are damaged or absent, stabilized designs with conforming surfaces have been found to give reliable results in the medium term. The more recent incongruous designs introduced in the 1980s, which may allow retention of the posterior cruciate ligament, must be under suspicion because of high wear rates associated with small contact areas and high contact stresses; some catastrophic failures have recently been reported. When all the ligaments are intact, as they are in about 50% of osteoarthritic knees requiring arthroplasty, medial unicompartmental arthroplasty with a fully conforming mobile meniscal prosthesis is a conservative option which has given reliable results in the medium term.

Acknowledgments. The work described in this chapter was supported by the Arthritis and Rheumatism Council and by Biomet Ltd. Mr. Andrew Carr FRCS provided Fig. 15.

References

1. Maquet P (1989) On the movement of animals. Springer, Berlin. Translation of: Borelli GA (1679) Di Motu Animalium
2. Kurosawa H, Walker PS, Abe S, Garg A, Hunter T (1985) Geometry and motion of the knee for implant and orthotic design. J Biomech 18(7):487–99
3. Daniel DM, Stone ML (1990) Case studies. In: Daniel DM, Akeson WH, O'Connor JJ (ed) Knee Ligaments: Structure, Function, Injury and Repair, Raven, New York, pp 31–55
4. O'Connor JJ, Shercliff TL, Biden E, Goodfellow JW (1989) The geometry of the knee in the sagittal plane. Proc Inst Mech Eng Part H, J Engng Med 203:223–33
5. O'Connor J, Shercliff T, FitzPatrick D, Bradley J, Daniel D, Biden E, Goodfellow J (1990) Geometry of the knee. In: Daniel DM, Akeson WH, O'Connor JJ (eds) Knee ligaments: Structure, function, injury, and repair. Raven, New York, pp 163–200
6. Weber WE, Weber EFW (1836) Mechanik der menschlichen Gehwerkzeuge. In: Dietrichschen Buchhandlung, Göttingen
7. Baratta R, Solomonow M, Zhou BH, Letson D, Chuinard R, d'Ambrosia R (1988) The role of the antagonist musculature in maintaining knee stability. Am J Sports Med 16(2):113–22
8. Morrison JB (1968) Bioengineering analysis of force actions transmitted by the knee joint. Biomed Eng 90:164–70
9. Morrison JB (1970) The mechanics of the knee joint in relation to normal walking. J Biomech 3:51–61
10. Harrington IJ (1976) A bioengineering analysis of force actions at the knee in normal and pathological gait. Biomed Eng 11:167–72
11. Collins JJ, O'Connor JJ (1991) Muscle-ligament interactions at the knee during walking. Proc Inst Mech Eng Part H, J Engng Med 205:11–18
12. Shiers LGP (1954) Arthroplasty of the knee — preliminary report of a new method. J Bone Joint Surg [Br] 36:553–60
13. Waldius B (1957) Arthroplasty of the knee using an endoprosthesis. Acta Orthop Scand [Suppl] 24
14. Lettin AW, Ware HS, Morris RW (1991) Survivor analysis and confidence limits. J Bone Joint Surg [Br] 73(5):729–31
15. Heinert K, Engelbrecht E (1988) Langzeitvergleich der Knie-endoprothesensysteme "St. Georg". Chirurg 59:755
16. Scuderi GR, Insall JN, Windsor RE, Moran MC (1989) Survivorship of cemented total knee replacements. J Bone Joint Surg [Br] 71(5):798–803
17. Engelbrecht E, Heinert K (1988) Experience with a surface and total knee replacement: Further development of the model St. Georg. In: Niwa S, Paul JP, Yamamoto S (eds) Total knee replacement. Springer Verlag, Tokyo, pp 257–275
18. Freeman MAR, Swanson SAV, Zahir A (1972) Total replacement of the knee using metal-polyethylene two-part prosthesis. Proc R Soc Med 65:374–5
19. Insall J, Scott WN, Ranawat CS (1979) The Total Condylar knee prosthesis, a report of 220 cases. J Bone Joint Surg [Am] 61(2):173–80

20. Attenborough CG (1978) The Attenborough total knee replacement. J Bone Joint Surg [Br] 60(3):320–6
21. Deane G (1974) A new concept in knee joint design. In: International Congress Series, no. 324. Excerpta Medica, Amsterdam, pp 244–247
22. Mathews LS, Sonstegard DA, Kaufer H (1973) The Spherocentric knee. Clin Orthop 94:234–41
23. Insall JN, Lachiewicz PF, Burstein AH (1982) The posterior stabilized condylar prosthesis: A modification of the Total Condylar design: 2–4 year clinical experience. J Bone Joint Surg [Am] 64(9):1317–23
24. Ewald FC, Jacobs MA, Miegel RE, Walker PS, Poss R, Sledge CB (1984) Kinematic total knee replacement. J Bone Joint Surg [Am] 66(7):1032–40
25. Hungerford DS, Kenna RV (1983) Preliminary experience with a porous coated total knee replacement used without cement. Clin Orthop 176:95–107
26. Ritter MA, Keating EM, Faris PM (1989) Design features and clinical results of the anatomic graduated components (AGC) total knee replacement. Contemp Orthop 19(6):641–7
27. Rosenberg AG, Barden R, Galante JO (1989) A comparison of cemented and cementless fixation with the Miller-Galante total knee arthroplasty. Orthop Clin North Am 20(1):97–111
28. Miller J, Johnson JA (1990) Anthropometric aspects of knee prosthesis design. In: Coombs R, Gristina A, Hungerford D (eds) Joint replacement, state of the art. Orthotext, London, pp 189–91
29. Bowden FP, Tabor D (1964) Friction and lubrication of solids, part II. Oxford University Press, London
30. Briscoe BJ, Pooley CM, Tabor D (1975) Friction and transfer of some polymers in unlubricated sliding. In: Lee L-H (ed) Advances in polymer friction and wear. Plenum, New York, pp 191–202
31. Landy M, Walker PS (1988) Wear of ultra-high-molecular-weight polyethylene components of 90 retrieved knee prostheses. J Arthroplasty, Suppl:73–85
32. Wright TM, Bartel DL (1986) The problem of surface damage in polyethylene total knee components. Clin Orthop 205:65–74
33. Dowling JM, Atkinson JR, Dowson D, Charnley J (1978) The characteristics of acetabular cups worn in the human body. J Bone Joint Surg [Br] 60(3):375–82
34. Wroblewski BM (1985) Direction and rate of socket wear in Charnley low friction arthroplasty. J Bone Joint Surg [Br] 67:757–61
35. Rose RM, Nusbaum HJ, Schneider H, Ries SB, Paul I, Crugnola A, Simon SR, Radin EL (1980) On the true wear rate of ultra-high-molecular-weight polyethylene in the total hip prosthesis. J Bone Joint Surg [Am] 62(4):537–49
36. Rose RM, Cimono WR, Ellis E, Crugnola AM (1982) Exploratory investigations on the structure dependence of the wear resistance of polyethylene. Wear 77:89–104
37. Rose RM, Goldfarb EV, Ellis E, Crugnola AM (1983) On the pressure dependence of the wear of ultra-high-molecular-weight polyethylene. Wear 92:99–111
38. Harris WH, Schiller AL, Scholler JM, Freiberg RA, Scott R (1976) Extensive localised bone resorption in the femur following total hip replacement. J Bone Joint Surg [Am] 58:612–8
39. Jasty MJ, Floyd WE, Schiller AL, Goldring SR, Harris WH (1986) Localised osteolysis in stable non-septic total hip replacement. J Bone Joint Surg [Am] 68:912–9
40. Nolan JF, Buckhill TM (1992) Aggressive granulomatosis from polyethylene failure in an uncemented knee replacement. J Bone Joint Surg [Br] 74(1):23–4

41. Goodfellow JW, O'Connor JJ (1978) The mechanics of the knee and prosthesis design. J Bone Joint Surg [Br] 60:358–69
42. Black J (1978) The future of polyethylene. J Bone Joint Surg [Br] 60(3):305–6
43. Rostoker W, Galante JO (1979) Contact pressure dependence of wear rates of ultra-high-molecular-weight polyethene. J Biomed Mater Res 13:957–64
44. Burwell JT, Strang CD (1952) J Appl Phys, A 212:470
45. Gunston RH, MacKenzie RI (1976) Complications of polycentric knee arthroplasty. Clin Orthop 120:11–7
46. Marmor L (1976) The Modular (Marmor) knee — case report with a minimum follow-up of 2 years. Clin Orthop 120:86–94
47. Williams EA, Hargadon EJ, Davies DRA (1979) Late failure of the Manchester prosthesis. J Bone Joint Surg [Br] 61(4):451–4
48. Engh GA (1988) Failure of the polyethylene bearing surface of a total knee replacement within four years: A case report. J Bone Joint Surg [Am] 70:1093–6
49. Christensen OM, Christiansen TG, Johansen T (1990) Polyethylene failure in a PCA unicompartmental prosthesis. Acta Orthop Scand 61:578–9
50. Lindstrand A, Ryd L, Stenström A (1990) Polyethylene failure in two total knees: wear of thin, metal-backed PCA tibial components. Acta Orthop Scand 61:575–7
51. Engh GA, Kimberly AD, Hanes CK (1992) Polyethylene wear of metal-backed tibial components in total and unicompartmental knee prostheses. J Bone Joint Surg [Br] 74(1):9–17
52. Jones SMG, Pinder IM, Moran CG, Malcolm AJ (1992) Polyethylene wear in uncemented knee replacements. J Bone Joint Surg [Br] 74(1):18–22
53. Scott RD (1982) Duopatellar total knee replacement. Orthop Clin North Am 13:89
54. Bayley JC, Scott RD, Ewald FC, Holmes GB (1988) Failure of the metal-backed patellar component after total knee replacement. J Bone Joint Surg [Am] 70(5):668–74
55. Windsor RE, Scuderi GR, Insall JN (1989) Patellar fractures in total knee arthoplasty. J Arthroplasty 4:S63–7
56. Merkow RL, Soudry M, Insall JN (1985) Patellar dislocation following total knee arthoplasty. J Bone Joint Surg [Am] 67:1321
57. Miller RK, O'Connor JJ, Goodfellow JN (to be published) Patello-femoral forces after knee replacement — an in vitro study
58. Jeffrey RS, Morris RW, Denham RA (1991) Coronal alignment after total knee replacement. J Bone Joint Surg [Br] 73(5):709–14
59. Lotke P, Ecker ML (1977) Influence of positioning of prosthesis in total knee replacement. J Bone Joint Surg [Am] 59(1):77–9
60. Bargren JH, Blaha JD, Freeman MAR (1983) Alignment in total knee arthroplasty. Correlated biomechanical and clinical observations. Clin Orthop 173:178
61. Johnson F, Leitl S, Waugh W (1980) The distribution of load across the knee — a comparison of static and dynamic measurements. J Bone Joint Surg [Am] 62:346–349
62. Hsu RWW, Himeno S, Coventry MB, Chao EYS (1990) Normal axial alignment of lower extremity and load-bearing distribution at the knee. Clin Orthop 255:215–27
63. O'Connor J, Biden E, Bradley J, FitzPatrick D, Young S, Kershaw C, Daniel D, Goodfellow J (1990) The muscle-stabilized knee. In: Daniel DM, Akeson WH, O'Connor JJ (eds) Knee ligaments: Structure, function, injury, and repair. Raven, New York, pp 239–78
64. Brantigan OC, Voshell AF (1941) The mechanics of the ligaments and menisci of the knee joint. J Bone Joint Surg [Am] 23:44–66

65. Brantigan OC, Voshell AF (1943) The tibial collateral ligament: Its function, its bursae, and its relation to the medial meniscus. J Bone Joint Surg [Am] 25(1): 121–31
66. Wang C-J, Walker PS, Wolf B (1973) The effects of flexion and rotation on the length patterns of the ligaments of the knee. J Biomech 6:587–96
67. Kapandji IA (1987) The Lower Limb, 5th edn. Churchill Livingstone, London (The Physiology of the Joints, Vol 2)
68. Knutson K, Lindstrand A, Lidgren L (1986) Survival of knee arthroplasties. A nationwide multicenter investigation of 8000 cases. J Bone Joint Surg [Br] 68(5):795–9
69. Shrive NG (1973) The weight-bearing role of the menisci of the knee. J Bone Joint Surg [Br] 56:381
70. Shrive NG, O'Connor JJ, Goodfellow JW (1978) Load-bearing in the knee joint. Clin Orthop 131:279–287
71. Seedhom BB, Dowson D, Wright V (1973) Function of the menisci, a preliminary study. J Bone Joint Surg [Br] 56:381
72. Seedhom BB, Dowson D, Wright V (1974) Functions of the menisci, a preliminary study. In: International Congress Series, no. 324. Excerpta Medica, Amsterdam
73. Walker PS, Erkmann MJ (1975) The role of the menisci in force transmission across the knee. Clin Orthop 109:184–92
74. Bradley J, Goodfellow JW, O'Connor JJ (1987) A radiographic study of bearing movement in Unicompartmental Oxford knee replacement. J Bone Joint Surg [Br] 69(4):598–601
75. O'Connor JJ, Goodfellow JW, Bradley JA (1988) Quadriceps forces following meniscal knee arthroplasty — An in vitro study. Trans Orthop Res Soc p 357
76. Goodfellow JW, Kershaw CJ, Benson MKD'A, O'Connor JJ (1988) The Oxford knee for unicompartmental osteoarthritis. J Bone Joint Surg [Br] 70:692–701
77. Goodfellow JW, O'Connor J (1986) Clinical results of the Oxford knee. Clin Orthop 205:21–42
78. Carr A, Keyes G, Miller R, O'Connor J, Goodfellow J (to be published) Medial unicompartmental arthroplasty: A survival analysis of the Oxford meniscal knee
79. Argenson J-N, O'Connor JJ (1992) Polyethylene wear in meniscal knee replacement: A 1–9 year retrieval analysis of the Oxford knee. J Bone Joint Surg [Br] 74:228–232

Gait

Current Trends in Gait Analysis: Advanced Techniques for Data Acquisition and Analysis

Tomokazu Hattori[1]

Summary. For kinematic gait measurement, a video device was developed using television techniques and optoreflective markers, and a personal-computer-based three-dimensional system was completed using direct linear transformation (DLT) method for three-dimensional calibration. Accuracy of the three-dimensional system was examined, and root mean square (RMS) errors of 3.3 mm, 2.6 mm, 1.4 mm in the X, Y, and Z axes were obtained with a camera angle of 60° and camera distance of 3 m. For kinetic measurement, a twin long force plate system was developed, which can measure the triaxial force components for several steps. The frequency response of the long force plate was investigated, and a finite impulse response (FIR) filter with a cutoff frequency of 10 Hz was designed and applied for an automatic data analysis program. Basic knowledge on gait analysis expressly stated in the program allowed detection of gait events and particular peak values, and calculation of time factors and stride dimensions. This kinetic measurement system was used on 95 normal female subjects, and mean values and standard ranges of those parameters were obtained in each age group.

Introduction

In clinical gait analysis in orthopedics, the main interest is assessment of gait disorders directly related to bone and joint diseases. However, conventional methods of data acquisition and data analysis are time-consuming. Recently, many advanced techniques on gait analysis have been developed for the clinical field with the increase in high performance personal computer systems, such as automated data acquisition, signal processing, and artificial intelligence (AI) techniques in computer programming.

[1] Department of Orthopaedic Surgery, Aichi Medical University, 21 Yazako-Karimata, Nagakute, Aichi 480-11, Japan

We have developed a video device for kinematic measurement which automatically measures two-dimensional coordinates of optical reflective markers, and have set up a three-dimensional gait analysis system with two video devices and twin long force plates. This report mainly concerns the technical details of our gait analysis system from on-line data acquisition to data analysis.

Kinematic Measurement

Since the late 1960s, several motion analysis systems using television techniques or photoelectro devices have been developed which can measure the location of optical light markers attached to body segments. One of the most applicable systems was developed at the University of Strathclyde in 1976, and this is now in use in leading biomechanics and rehabilitation engineering institutes worldwide. Recently, new developments in electriconics have provided a new environment for developing automatic motion analysis systems with a personal computer. We have developed a video device for automatic motion analysis using a charge coupled device (CCD) black and white (B/W) video camera and video signal processing unit with built-in memory in order to construct a compact system for clinical gait assessment.

Video Device

The principle of light marker detection followed the Strathclyde TV system reported by Jarrett [1] and Andrews [2] (Figs. 1, 2). The main developments with this video device are the built-in memory and its interface technique. The device consists of a CCD B/W video camera with a 8.5-mm wide-angle lens and optical Infrared (IR) pass filter, an IR strobe using high intensity IR LEDs

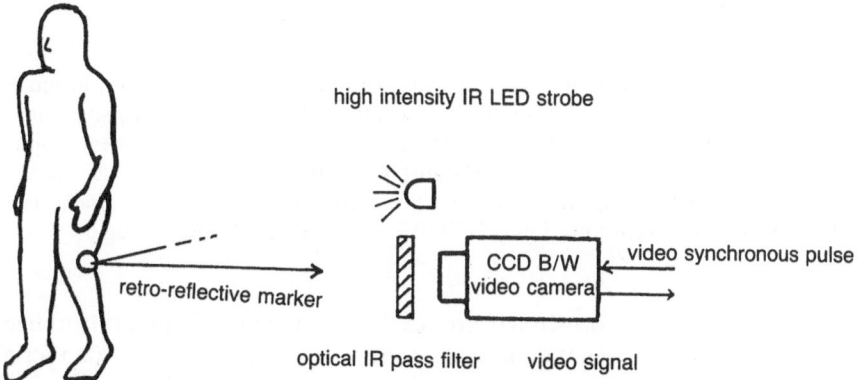

Fig. 1. Obtaining a light marker image. Infrared strobe illumination, a retro-reflective spherical marker, and optical IR filter provide a high contrast still image for video digitizing. *VD*, vertical synchronized pulse; *HD*, horizontal synchronized pulse

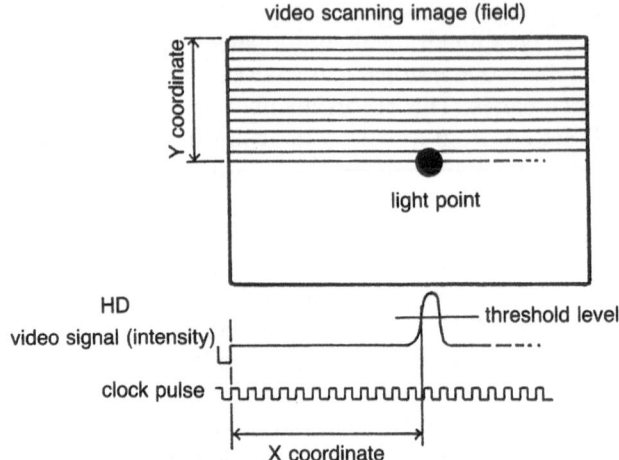

Fig. 2. Video digitizing using a television technique. X and Y coordinates are given by counting the clock pulse and horizontal synchronous pulse (*HD*)

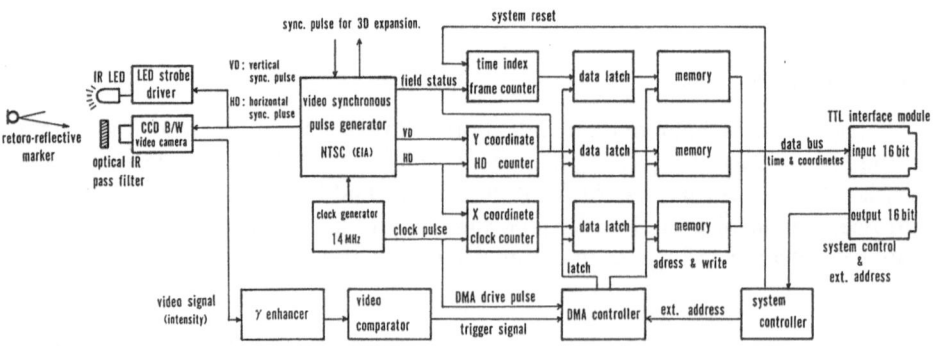

Fig. 3. Block circuit diagram of video locater with built-in DMA memory, including an assortment of circuit modules for camera control, video signal processing (analog), pulse counter, memory, direct memory access controller, and system controller (digital). *Time index*, counting field status pulse from system reset; *X coordinate*, counting clock pulse from horizontal synchronized pulse (*HD*); *Y coordinate*, cunting *HD* from vertical synchronized pulse (*VD*); *TTL*, transistor transistor logic

(940 nm) and a strobe driver synchronized with the video frame, retroreflective markers, a signal processing unit called the video locator, and a video monitor [3, 4].

The video locator consists of an assortment of circuits: the clock generator, video synchronous pulse generator, gamma enhancer, video comparator with threshold adjuster, frame counter, X and Y coordinate counter, 128-kb memory with data latch, direct memory access (DMA) controller, and system controller. These are assembled in two circuit boards separated into an analog part and digital part. Data bus, address bus, and system control lines are connected to the transistor transistor logic (TTL) interface modules of a personal computer (Fig. 3).

The video synchronous pulse generator provides the field status pulse, vertical synchronous pulse (VD), and horizontal synchronous pulse (HD) according to the Electronics Industries Association (EIA) standard. Vertical and horizontal synchronous pulses are supplied to the video camera for two-dimensional scanning of the image, and a vertical synchronous pulse is also supplied to the strobe driver for the synchronous IR ray projection. At the same time, these synchronous pulses, the field status pulse, and clock pulse are supplied into a frame counter (time index), HD counter (Y coordinate), and clock counter (X coordinate) for digital conversion of the time and video scanning position.

The IR rays projected from the LED strobe are reflected by the retro-reflective markers on the anatomical landmark, and the camera can take an objective image of those light markers. Visual light is cut off from the image by the optical IR pass filter (850 nm) to reduce the background noise, and the contrast of the image is improved by the gamma enhancer in order to get stable video digitizing. When the intensity of the video signal which represents the brightness of the image exceeds a threshold level, the video comparator discriminates the light point and gives a trigger pulse for the DMA controller. Immediately, the digital data of time index and X-Y scanning position are kept at the data latch gate, then stored into the built-in memory. In this way, the location of the optical light makers on the video image can be automatically obtained as two-dimensional coordinates of X and Y. This data acquisition sequence is repeated until memory saturation or an external stop signal. After measurement, the system controller allows external memory access from computer through the TTL interface modules.

In the specification, the sampling frequency of this video device is 30 Hz according to the EIA standard. It seems that 30 Hz is fast enough for gait analysis, but not for jumping and running. The resolution of the camera module is 1/756 in the horizontal axis and 1/486 in the vertical axis. The measurement error was observed specially at a diagonal corner in use of such a wide-angle lens, which is mainly caused by symmetric lens distortion. However, those errors in the X and Y axes are directly proportional to $X(X^2 + Y^2)$ and $Y(X^2 + Y^2)$ in lens distortion theory, so it can be easily corrected (Fig. 4).

This video device can be combined with any personal computer by TTL interface modules. Video digitizing and fast data transfer are provided by internal circuits such as the built-in memory and DMA controller. Thus it is easy to build up a target system for practical measurement that has a flexibility for software, without any specific computer system. Also the computer system can control or measure other equipment during kinematic measurement, because the video digitizing and data storage sequence is independent from the computer system (Fig. 5).

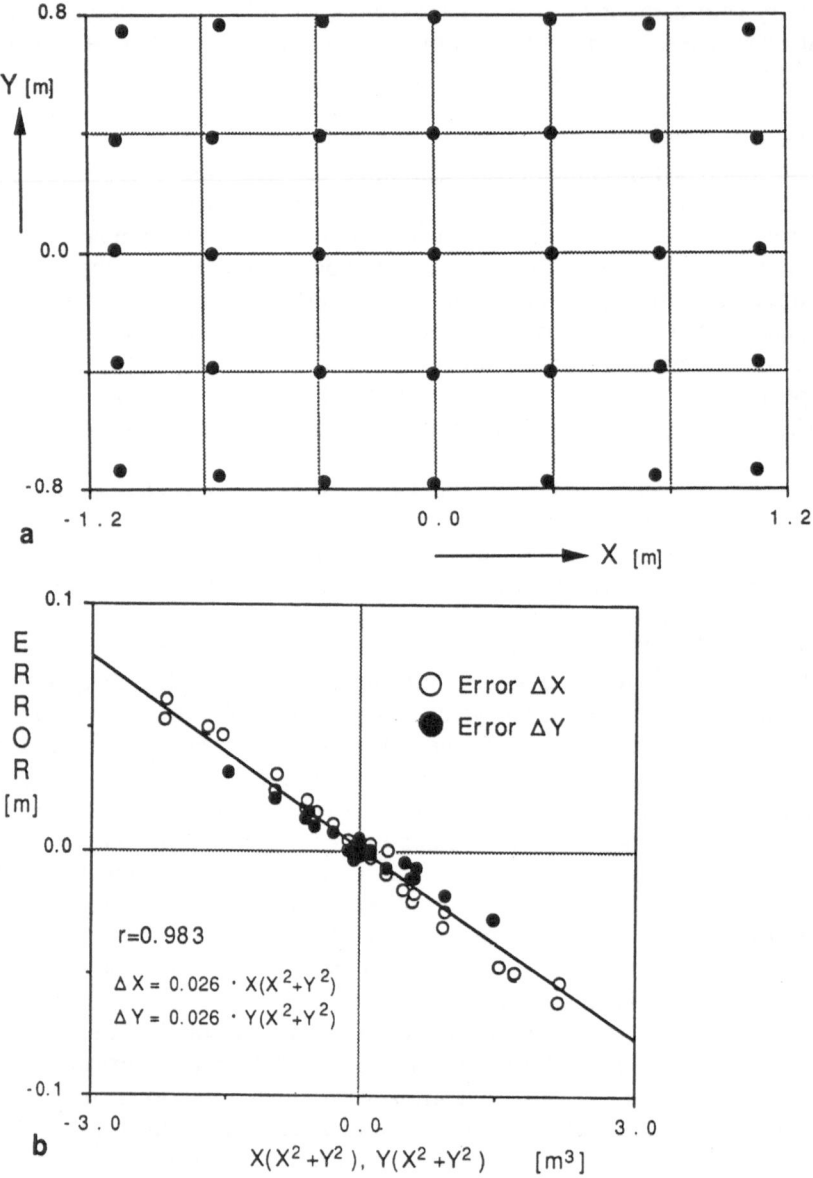

Fig. 4a,b. Symmetrical erros in 2-dimensional measurement. **a** Barrel-shaped distortion of the image due to lens distortion. Grid intersections are expected locations, error is expressed as a distance between the intersection and measured point, which is here exaggerated twice. **b** Colinearity of symmetrical error in proportion to $X(X^2 + Y^2)$ and $Y(X^2 + Y^2)$ when using an 8.5-mm wide lends. *Open circles*, error ΔX; *solid circles*, error ΔY

Fig. 5. Stick diagram with force vector in normal subject. Additional kinetic measurement was performed using an AD converter synchronized with a video frame pulse

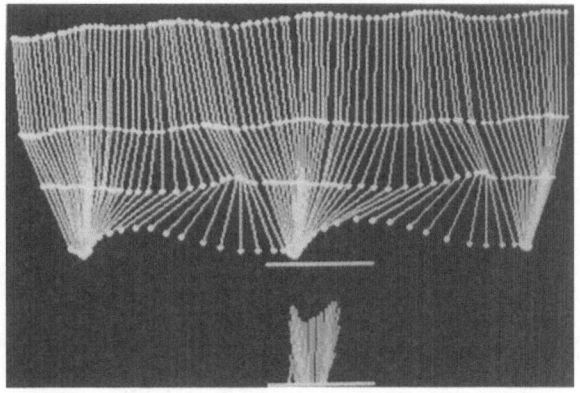

Three-Dimensional Measurement

Current trends in kinematic study are toward three-dimensional measurement with television systems or photoelectro devices, which reconstruct three-dimensional spatial coordinates using multidirectional two-dimensional measurements. However, this requires extremely complicated mathematical solutions for the spatial geometry and optical projection system.

If the cameras can be set up in an exact location with no rotations around the camera light axis and transverse axis, the projective centers and intersectional angle of the camera light axes are easily obtained, and subsequently three-dimensional calibration can be achieved by trigonometry alone. However, it is quite difficult to obtain such an ideal setup in practice.

The direct linear transformation (DLT) method developed by Abdel and Karara [5] and Marzan and Karara [6] is widely accepted for three-dimensional calibration with a free camera setup in photogrammetry, although it needs use of the least square method for solution of simultaneous equations with 11 unknown parameters [7–18]. The basic theoretical concept is that the photo-

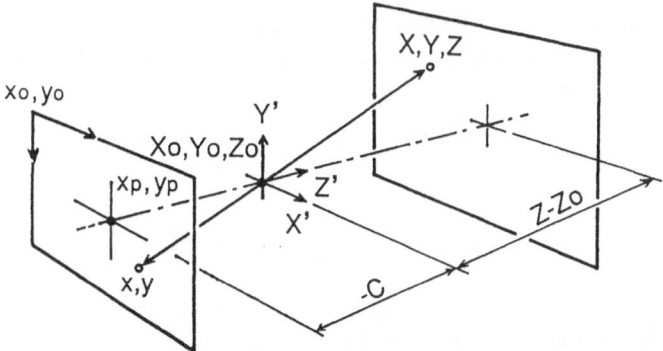

Fig. 6. Theoretical concept of direct linear transformation. The photograph is a central projection of the objective space to the perfect plane

graph, being a perfect plane, is a central projection of the object space, which is the result of projective transformation with the colinearity of the image point (x, y), the projection center (X_o, Y_o, Z_o) and the object point (X, Y, Z) (Fig. 6). The relationship was expressed with triaxial camera rotations (a rotation matrix of $m_{11}-m_{33}$), a principle point of the two-dimensional image (x_p, y_p), and distance (c) between the projective center and the principle point:

$$\begin{bmatrix} \bar{x} - x_p \\ \bar{y} - y_p \\ -c \end{bmatrix} = \lambda \begin{bmatrix} m_{11} & m_{12} & m_{13} \\ m_{21} & m_{22} & m_{23} \\ m_{31} & m_{32} & m_{33} \end{bmatrix} \begin{bmatrix} X - X_o \\ Y - Y_o \\ Z - Z_o \end{bmatrix} \tag{1}$$

where $\bar{x}, \bar{y}$ = refined photo coordinates of a point
            x_p, y_p = photo coordinates of the principle point
            c = camera principle distance
            λ = scale factor
            m_{ij} = rotation matrix due to tri-axial camera rotation
    X, Y, Z = object space coordinates of the point
X_o, Y_o, Z_o = object space coordinates of projection center

$$\bar{x} - x_p = \lambda_x(x + \Delta x - x_o)$$
$$\bar{y} - y_p = \lambda_y(y + \Delta y - y_o) \tag{2}$$

where λ_x, λ_y = scale factors which allow for a different scale in the two axes
       x_o, y_o = coordinates of the principle point referred to the comparator coordinate system
       x, y = photo coordinates of a point
    $\Delta x, \Delta y$ = systematic error in coordinates

Through some arrangements by substitution, Eq. 3 was obtained:

$$x + \Delta x - x_o = -c_x \frac{m_{11}(X - X_o) + m_{12}(Y - Y_o) + m_{13}(Z - Z_o)}{m_{31}(X - X_o) + m_{32}(Y - Y_o) + m_{33}(Z - Z_o)}$$

$$y + \Delta y - y_o = -c_y \frac{m_{21}(X - X_o) + m_{22}(Y - Y_o) + m_{23}(Z - Z_o)}{m_{31}(X - X_o) + m_{32}(Y - Y_o) + m_{33}(Z - Z_o)} \tag{3}$$

where c_x, c_y = scale factors which allow for a different scale. These were also simplified with 11 summarized parameters, the so-called DLT 11 parameters, and systematic error($\Delta x, \Delta y$) due to lens distortion, as in Eq. 4:

$$x + \Delta x = \frac{L_1 X + L_2 Y + L_3 Z + L_4}{L_9 X + L_{10} Y + L_{11} Z + 1}$$

$$y + \Delta y = \frac{L_5 X + L_6 Y + L_7 Z + L_8}{L_9 X + L_{10} Y + L_{11} Z + 1} \tag{4}$$

where Li = DLT parameters. Furthermore, Δx and Δy can be corrected for symmetrical and asymmetrical lens distortion as in Eq. 5.

$$\Delta x = x'(K_1 r^2 + K_2 r^4 + K_3 r^6 + \ldots) + P_1(r^2 + 2x'^2) + 2P_2 x'y'$$

$$\Delta y = y'(K_1 r^2 + K_2 r^4 + K_3 r^6 + \ldots) + P_1(r^2 + 2y'^2) + 2P_2 x'y' \tag{5}$$

where $x' = x - x_o$
$\quad\quad y' = y - y_o$
$\quad\quad r^2 = x'^2 + y'^2$
$\quad\quad K_i$ = coefficient of symmetrical lens distortion
$\quad\quad P_i$ = coefficient of asymmetrical lens distortion

These are the basic formulas derived by Abdel and Karara [5] for the DLT method of solving the colinearity condition, which was originally developed for photosurveying in architectural and civil engineering.

This DLT three-dimensional calibration method is actually performed by two steps involving some error corrections due to systematic lens distortion and random error in the measurement. Firstly, several object points with known three-dimensional coordinates (X_i, Y_i, Z_i, where i is point number), the so-called control points, are prepared in objective space. Then multidirectional images of those control points are taken, and two-dimensional coordinates (x_i, y_i) of every control point are obtained in each observed image. Putting known coordinates (X_i, Y_i, Z_i) and observed coordinates (x_i, y_i) into Eq. 4, simultaneous equations referred to each control point can be made with the 11 unknown parameters (L_1–L_{11}) for each individual image. Solving by the least square method, the DLT 11 parameters and coefficients of lens distortion can be obtained mathematically.

Secondly, multidirectional images of objective points are taken in the same configuration, and two-dimensional coordinates (x_i, y_i) are obtained for each point. To find the unknown three dimensional coordinates (X_i, Y_i, Z_i) of each objective point, a set of observed two-dimensional coordinates (x_i, y_i) on multidirectional images is put into Eq. 4 with the DLT 11 parameters previously obtained and the three unknown parameters of (X_i, Y_i, Z_i). Again by solving the simultaneous equations by the least square method, the three-dimensional coordinates (X_i, Y_i, Z_i) of the objective point can be obtained. This procedure should be repeated for each objective point.

Considering error components due to lens distortion and its correction, it is possible to set some function model of error components, using Eq. 5 in the DLT equations, and solve their unknown coefficients through the least square method. In this process, the control points have to be dispersed all around the image as the master references of lens distortion. It is hard to set such dispersed control points in objective space for each calibration time. Thus error correction should be performed on two-dimensional coordinates prior to the DLT three-dimensional calibration with only 11 parameters, to minimize the number of control points and processing time for the least square method which is repeated for every objective point in all multidirectional image pairs throughout the walking.

The three-dimensional kinematic measurement system was built up from two sets of the video device, and its accuracy with 25 control points was examined in several camera setups. The 25 control points were precisely set into the objective space using a calibration wire frame, and the setting error was less than 1 mm in each axis. Then two video cameras were placed at varying

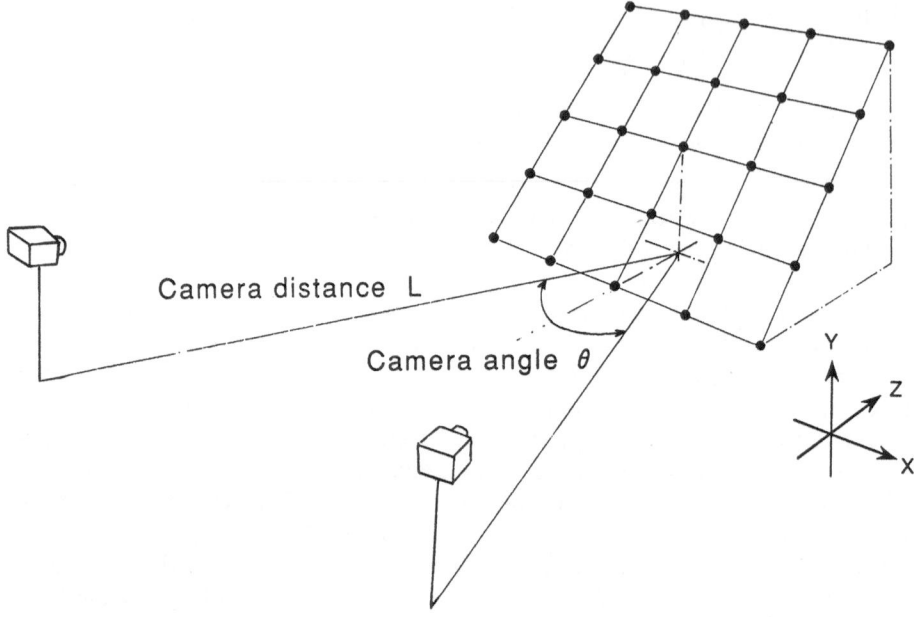

Fig. 7. Examination of practical accuracy in a 3-dimensional measurement. Varying camera distance and intersectional angle of light axes

Table 1. Errors in 3-dimensional measurement by the DLT calibration method will 11 parameters, RMS error, and maximum error (in parentheses). Symmetrical lens distortion was corrected in 2-dimensional measurement prior to 3 dimensional calibration

Camera distance L (m)		Camera angle θ		
		30°	60°	90°
2.5	X	—	5.3 (12.2)	—
	Y	—	4.3 (9.4)	—
	Z	—	2.4 (4.2)	—
3.0	X	5.1 (12.4)	3.3 (6.6)	6.2 (16.1)
	Y	13.2 (54.3)	2.6 (6.7)	3.7 (10.3)
	Z	7.7 (31.5)	1.4 (3.7)	2.0 (5.7)
3.5	X	—	12.6 (28.3)	—
	Y	—	18.4 (53.2)	—
	Z	—	10.6 (30.4)	—

distances of 2.5–3.5 m from the objective center and the intersectional angle between the camera light axes was varied to 30°, 60°, and 90° (Fig. 7). The practical error was evaluated by the root mean square (RMS) errors and maximum errors, and it was small enough for general gait analysis (Table 1). It is well known that the accuracy of the DLT calibration is directly related to the intersectional angle up to 90°, and negatively related to camera distance. Consequently there is a negative relationship between the accuracy and the size of objective space.

Kinetic Measurement

Measuring external force actions applied on the floor during walking, the so-called ground reaction force, is a gross assessment of gait performance. Several systems have been developed and are widely used in clinical institutions as standard equipment for gait analysis, and a number of studies on various diseases have been reported. The basic approach is waveform analysis of triaxial peak values of the vertical, fore-aft, and bilateral force components. For instance, the vertical force component has a well-known biphasic pattern which consists of two peaks on deceleration and acceleration periods in normal gait, and those two peak values usually exceed the body weight of subject. Where there is either no biphasic pattern or peak values of less than the body weight, there is something unusual in the gait. Furthermore, asymmetry of the waveform suggests some disorder of bipedal human gait. It is said that the ground reaction force is the overall result of movement of the whole body; not only the lower extremities but also other body segments such as the trunk and upper extremities are involved. Thus the waveform analysis grossly describes gait abnormality in whole body movement. However, in some cases results contradictory to clinical observation are obtained, showing apparently normal wavefrom in force measurement with abnormal posture or trunk motion as a compensation for pain or muscle weakness.

If we can measure bipedal ground reaction forces separately in several steps, we can obtain time factors and stride dimensions such as gait cycle, double supporting time, and step length. This information can be more useful for assessing gait and resolving the contradiction concerning the transition behavior of weight-bearing in bipedal walking.

Force Plate System

We have developed a twin force plate system using a pair of Kistler long plates (Type Z12888 2.0(L) × 0.4(W) × 0.2(H); Kistler Instrument, Switzerland) a and personal computer system (PC9801VX, NEC, Tokyo), which can measure more than three steps in ordinary step length and calculate those time factors and stride dimensions by automatic data analysis. A 12-bit 16-ch A/D converter module was used for on-line data acquisition with a sampling frequency of 200 Hz for gait study. Digital filtering was performed prior to data analysis because considerable vibration noise was observed. Generally, long plates have poor frequency response, losing rigidity in proportion to the span length. Accordingly plate vibration is easily caused by the stepping impact and continues for a long time. Additionally force transducers used in the force plate are very sensitive. Consequently the force plate measures not only the applied

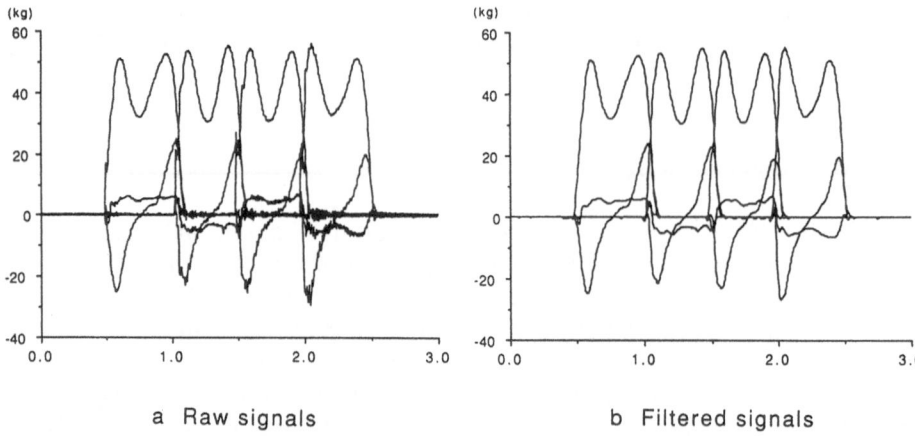

a Raw signals   b Filtered signals

Fig. 8a,b. Ground reaction force components. **a** Raw signals with vibration noise. **b** Filtered signal by an FIR digital filter with a cutoff frequency of 10 Hz

force but also its own vibration, which seriously interferes with data analysis (Fig. 8). However, in this computerized system, digital signal processing techniques can be applied to investigate and remove the vibration noise characterized by its frequency.

Signal Processing

Fast fourier Transform (FFT) is a well-known method for spectral analysis to estimate frequency distribution, but it results in broadhand frequency components. In recent years, several high resolution spectral estimation techniques have been developed to resolve narrowband frequency components. In this long plate study, the maximum entropy method (MEM) [19] with an autoregressive model was performed to investigate vibration noise and determine the cutoff frequency for digital filtering. According to the results, the main frequency components of triaxial ground reaction forces were below 10 Hz, and vibration noise components were observed between 40–60 Hz. Thus a cutoff frequency of 10 Hz was determined (Fig. 9).

Regarding signal filtering, many kinds of digital filters have been described in digital signal processing and various filter-designing software are commercially available for personal computer systems [19–22] Their theoretical derivation is quite complicated, although only a little background knowledge is required for practical application. The finite impulse response (FIR) filter is one of the most popular and applicable filters with simple computing and linear phase shift. The computing procedure is very similar to an average smoothing method with particular coefficients, but it has a fine frequency response and capability for specific designs such as low pass, high pass and band pass filtering. In this study, an equilipple FIR lowpass filter with a cutoff frequency of 10 Hz was

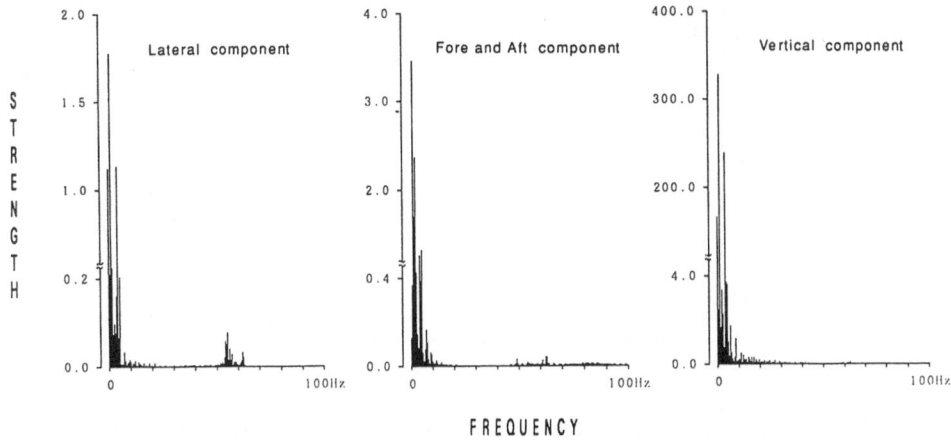

Fig. 9. Power spectrum analysis by maximum entropy method. Amplitude characteristics due to signal frequency were investigated by the maximum entropy method with an autoregressive model

Table 2. Coefficients for the FIR low pass filter designed by equiripple techniques. The cutoff frequency was 10 Hz; filter length, 25 taps; and sampling frequency, 200 Hz

$$X'_n = \sum_{K=-N}^{N} C_K X_{n-k} = C_{-12} X_{n-12} + \ldots + C_0 X_n + \ldots + C_{12} X_{n+12}$$

$C(-12), C(12) =$	0.0008085120352916
$C(-11), C(11) =$	0.0025434035342187
$C(-10), C(10) =$	0.0041539776138961
$C(-9), C(9) =$	0.0027216307353228
$C(-8), C(8) =$	-0.0047347121872008
$C(-7), C(7) =$	-0.0174989178776741
$C(-6), C(6) =$	-0.0279651936143637
$C(-5), C(5) =$	-0.0227684658020735
$C(-4), C(4) =$	0.0102347135543823
$C(-3), C(3) =$	0.0723628029227257
$C(-2), C(2) =$	0.1486791968345642
$C(-1), C(1) =$	0.2122019976377487
$C(0) \quad =$	0.2369469255208969

X'_n, filtered data; N, $\frac{1}{2}$(Filter length $-$ 1); C_k, filter coefficients; X_{n-k}, raw data; n, data index; K, offset index

designed from the result of previous spectra analysis, and 25 symmetrical coefficients were obtained (Table 2, Fig. 10). Through this FIR low pass filter, vibration noise was easily removed, and the refined triaxial ground reaction forces were then ready for automatic data analysis (Fig. 8).

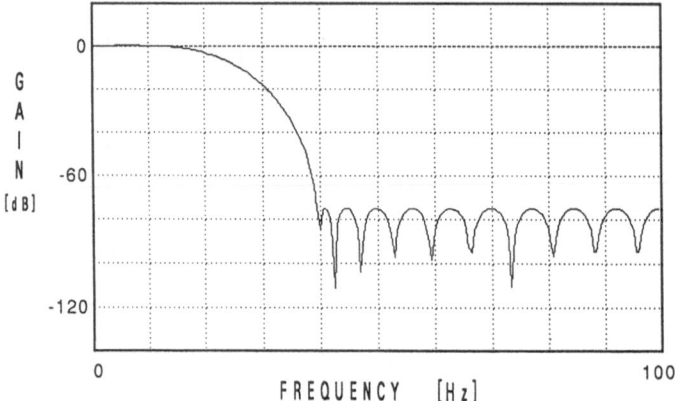

Fig. 10. Frequency response of the designed filter. Magnitude (attenuation) of the transfer function was demonstrated as a gain in dB, and the pass band edge was 10 Hz, the stop band edge, 40 Hz

Automatic Data Analysis

A number of parameters can be obtained with this bipedal measurement, including force peak values in each walking phase and their existing time, gait cycle and its composition, step and stride dimensions, and walking velocity and force vector angles as secondary products. These parameters have been clearly defined and used in previous gait studies using ordinary force plates, cinemato- graphy, foot switches, and foot prints [23–25]. An automatic analysis program was developed using technical knowledge of gait analysis to perform a large number of gait assessments, which can automatically detect gait events and calculate various parameters.

Firstly gait events were defined by the state of the ground reaction forces, and then the gait cycle and its composition were calculated from those gait events as follows. Heel contact (HC) is the time over which the vertical force has just exceeded the threshold value for probable random error, and toe off (TO) is the time over which the vertical force has vanished below the threshold value after heel contact. Consequently the gait cycle (CYC) is the interval of heel contacts on the same side, and cadence (CDN) can be obtained from an inverse function of gait cycle. Also, the stance phase (STN) and swing phase (SWG) are expressed as the periods from heel contact to toe off, and toe off to next heel contact on the same side. In the combination of vertical and fore-aft force components, mid-stance (MS) is defined here to mean the time with no fore-aft force in the stance phase, where the center of body mass seems to be just above the center of pressure with no deceleration and acceleration ahead. From this point of view, deceleration (DEC) and acceleration times (ACC) are expressed as the period from heel contact to mid-stance, and the period from mid-stance to toe off. In the bipedal loading state, gait events should be extensively described as attentive heel contact, opposite heel contact, and so

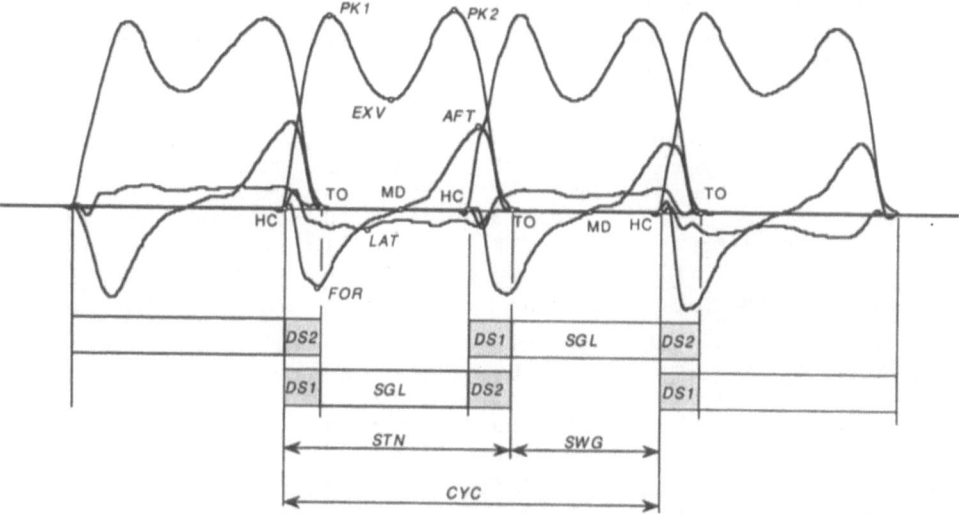

Fig. 11. Gait events and parameters. Gait events are defined by the state of the ground reaction force, and gait parameters are derived from those events which are shown in italics. *PK1*, *PK2*, peak values; *EXV*, extreme value; *FOR*, *AFT*, fore and aft peak values in fore-aft component; *HC*, heel contact; *TO*, toe off; *MD*, mid-stance; *LAT*, maximum value in the lateral component; *DS1*, *DS2*, double supporting times; *SGL*, single supporting time; *STN*, stance phase; *SWG*, swing phase; *CYC*, gait cycle

on. Thus double supporting times (DS1, DS2) are described as the period from attentive heel contact to opposite toe off and the period from opposite heel contact to attentive toe off. Single supporting time (SGL) is also expressed as the period from opposite toe off to opposite heel contact in the attentive stance phase. These compositions of the gait cycle were standardized by a percentile of the gait cycle. Subsequent waveform analysis was performed with logical references to gait events and walking phases, such as three components of lateral, fore-aft and vertical forces at each gait event. Also conventional parameters were obtained such as two peak values (PK1, PK2) and an extreme value (EXV) in the biphasic pattern of the vertical component, fore and aft peak values (FOR, AFT) in the fore-aft component, and maximum value in the lateral component (LAT). All force values were also standardized by a percentile of body weight (Fig. 11).

Step and stride dimensions were obtained by calculating the center of pressure using the moment equations of applied forces and the span length of the force plate. Step length (STP) was expressed here as the displacement of the center of pressure from the previous opposite mid-stance to attentive mid-stance taking place at the previous opposite heel contact and attentive heel contact, because the calculation possibly produces considerable error due to a small vertical force as at heel contact. Stride length (STR) was expressed as the displacement from previous mid-stance to present mid-stance on the attentive side. Then these parameters were standardized by the height of the subject.

Table 3. Mean values and standard deviations of various gait parameters in 95 normal female subjects. Most of the parameters are standardized by gait cycle (*cyc%*), body weight (*BW%*), and body height (*HT%*)

Age group	n	Age (years)	Height (cm)	Weight (kg)	CYC (s)	CDN (st/min)
20–30	25	25.0 ± 2.5	160.3 ± 4.8	52.7 ± 6.5	0.98 ± 0.06	123.6 ± 7.7
30–40	11	33.6 ± 2.6	156.0 ± 4.7	49.5 ± 3.3	0.97 ± 0.04	124.5 ± 5.4
40–50	17	45.1 ± 3.0	155.9 ± 5.2	55.6 ± 4.4	0.95 ± 0.05	126.6 ± 6.1
50–60	27	54.7 ± 3.0	153.0 ± 4.3	54.8 ± 7.0	0.99 ± 0.06	122.4 ± 6.6
60–70	12	63.1 ± 2.5	151.8 ± 3.8	51.2 ± 5.1	0.97 ± 0.05	124.4 ± 6.2
70–	3	71.7 ± 1.7	147.2 ± 9.3	54.6 ± 18.6	0.91 ± 0.06	133.0 ± 8.3

Age group	SWG (cyc%)	STN (cyc%)	DS1 (cyc%)	SGL (cyc%)	DS2 (cyc%)
20–30	38.5 ± 1.0	61.5 ± 1.0	11.5 ± 1.0	38.5 ± 1.0	11.6 ± 1.0
30–40	38.4 ± 0.9	61.6 ± 0.9	11.6 ± 0.9	38.4 ± 0.9	11.6 ± 0.8
40–50	38.8 ± 1.4	61.2 ± 1.4	11.2 ± 1.2	38.8 ± 1.3	11.2 ± 1.1
50–60	38.5 ± 1.3	61.5 ± 1.3	11.5 ± 1.1	38.4 ± 1.2	11.6 ± 1.1
60–70	37.6 ± 0.9	62.4 ± 0.9	12.4 ± 0.8	37.6 ± 1.0	12.4 ± 0.8
70–	37.5 ± 1.2	62.5 ± 1.2	12.5 ± 1.1	37.4 ± 1.3	12.7 ± 1.1

Age group	DEC (cyc%)	ACC (cyc%)	LAT (BW%)	FOR (BW%)	AFT (BW%)
20–30	32.6 ± 2.0	28.9 ± 1.8	7.5 ± 1.3	21.4 ± 2.9	22.2 ± 2.4
30–40	32.3 ± 2.3	28.3 ± 2.2	7.3 ± 0.9	19.8 ± 3.5	20.8 ± 2.6
40–50	33.6 ± 1.9	27.6 ± 2.2	7.4 ± 1.0	20.2 ± 3.8	20.9 ± 2.8
50–60	33.6 ± 1.9	27.9 ± 2.2	7.7 ± 1.3	18.8 ± 3.2	20.3 ± 2.8
60–70	33.9 ± 2.4	28.5 ± 2.2	7.4 ± 1.0	16.6 ± 2.5	18.5 ± 2.4
70–	32.1 ± 2.8	30.4 ± 1.7	6.7 ± 0.8	13.1 ± 1.4	14.7 ± 1.2

Age group	PK1 (BW%)	EXV (BW%)	PK2 (BW%)	STP (HT%)	STR (HT%)
20–30	114.9 ± 7.0	72.8 ± 4.9	112.5 ± 5.6	39.8 ± 2.6	78.7 ± 3.7
30–40	109.9 ± 7.8	76.1 ± 7.0	112.5 ± 4.0	40.8 ± 2.5	80.7 ± 4.4
40–50	113.1 ± 7.8	76.4 ± 6.2	109.9 ± 7.1	39.7 ± 2.7	78.5 ± 4.6
50–60	111.0 ± 6.0	78.7 ± 4.8	107.6 ± 5.7	38.9 ± 2.7	77.1 ± 4.8
60–70	108.0 ± 5.7	79.8 ± 4.9	107.2 ± 4.2	37.4 ± 3.5	74.3 ± 6.2
70–	105.0 ± 4.2	84.8 ± 1.5	103.1 ± 7.9	31.9 ± 1.7	64.0 ± 3.5

Age group	VEL (m/s)	GRD (BW%/%)	VQS (°)	VQE (°)	VQM (°)
20–30	1.28 ± 0.10	9.2 ± 1.1	−9.9 ± 1.3	10.6 ± 1.3	3.2 ± 0.7
30–40	1.29 ± 0.09	9.3 ± 1.2	−9.5 ± 1.3	10.9 ± 1.2	3.1 ± 0.8
40–50	1.28 ± 0.13	10.0 ± 1.5	−9.6 ± 1.8	11.2 ± 1.6	3.1 ± 0.5
50–60	1.18 ± 0.10	9.6 ± 1.2	−8.9 ± 1.4	10.7 ± 1.7	3.2 ± 0.8
60–70	1.14 ± 0.13	8.6 ± 0.9	−7.7 ± 1.1	9.5 ± 1.8	3.3 ± 0.4
70–	1.00 ± 0.02	8.2 ± 0.6	−6.4 ± 0.9	7.9 ± 0.8	3.4 ± 0.5

CYC, gait cycle; CDN, cadence; SWG, swing phase; STN, stance phase; DS1, DS2, double supporting times; SGL, single supporting time; DEC, ACC, deceleration and acceleration times; LAT, maximum value in the lateral component; FOR, AFT, fore and aft peak values in fore-aft component; PK1, PK2, peak values; EXV, extreme value; STP, step length; STR, stride length; VEL, walking velocity; GRD, gradient of the vertical force component in the first double support time; VQS, VQE, VQM, angles of resultant force vectors in the sagttlal plane at the beginning and of single support and end in the frontal plane at mid-stance

Step width can be obtained in the same manner, although it is difficult to use this parameter as no step across two plates is allowed in this twin long plate system, and the subject is thus forced to take slightly wide steps consciously.

Many second products were obtained through combination of the parameters mentioned above, including walking velocity (VEL), gradient of the vertical force component in the first double support time (GRD), and angles of the resultant force vectors in the saggital plane at the beginning and end of single support (VQS, VQE) and in the frontal plane at mid-stance (VQM). Some of these derivative parameters have been proposed in the fundamental concepts of kinetics and kinematics, but some others were created for the convenience of data analysis, and have no direct translation for actual gait performance because the ground reaction force is the overall result of whole body movement. However, they suggest something unusual in comparison with the contralateral side or their standard ranges on normal subjects. Finally practical results on 95 normal female subjects were obtained using this twin plate system. The mean values and standard deviations may allow an estimate of the degree of abnormality by referring to statistical probability (Table 3).

Discussion

Gait analysis has a long history and has gone through cyclic trends corresponding to new developments in equipment, such as cinematography, force plates, electrogoniometers, and so on. In research-oriented studies, gait has been investigated and the complex mechanisms explained. In the clinical field, gait analysis has been performed as a functional assessment of gait disorders for the purpose of investigating impediments and progress in a case. There have been many studies on joint diseases and artificial joint replacements in the lower extremities, which examined pre- and post-operative gait performance in comparison with normal subjects. However, in previous studies aiming at the movement of only the lower extremities and the ground reaction force, if a patient had a severely impeded gait because of intolerable pain, muscle weakness or joint immovability, unusual movements and force actions can be measured and represented by gait parameters using those facilities. However, a patient with exaggerated limping sometimes shows apparently normal results because of the compensatory movements. In effect, there is no clear borderline between normal and abnormal gait, and no exact interpretation for each parameter because of the wide variation in normal gait and various compensatory movements in clinical cases. So the questions remain as to how to objectively define normal gait and how to evaluate and classify gait disorders.

In recent gait studies, three-dimensional kinematic measurement systems using television techniques or photoelectro devices are commonly used in combination with DLT three-dimensional calibration methods with free camera setup and no parallax error. Furthermore, synchronous kinetic measurement is preferably performed by the force plate system, and digital signal filtering is necessarily performed prior to data analysis instead of implicit noise cancellation manually. Thus both kinematic and kinetic gait parameters can be obtained from a refined data series in practice using the integrated system. Further investigation should move on to combinatorial analysis of these abnormal parameters, referring to the mean values and standard ranges. However not all of these parameters are useful, and use of some can result in a flood of information or combinatorial "explosion" at this stage. Therefore the parameters should be carefully selected with regard to their significance for the purpose of the study.

On the other hand, AI techniques have been developed and widely introduced to information processing, including knowledge-based expert systems for automatic diagnosis, decision making and classification [26–28]. In gait analysis and functional electro-stimulation for walking aids, several expert systems have already been developed for classifying gait disorders and automatic gait event detection [29, 30]. Expert systems have considerable knowledge and information in specific fields expressly stated in the computer software. There are two basic types of expert system, rule-based systems and induction systems. A rule-based system uses production rules for representing knowledge. Automatic data analysis on the ground reaction force, as described above, is a kind of rule-based expert system with definite knowledge on the loading status of bipedal walking. IF-THEN statements are often used as conditional rules to detect gait events in the program. The induction type of expert system has rules inferred from a series of examples with attributes or conditions and specific results. In other words, results are derived from the various combinations of attributes. This is originally the same concept as the combinatorial analysis of gait parameters, and most applicable for classifying gait disorders when a sufficient number of examples are obtained by the integrated system previously described.

Conclusion

A personal-computer-based kinematic gait measurement system was developed using advanced electronics technology, which can accurately measure three-dimensional coordinates of optoreflective markers on body segments using television techniques and direct linear transformation for three-dimensional calibration. A kinetic measurement system with automatic data analysis was also developed using twin long force plates, which can measure triaxial force components for several steps, and various gait parameters are calculated by automatic data analysis software with digital filtering. Both rule-based and induction expert system are promising approaches to solve further problems in analysis and evaluation of gait performance.

Acknowledgement. This study was greatly supported by Prof. Shigeo Niwa, head of the Department of Orthopedic Surgery, Aichi Medical University, who is strongly interested in Biomechanics.

References

1. Jarrett MO (1976) A television/computer system for human locomotion analysis. PhD thesis, University of Strathclyde, Glasgow
2. Andrews BJ (1983) On-line locomotion analysis using television techniques. PhD thesis, University of Strathclyde, Glasgow
3. Hattori T, Hirose S, Sawai K, Mitsui T, Niwa S (1990) Video device for an automatic motion analysis system. Proceedings of the 7th Meeting of the European Society of Biomechanics, Aarhus, Denmark, July 8–11, p 44
4. Hattori T, Hirose S, Sawai K, Mitsui T, Niwa S (1991) Video device for an automatic motion analysis system. J Biomech 24(6):470
5. Abdel-Aziz YI, Karara HM (1971) Direct linear transformation from comparator coordinates into object space coordinates in close range photogrammetry. Symposium on close range photogrammetry, Urbana, Il January 26–29
6. Marzan GT, Karara HM (1975) A computer program for direct linear transformation solution of the colinearity condition, and some applications of it. Proceedings of the Symposium on Close Range Photogrammetric Systems, American Society of Photogrammetry, Champain, Il July 28–August 1
7. Philippens M (1980) Aspects of load acquisition and analysis in the index finger. Internal Report, Bio-Engineering Unit, University of Strathclyde, Glasgow
8. Shapiro R (1978) Direct linear transformation method for three-dimensional cinematography. J Res Quart 49(2):197–205
9. Woltring HJ (1980) Planar control in multi-camera calibration for 3-D gait studies. J Biomech 13:39–48
10. Miller NR, Shapiro R, McLaughlin T (1980) A technique for obtaining spatial kinematic parameters from cinematographic data. J Biomech 13:535–547
11. Dapena J, Harman EA, Miller JA (1982) Three-dimensional cinematography with control object of unknown shape. J Biomech 15:11–19
12. Dapena J (1985) Correction for three-dimensional cinematography with control object of unknown shape. J Biomech 18(2):163
13. Whittle MW (1982) Calibration and performance of a 3-dimensional television system for kinematic analysis. J Biomech 15(3):185–196
14. Fioretti S, Germani A, Leo T (1985) Stereometory in very close-range stereophotogrammetry with non-metric cameras for human movement analysis. J Biomech 18(11):831–842
15. Wood GA, Marshall RN (1986) The accuracy of DLT extrapolation in three-dimensional file analysis. J Biomech 19(9):781–785
16. Hatze H (1988) High-precision three-dimensional photogrammetric calibration and object space reconstruction using a modified DLTapproach. J Biomech 21(7): 533–538
17. Ghosh SK (1983) A close-range photogrammetric system for 3-D measurements and perspective diagramming in biomechanics. J Biomech 16(8):667–674
18. Van Gheluwe B (1974) A new three-dimensional filming technique involving simplified alignment and mêasurement procedures. Biomechanics IV:476–481. University Park Press, Baltimore
19. Minami S (1986) Waveform data processing for scientific measurement and instrumentation. CQ Press, Tokyo
20. DeFatta DJ, Lucas JG, Hodgkiss WS (1988) Digital signal processing: A system design approach. Wiley, New York

21. Bateman A, Yates W (1988) Digital signal processing design. Pitman, London
22. Ochi H (1990) Introduction of digital filter design. CQ Press, Tokyo
23. Murray MP, Drought AB, Kory RC (1964) Walking patterns of normal man. J Bone Joint Surg [Am] 46(2):335–360
24. Dagg AI (1977) Running, walking, and jumping. Wykeham, London
25. Galley PM, Forster AL (1982) Human movement. Churchill-Livingstone, Edinburgh
26. Hart A (1989) Knowledge acquisition for expert systems. Kogan Page, London
27. James M (1986) Basic artificial intelligence. Butterworth Borough Green, Sevenoaks, UK
28. Frenzel LE (1987) Understanding expert systems. Howard W. Sams, Indiana
29. Dzierzanowski JM, Bourne JR, Shiavi R, Sandell HSN, Guy D (1985) Gaitspert: An expert system for the evaluation of abnormal human locomotion arising from stroke. IEEE Trans Biomed Eng 32:935–942
30. Kirkwood CA, Andrews BJ, Mowforth P (1989) Automatic detection of gait events: A case study using inductive learning techniques. J Biomed Eng 11:511–516.

Knee Joint Force in Normal and Pathological Gait

IAN J. HARRINGTON[1]

Summary. Knee joint loading in normal adults is characterized by three distinct joint force peaks corresponding to force actions generated in the hamstring, quadriceps and gastrocnemius muscles. These peaks tend to be absent during load transmission in pathological knees. The resultant knee joint force is transmitted primarily through the medial joint compartment in normal knees but in pathological joints, particularly those with varus-valgus angulation, its force (the centre of joint pressure) is frequently unpredictable and there are no distinct peaks. The magnitude of the knee joint bearing force tends to increase with increasing angular deformity but does not exceed normal joint loading values. Joint force transmission in prosthetic knees does not follow a normal loading pattern but is more typical of pre-operative loading profiles.

Key words. Joint force — Force actions — Centre of pressure — Moments

Introduction

Human locomotion is an extremely complex phenomenon. The fact that walking is a three-dimensional activity creates major difficulties in the analysis of force actions transmitted between limb segments. Interest in the study of muscular control of locomotion was stimulated by the Weber brothers [1] who claimed that during the swing phase of walking, muscular control was not necessary and the motion of the leg occurred much like a simple pendulum. Further contributions to study of human gait were made by Marey [2] Braune and Fischer et al. [3] and Bernstein [4]. Elftman [5] studied the distribution of pressure in the human foot, the function of the arms in walking, the rotation of

---
[1] Toronto East General and Orthopaedic Hospital, 825 Coxwell Ave., Toronto, Ontario, Canada

121

the body, and the function of the muscles in walking. Pauwels [6] calculated the magnitude and direction of the resultant force transmitted between the femoral head and the acetabulum for an individual standing on one leg. He also extended this type of static two-dimensional frontal plane analysis to formulate a rationale for osteotomy in the hip for the treatment of osteoarthritis.

Bresler and Frankel [7] used a force plate developed by Cunningham and Brown [8] to measure ground-to-foot forces and a three-dimensional analysis to position the limb segments in space, and calculated the external forces and moments acting at the ankle, knee and hip in four young male subjects. They did not, however, attempt to calculate force actions occurring in muscles, ligaments or at the bearing surfaces. They reported that the bulk of their computations involved determining gravitational and inertial forces acting on limb segments. They concluded that since the contribution of these forces to total moments was small, an accurate assessment of force actions could be achieved by considering the ground-to-foot forces only during the stance phase of gait. Bresler and Frankel's work has become the basis for virtually all subsequent studies related to force transmission involving the lower extremity.

A direct determination of joint force at the hip was reported by Rydell [9]. A modified Austin-Moore prosthesis capable of measuring force at the hip was surgically implanted in two patients.

At the same time, Paul [10] at the University of Strathclyde, used a non-surgical three-dimensional experimental technique to measure the variation of ground-to-foot force action correlated with cine photographic records of the subjects' gait to calculate force transmission at the hip. This was an extension of the basic technique developed previously by Bresler and Frankel. Gravitational and inertial forces were included in the computations. It is interesting to note that the phasic relationship of Paul's and Rydell's work was comparable but the magnitude of force transmitted at the hip as determined by Rydell was less than that calculated by Paul. Paul's subjects were young healthy adults, whereas Rydell's patients were elderly and recovering from hip surgery.

Morrison [11] used essentially the same technique as Paul to calculate forces transmitted at the normal knee. Morrison's and Paul's results were based on the assessment of young, healthy individuals without gait abnormalities.

Harrington [12] used Morrison's technique but eliminated gravitational and inertial forces from the computations to calculate joint force transmission at the knee for both normal individuals and those with various pathological conditions affecting gait. Much of the data presented in this paper is based on that work.

Experimental Technique

All tests are conducted using a walk path along which is centred a force plate. The force plate measures three forces, vertical, fore and aft, medio-lateral, and three moments relative to a set of orthogonal axes intersecting at the centre of the plate. The forces and moments define the centre of foot pressure and the magnitude and direction of the resultant force between the ground and foot. In our studies, skin markers were placed on each subject at the hip, knee, and ankle to determine the centres of joint rotation, and each subject was photographed from the front and side by two synchronised 16 mm cameras at 50 frames per second. In this way, a continuous record of the variation of ground to foot force at 0.02 second intervals correlated with the position of the limb in space was obtained from synchronized force plate and cine photographic measurements. This information was then processed to obtain the resultant external forces and moments about the knee joint.

The external force system acting on the tibia when inertial and gravitational forces are ignored consists of the three ground to foot forces, $(F_x F_y F_z)$ and moments or torques $(M_x M_y M_z)$ due to these forces, acting relative to the centre of rotation of the knee joint. F_x is the fore and aft shear force acting at the foot in the line of progression. F_y is the vertical component and F_z is the side-to-side ground-to-foot force. The effect of M_z is to either flex or extend the knee. Similarly, M_x tends to abduct or adduct the tibia relative to the thigh and M_y tends to rotate it or twist it about its long axis.

Force Model Analysis

For equilibrium, the external moments and forces at the knee must be balanced by force actions occurring in muscles and ligaments spanning the joint as well as by the articular bearing surfaces. In general, the moments are transmitted by tensions developed in the muscles and ligaments. A simplified force model of the knee is shown in Fig. 1 which includes the three major muscle groups — the quadriceps (Q), Hamstrings (H) and Gastrocnemius (G), the four main ligaments affecting the knee — the anterior (ACL) and posterior cruciate (PCL) ligaments respectively and the medial and lateral collaterals (MCL) and

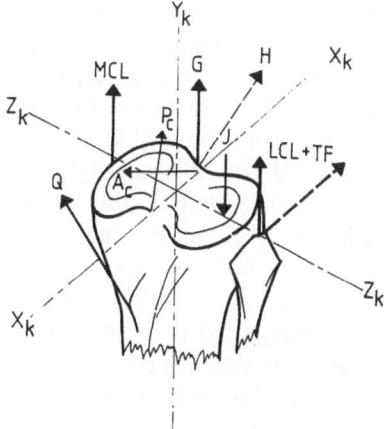

Fig. 1. Simplified knee force model. Of Quadriceps (*Q*), hamstring (*H*), and gastrocnemius muscle (*G*) force actions, anterior (*AC*) and posterior (*PC*) cruciate forces, medial collateral (*MCL*) and lateral collateral (*LCL*) ligament forces, force action in the tensor fascialata (*TF*), and resultant joint bearing force (*J*)

(LCL), the reactive joint bearing force (J) and the location of the centre of joint pressure (Zo). There are nine unknowns in the force model so that a solution is not possible unless certain assumptions are made. For example, it is assumed that each muscle group has one main function acting as one unit with no antagonistic or agonistic activity; and that this activity can be represented by a single line that can be defined anatomically. Its direction is determined from anatomical considerations and angular relationships of limb segments determined from cine film analysis. It is also assumed that when the externally applied moment is tending to flex the knee, this will be balanced by an equal and opposite force action in the quadriceps muscle group and that in this situation no activity occurs in the knee flexors, (i.e., the hamstrings and gastrocnemii). The reverse applies when the knee is being extended.

From Morrison's analysis, the cruciate ligaments are supposed to act only one at a time and because of their specific anatomical arrangement, are assumed to exclusively resist antero-posterior displacements of the tibia relative to the femur. A tensile force in one ligament implies no force action in the other. Morrison did not include medio-lateral shear forces in the calculation of cruciate ligament force actions, however it is evident from anatomical considerations that these ligaments are able to resist medio-lateral shear and can be accounted for in the analysis [12]. The cruciates are considered not to transmit moments in either the sagittal or frontal planes. They do, however, in combination with the collateral ligaments, transmit torque about the Y-axis. This specific function, however, is not accounted for in the analysis. Mechanically, the collateral ligaments are anatomically better situated to resist torque and abduction-adduction moments in the frontal plane than are the cruciates.

The main function of the collateral ligaments is to resist abduction and adduction moments about the knee. When there is a tendency to adduct the tibia relative to the thigh, rotation will be resisted by a tensile force in the lateral collateral ligament and by the joint force acting on the medial tibial condyle. If the externally applied moment tends to abduct the tibia, equilibrium will occur when a tensile force occurs in the medial collateral ligament and the joint force is centred over the lateral tibial condyle. As the adducting moment increases in magnitude, the resultant joint force and its point of application, the centre of joint pressure (Zo), will shift towards the medial side of the knee; whereas the opposite situation occurs when the knee is subjected to an abducting moment (Fig. 2a,b). The limiting value of (Zo) is W/4, i.e., the center of the tibial condyle. Depending on the loading situation, total joint force transmitted by the articular surfaces of the knee may be concentrated entirely in either the medial or lateral compartment as a one-point bearing contact, (100% of total load) or shared as a two-point contact in both tibial condyles. In general, there are three solutions for (Zo). If, from the computations, (Zo) is greater than W/4 then the joint force (J) will be located entirely in either the lateral or medial condyle depending on whether the knee is being abducted or adducted, respectively. For equilibrium there will be a force action in one or the other of the collateral ligaments (MCL, and/or

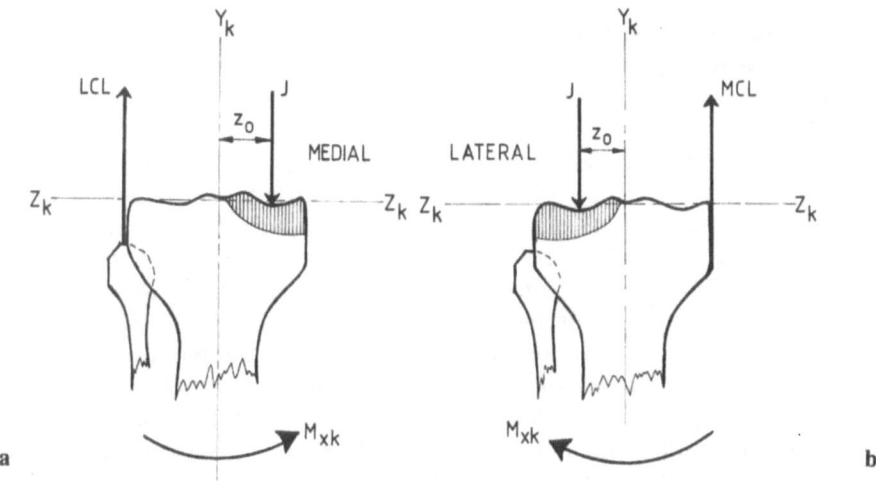

Fig. 2a,b. a Adduction knee moment. When the knee is subjected to an adduction moment (M_{xk}) the joint bearing force (J) is located in the medial compartment at the distance Z_0 from the knee centre (centre of joint pressure) and the lateral collateral ligament (LCL) is loaded. **b** Abduction moment. When the knee is subjected to an abduction moment (M_{xk}), the resultant joint bearing force (J) is located in the lateral compartment and the medial collateral ligament (MCL) is loaded

LCL). If (Zo) is calculated to be less than W/4 then there is no force transmitted by the collateral ligaments and the total joint load (J) will be distributed proportionately in the medial and lateral compartments as (Jm) and (Jl) respectively (Fig. 3). With these simplifying assumptions, a valid solution can be made for the force actions transmitted by the various muscle groups and ligaments spanning the knee joint as well as the joint articular bearing force, for all possible loading conditions during the stance and swing phase of the gait cycle.

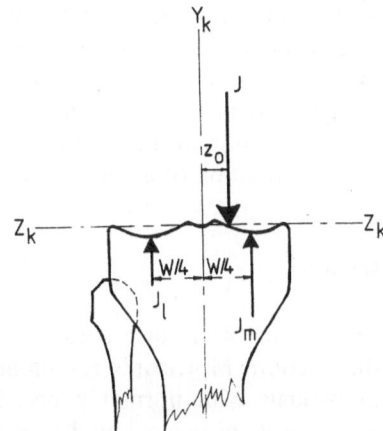

Fig. 3. Centre of joint pressure (Z_0) and load distribution. The limiting value of Z_0 is $W/4$. Total joint load J is distributed proportionately as J_m and J_l in the medial and lateral joint compartments, respectively

Gravitational and inertia forces have been omitted from the analysis. This is in accordance with Bresler and Frankel's work which showed that the contribution of these forces to the sagittal plane moments for the stance phase of gait in normal individuals is negligible. It is anticipated that their effect would be even less for subjects walking at reduced speeds due to pathological gait.

Nissan [13] reviewed the basic assumptions required for a solution of force transmission at the knee using this technique and concluded that the assumptions used to define the lines of actions of muscles and ligaments was a good approximation. The exact coordinates of the ligament insertions were found to be unimportant. The point of application of the inter-condylar forces however proved to be important, and the same applied to the definition of knee centre. He concluded that the sensitivity of the results to the exact location of the knee centre and the point of application requires an accurate description and definition to be used in the literature in order to enable comparison of results. Nissan also pointed out that it is not possible to calculate agonistic and antagonistic muscle activity even though it is well known that this exists. Most investigators have assumed it to be secondary and negligible in order to enable a single solution to the equilibrium equations. Naturally, any antagonistic muscle action would give rise to considerable increases in load action above those calculated in this way. Experimental evidence from E.M.G. readings in normal individuals, however, indicate that the only major areas of antagonistic muscle action in the walking cycle are in the region of 10% of the gait cycle before and after heel strike and in general the maximum values of joint loads do not occur at this time. In pathological gait, particularly where there is isotonic contraction of muscle fibres (spasticity) such as in patients with cerebral palsy, stroke and Parkinson's disease etc., this effect may be further magnified. The resultant increased force transmission across joint bearing surfaces for these patients cannot be calculated by the force plate cine photographic technique. Paul has shown that the errors in the calculation of the forces transmitted by internal body structures during locomotion, taking the most pessimistic view, may amount to 50% of the value at any instant. The sources of error relate to the programme for analysing the force plate and displacement data and the programme for computation of muscle and joint force. Nevertheless, it is contended that unless and until transducers can be designed to measure the loads transmitted by body structures without interfering with normal function, values calculated in this way form the best guide to interpretation of a very complicated area.

Results

The results of my own studies as well as those of Morrison's are presented in this section. Morrison's results are based on the assessment of young healthy individuals with normal knees. My results are based on the bilateral limb assessments of normal adults, as well as some normal children, and a group of subjects of varying ages with abnormal gait due to a variety of problems

Fig. 4. *Normal adult joint force* exhibits three distinct loading peaks *a*, *b*, and *c* which are characteristic of normal joint loading. Proportionately, most of the joint force is transmitted through the medial part of the knee (J_m)

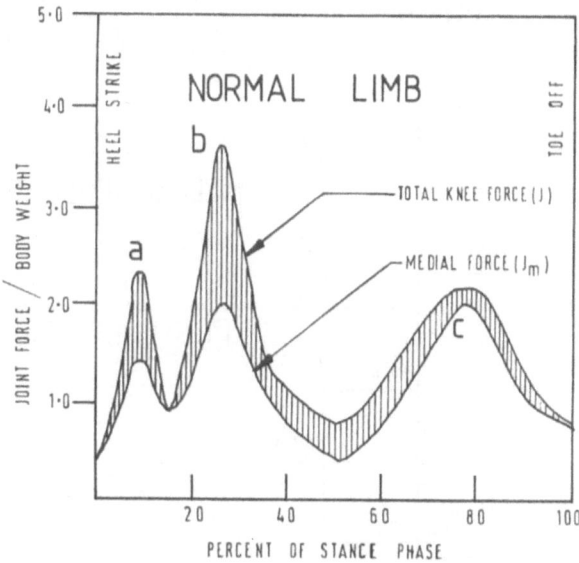

including osteoarthritis and rheumatoid arthritis of the knee, and paralysis due to polio. Joint loading has also been calculated for patients with angular deformity, i.e., varus, valgus and flexion deformity of the knee under conditions of dynamic and static loading. The gait and joint loading characteristics of a few patients who have had total knee replacements are also discussed.

Force Actions in Normal Subjects During Level Walking

Joint Force Profiles

The magnitude and phasic relationship of force actions transmitted at the muscles and ligaments of the knee joint and bearing force in normal individuals for the activity of level walking, although not identical, compare favourably with the results obtained by Morrison. A normal loading cycle during the stance phase of gait is characterized by three peak loads in Fig. 4, "a", "b", and "c" which correspond to hamstring (H), quadriceps (Q) and gastrocnemius (G) contraction respectively. The overall magnitude of loading is 3–5 times body weight. The average maximum bearing force transmitted at the knee is 3.5 times body weight.

Ligaments

Cruciate Ligaments. Cruciate ligament force varied in magnitude from 150–600 Newtons. For most subjects, the anterior cruciate ligament (ACL) transmits load during the early part of the stance phase and the posterior cruciate (PCL) towards the end of the cycle. The posterior cruciate ligament tends to have higher force values on average than the anterior. When the medio-lateral shear

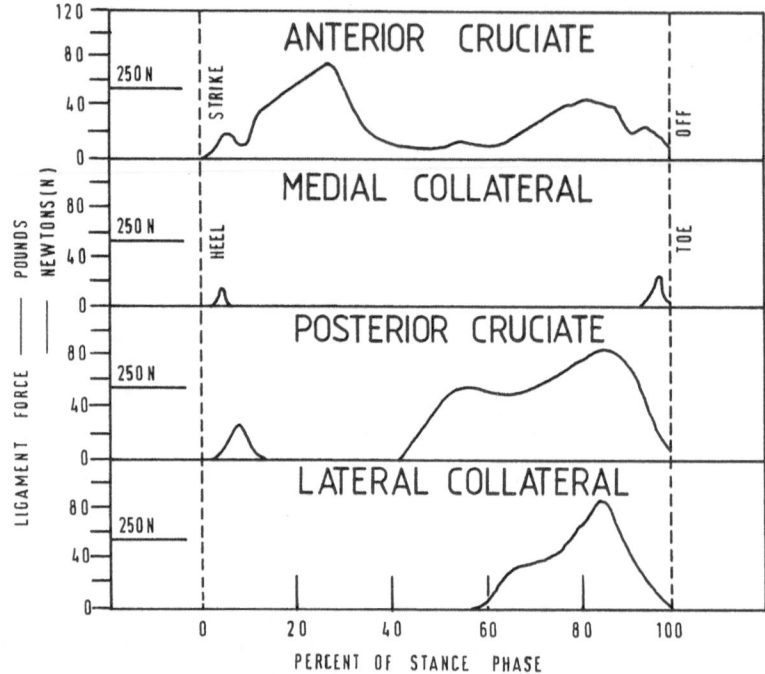

Fig. 5. Forces transmitted by the cruciate and collateral ligaments of the knee in normal adults

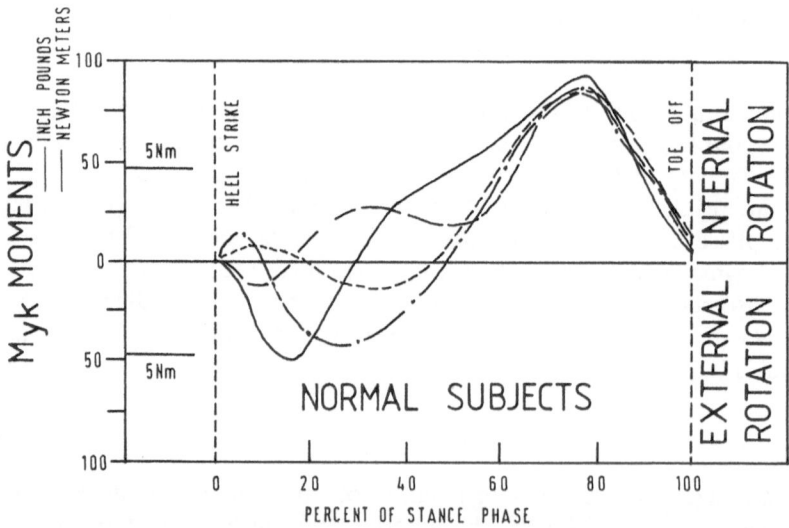

Fig. 6. Tibial shaft torque. The tibia is subjected to a torque that tends to externally rotate the knee during the early part of stance and to internally rotate it toward the end of the stance phase of gait

force component is included in the calculations for cruciate ligament force actions the anterior cruciate (ACL) is active throughout the stance phase of gait. The posterior cruciate ligament, however, transmits the greater load.

Collateral Ligaments. Almost no force is transmitted by the medial collateral ligament (MCL) during weight bearing. The lateral collateral ligament (LCL), however, transmits load for a major portion of the stance phase (Fig. 5). Anatomically, this is a relatively small ligament. It is likely that the lateral collateral ligament shares force transmission with larger laterally placed structures such as the tensor fascia lata, ileo-tibial tract, and the biceps femoris muscle. The maximum force calculated for this ligament was 625 N.

Muscle Forces

Maximum values of 1350–1800 N of force were computed for the three major muscle groups. The phasic relationship of the major muscle groups, (Q), (H), and (G) correspond to the "a", "b", and "c" peak force transmissions across the knee bearing surface.

Knee Joint Torque

During the early stance phase of gait, the knee is subjected to a torque, M_{yk}, that tends to externally rotate the shank relative to the femur. The applied torque changes during the later part of the stance phase to internally rotate the shank. Generally, the magnitude of the tibial torques is not high — maximum 9 Newton Meters (Fig. 6).

Force Actions in Normal Subjects While Ascending and Decending Stairs

Morrison has shown that forces transmitted by ligaments, muscles and the knee bearing surfaces tend to increase for the activity of ascending and decending a ramp or stairs. The Quadriceps muscle (Q) is active throughout. When walking down stairs the function of the quadriceps is to control flexion of the knee whereas while ascending stairs it acts to extend the knee. In both situations muscle force is increased, particularly while walking down stairs. The bearing force (J) is also increased. Force in the posterior cruciate (PCL) and lateral collateral (LCL) ligaments is greatest while going up stairs whereas force in the anterior cruciate ligament (ACL) increases while decending stairs. The function of the cruciates in relation to stair climbing may have importance for surface replacement artificial knee designs. Andriacchi et al. [14] has suggested that the presence and function of the posterior cruciate ligament (PCL) may affect the patients' ability to descend stairs. He showed that patients with posterior cruciate sacrificing prostheses had a reduced descent velocity as well as a reduced range of knee motion. As the knee flexes while the patient descends stairs, the posterior cruciate ligament is in a position to resist the forward thrust of the femur on the tibia (Fig. 7). In the absence of the ligament, the joint must be stabilized by constraints built into the prosthesis

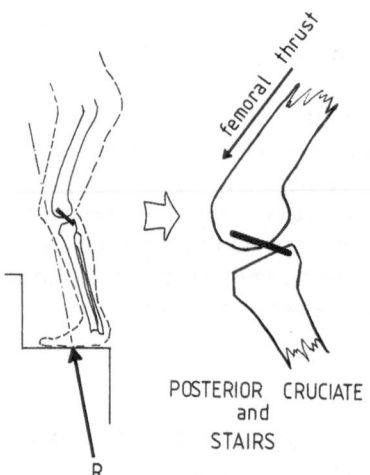

Fig. 7. Posterior cruciate ligament function and stair use. The posterior cruciate ligament is best situated to resist the forward thrust of the femur while descending stairs

femoral thrust

POSTERIOR CRUCIATE
and
STAIRS

R

since the muscles and other soft tissues spanning the joint are not in a mechanically advantageous position to do so. Differences in design however did not influence gait during level walking.

Children's Force Actions

Force transmission at the knee in children was similar to that of adults but with higher magnitude of the "a" peak (Fig. 8). This was interpreted as due to a more vigorous heel strike resulting in a greater extension moment at the knee resisted by a force action in the hamstrings. Collateral and cruciate ligament

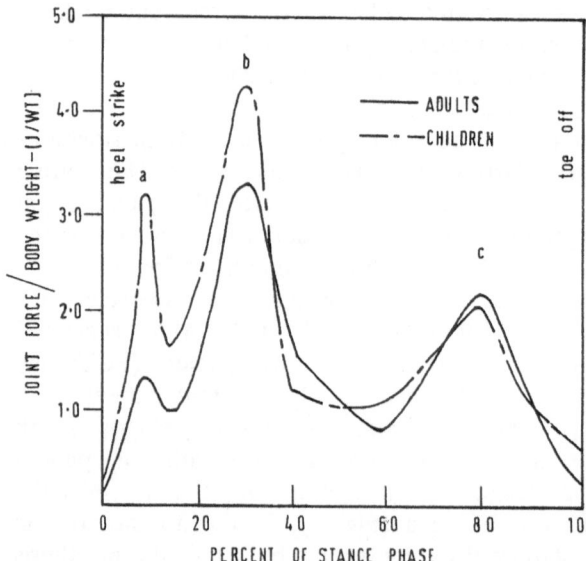

Fig. 8. Knee joint force in children and adults. Joint force transmission in children is characterized by a distinctive *a* peak

values for children compared well with adult values on a per weight basis and also for phasic relationships. Like adult cruciate forces, higher maximum values occurred in the posterior cruciate ligament (PCL).

Centre of Knee Joint Pressure

Morrison's studies have suggested that the centre of joint pressure (Zo) tends to be located in the medial compartment for normal adult knees. Prior to his work, it was generally assumed that because of the natural valgus obliquity of the femur, most of the knee joint force would be borne by the lateral condyle. Our results are identical to Morrison's and show that in normal adults for the activity of walking, the centre of joint pressure is concentrated mostly in the medial compartment of the knee throughout the stance phase of gait. Morrison showed that during the swing phase of gait the centre of pressure (Zo) shifted laterally. The lateral shift is entirely due to inertia forces. There are, however, individual variations in the location of the centre of pressure during stance, particularly in those adults who walk at a rapid rate and in some children. For these individuals, the centre of pressure is frequently calculated in the lateral compartment (Fig. 9).

Force Actions in Pathological Knees

Drillis [15], was probably the first to show a relationship between force plate data (ground-to-foot external force actions) and pathological gait. He showed that the slope of the vertical load curves for level walking for normal speed was

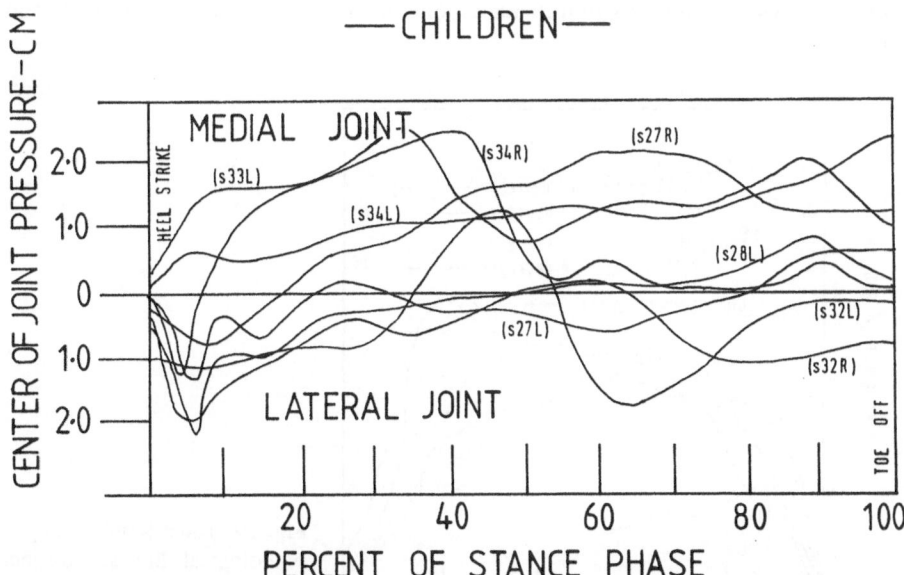

Fig. 9. Centre of joint pressure in children. Shift of the centre of pressure into the lateral knee compartment commonly occurs in children during the stance phase of gait

greater than that for pathological gait. This was interpreted as an attempt by the individual with pathological gait to protect the affected leg from sudden force change. Drillis did not attempt to calculate the forces transmitted by the muscles, ligaments, and bearing surfaces of the involved limb. More recent studies of pathological gait have also made similar observations from ground-to-foot force data. The interpretation of this material is difficult. It provides limited information since it does not account for force transmission by the ligaments and muscles spanning the affected joint and cannot give an accurate estimate of joint force.

Our results also show that Force actions in pathological knees differ strikingly from normals. At first glance, one might expect force actions to be increased in pathological limbs but for these individuals forces transmitted by muscles, ligament, and bearing surfaces were significantly reduced in every instance. Elimination of the "a", "b," and "c" force peaks was characteristic of the joint force bearing curves for these subjects, including those individuals who had knee replacements.

Polio Subjects

The joint force curves for the polio subjects illustrate many of the features typical for all pathological limbs tested (Fig. 10). The "a", "b," and "c" peaks tended to be absent or reduced in magnitude. For polio limbs, the resultant knee joint force was considerably lower than the calculated value for normal limbs and the magnitude approached body weight. The knee in these subjects was subjected to an abduction moment of low magnitude throughout stance. Some individuals also exhibited an extension moment pattern in the sagittal plane.

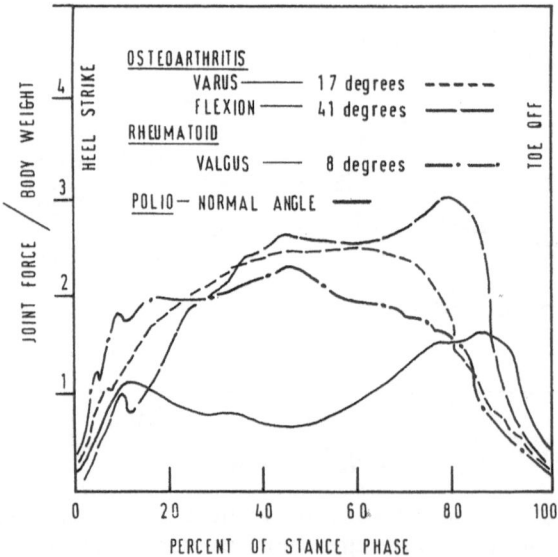

Fig. 10. Knee joint loading in pathological knees. Absence or blunting of the *a*, *b*, and *c* force peaks is characteristic of loading in pathological knees

Arthritic Subjects

Similarly shaped force curves were also noted for the arthritic patients. For most of the individuals in this group the "a", "b," and "c" force peaks were also eliminated. Maximum force tended to be transmitted during the midstance phase of the gait cycle. The only exceptions were those individuals who walked at increased speeds. The average maximum knee joint force for osteoarthritic patients was 2.5 times body weight, and 2.1 times body weight for rheumatoid arthritic patients.

Ligament and Muscle Forces

Ligament and muscle force actions in pathological limbs are lower than the calculated values for normal subjects. For most of the arthritic subjects, there was no force action computed for the MCL or the LCL.

Centre of Knee Joint Pressure

The location of the center of knee joint pressure (Zo) showed greater variability in pathological knees tending to oscillate about the midline of the knee in those individuals with normally aligned limbs and for those subjects who walked slowly. Shifts of the center of pressure into the medial or lateral compartment of the knee occurred with varus and valgus deformity respectively but not to the same degree as one would expect from static analysis.

Knee Joint Deformity

Varus Angulation

From a comparison of dynamic and static loading conditions in our results, it was shown for varus deformity that the magnitude of the total joint load was greater under conditions of dynamic loading for all subjects. In each case of static analysis, the total joint force, (100% of the joint load) was transmitted by the medial condyle, (one-point contact). The force in the lateral compartment was zero. For *dynamic analysis*, force was shared by both condyles, (two-point contact) in most patients, however, the point of application of the resultant joint force, (center of joint pressure) and therefore the major proportion of total load was located in the medial condyle. In some varus limbs tested dynamically, joint load was localised exclusively to the medial compartment throughout the stance phase of the gait cycle (one-point contact). For these individuals, 100 per cent of the knee joint force was carried by the medial condyle and zero pressure occurred laterally. The magnitude of joint force tended to increase with increasing angular deformity in both groups.

Valgus Angulation

For patients with valgus deformity, maximum joint bearing force was also greater for dynamic than for static methods. The point of application of joint load, (Zo), for all knees with valgus deformity assessed by the static method was located in the lateral compartment. Shared load, (two-point bearing contact) was calculated for some patients by static and dynamic methods. For the individuals tested dynamically, the centre of pressure did not remain in the lateral condyle throughout the stance phase of the gait cycle but shifted into the medial compartment for major portions of the stance phase and corresponded to the period of maximum load transmission. Joint force also tended to increase in both groups with increasing valgus angulation.

Flexion Deformity

For patients with flexion deformity, the major angulation for this group was in the sagittal plane. Each subject had a flexion contracture that was associated with minimum varus valgus angulation. The difference between the calculated values of joint load from static and dynamic analysis was larger than for the subjects with varus or valgus deformity. The centre of joint pressure was located in the medial compartment by static and dynamic analysis and two point bearing contact was calculated for each subject. It is interesting to note that significant force actions occur in the quadriceps muscle throughout the stance phase of the gait cycle for these patients. Total joint load was also noted to increase with increasing knee flexion deformity.

In each group (varus, valgus and flexion deformity), there was a general tendency for increase in magnitude of joint load with increasing angulation of the knee and the effect was most apparent in subjects with flexion deformity. This is because force actions in the major muscle groups cannot be accounted for in a static frontal plane analysis. Loading patterns seem also to be more predictable for varus than for valgus knees. A fairly large valgus angulation is required before the centre of pressure (Zo) shifts entirely into the lateral compartment, whereas concentration of joint load over the medial condyle can be expected in a joint with varus deformity even with a small amount of angulation since the normal tendency is to transmit force through the medial side of the knee. The average maximum joint loads were 2.1 times body weight for subjects with varus deformity, 2.4 times body weight for those with valgus deformity, and 2.7 times body weight for those with flexion deformity.

In each pathological group, including patients with significant angular deformity, force transmission at the knee was less than for normal individuals.

Discussion

The classic paper describing force transmission at the knee when a static analysis is used is that of Kettelkamp and Chao [16]. Their results indicate that load sharing is related to the angle between the tibial and femoral axes when a static frontal plane analysis is used to calculate force. Johnson et al. [17], however, has explained that in a static knee the vertical reaction from the floor passes through the centre of the knee and no horizontal force exists. The load

is shared equally between the lateral and medial compartments. In a varus
knee, again standing still, the load passes medial to the knee centre and the
load on the medial compartment is correspondingly greater. From dynamic
analysis at mid stance, which is the closest approximation to standing on one
leg, force platform measurements show that there is an additional horizontal
component to the floor reaction causing the resultant vector to be directed
medial to the knee. This produces an adduction moment. The effect will be to
increase the load in the medial plateau. Johnson has shown that the load on the
medial plateau for a varus knee rapidly approaches 100 per cent of total load
when calculated during conditions of dynamic analysis whereas for a valgus
deformity the load remains medial in the majority of individuals. Our results
are in complete agreement with this observation.

Where there is no medial lateral horizontal ground-to-foot force acting on
the free body of the shank, such as occurs when standing on one leg (static
analysis), it is evident that the location of the resultant knee load, (Zo) i.e.,
medial or lateral, will depend on the inclination of the shank in the frontal
plane (Fig. 11). During ambulation, however, it is the location of the centre of
gravity (c.g) of the body's mass above the knee that will determine whether the
knee will be adducted or abducted. In the normal limb for example the
mechanical axis is such that the hip, knee and ankle are in a straight line
whereas the angle between the femur and tibia is a valgus angulation of about 9
degrees. Even so, joint force is concentrated in the medial part of the joint.
This is basic to understanding the difference between static and dynamic
loading of the knee, i.e., whether there is one- or two-point bearing contact,
and the location of the centre of joint pressure. It is also basic to understanding
the differences in loading that occur in individuals with pathological gait.

This is best illustrated by recent studies in gait related to tibial osteotomy at
the University of Toronto. We have attempted to compare load transmission
at the knee before and following high tibial valgus osteotomy using force
plate cine-photographic techniques. The major proportion of total joint load,

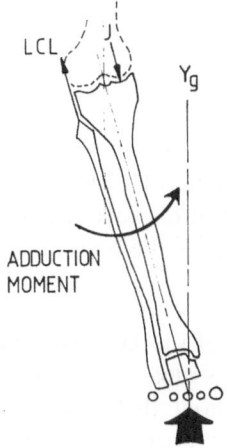
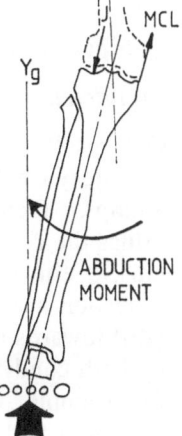

Fig. 11. Static loading of the
knee. When an individual stands
on one leg, there is no horizontal
ground-to-foot force therefore the
$M_x k$ abduction-adduction moments
will depend on the inclination of
the shank in the frontal plane

(86%) was transmitted by the medial knee compartment preoperatively. The maximum average total joint load for these patients was 2.5 times body weight and the mean preoperative varus angle was 8 degrees. Following surgery, (3 and 6 months postoperative) maximum joint load was reduced to 1.7 times body weight. The proportion of load carried by the medial side was lowered to 68% for a mean valgus angle of 9 degrees. Although this is a preliminary study it does suggest that even with osteotomy and a valgus alignment of the knee the major proportion of joint load is still transmitted by the medial compartment.

Quadriceps and Hamstring Force Actions in the Frontal Plane

Another important consideration is the contribution of the hamstring, (H) and quadriceps, (Q) muscle force actions in the production of adduction or abduction moments to the shank in the frontal plane. Morrison assumed that quadriceps function was confined to the sagittal plane and that its force action had no "z" component. The line of action of these muscle groups will tend to follow the inclination of the femur so that with a varus deformity the quadriceps and hamstrings will tend to apply an adduction moment about the knee which will increase the force transmitted by the medial femoral condyle and shift the centre of pressure (Zo) medially. The opposite will occur with a valgus knee, i.e., the tendency to produce an abduction moment of the knee and shift the centre of pressure (Zo) laterally. This effect will be more pronounced with valgus knees due to the natural valgus obliquity of the femur in a normal knee. The contribution of the quadriceps and hamstring force actions to moments in the frontal plane cannot be accounted for by static methods of analysis. It has also been neglected in most dynamic gait studies to date. Although the contribution of the hamstring and quadriceps moment in terms of magnitude will not be as great as the moment due to the medial lateral ground-to-foot shear force, it should not be neglected. It will certainly have an effect in determining the position of the centre of joint pressure.

Compensating Mechanisms

Trendelenburg Gait

Our results suggest that individuals can modify force transmission by compensatory mechanisms. Maquet [18] has observed that subjects who walk with a limp or Trendelenberg gait displace the trunk toward the affected limb. This brings the centre of gravity of the body mass in line with the joint so that less muscle effort is required to balance body weight. The overall effect is a reduction in the magnitude of joint force and a shift of the centre of pressure (Zo) towards the midline of the knee so that joint load is shared more evenly by both condyles. This was most apparent with the polio patients. It is noted, for example, that throughout the stance phase of the gait cycle, the shank was subjected to a net abducting moment rather than the normal adduction

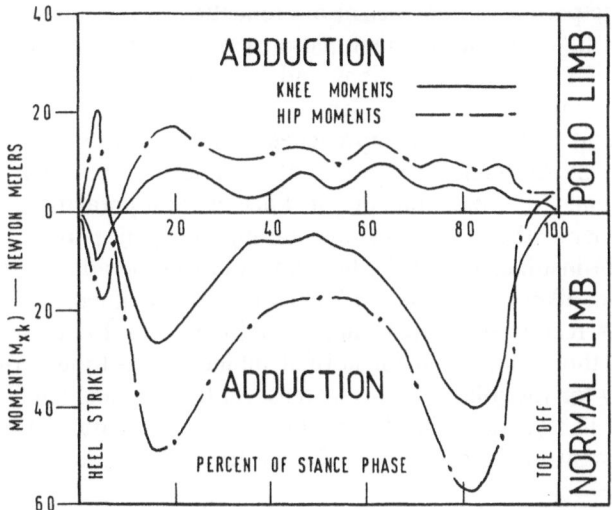

Fig. 12. $M_x k$ moments in polio patients. The knee is subjected to an abduction moment ($M_x k$) throughout the stance phase of gait in polio patients whereas the normal knee is subjected to an adduction moment

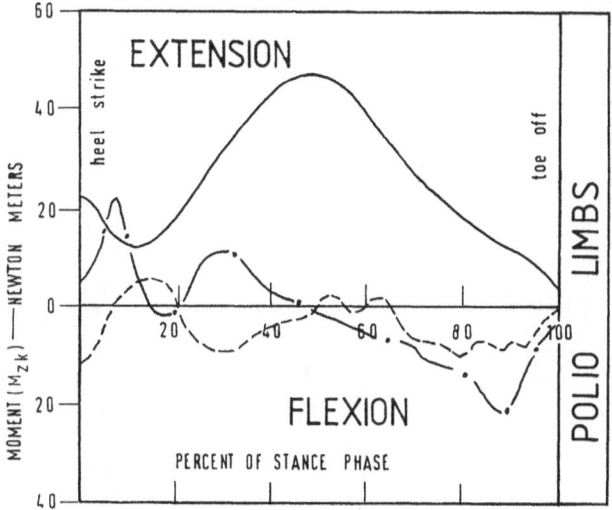

Fig. 13. $M_z k$ knee moments in polio patients. An extension moment ($M_z k$) provides knee stability for weight bearing

moment (Fig. 12). The overall effect is a shift of the centre of joint pressure to the lateral joint compartment. The centre of joint pressure, (Zo) was calculated to lie within the lateral compartment of the knee for each polio limb even though these subjects had normal angulation of the knee in the frontal plane. For these patients, there was also a reduction in sagittal plane moments (most showed an extensor moment pattern) and a corresponding decrease in magnitude of muscle force actions and joint bearing load (Fig. 13). The resultant knee joint force was considerably lower than the calculated value for normal limbs and in magnitude approached body weight. The "a", "b," and "c" joint force peaks tended to be absent or reduced. A similar mechanism probably accounts for unloading of the knee in non-paralytic patients with pathological knees; that is to say by altering the gait pattern the subject compensates for insufficiency of the limb and a functional adaptation develops. This mechanism likely accounts for unloading of the knee in subjects with deformity. Our results show that values for joint force for these individuals were less than those calculated for normal subjects even though overall there was a tendency for the joint force to increase with increasing angulation of the knee. It seems likely, then, that factors other than structural deformity of the limb are important in the estimation of force transmission at a weight-bearing joint.

Body Segment Displacement Patterns

The trendelenberg gait pattern or so-called lurch is clinically obvious. There is little doubt that significant displacements of segments of the body occur in the frontal plane as the centre of gravity of the individual moves over the affected joint. It is also important to appreciate that abnormal displacements of the body also occur in the sagittal plane with these subjects (Figs. 14a–c). In normal individuals, the center of gravity translates through a smooth,

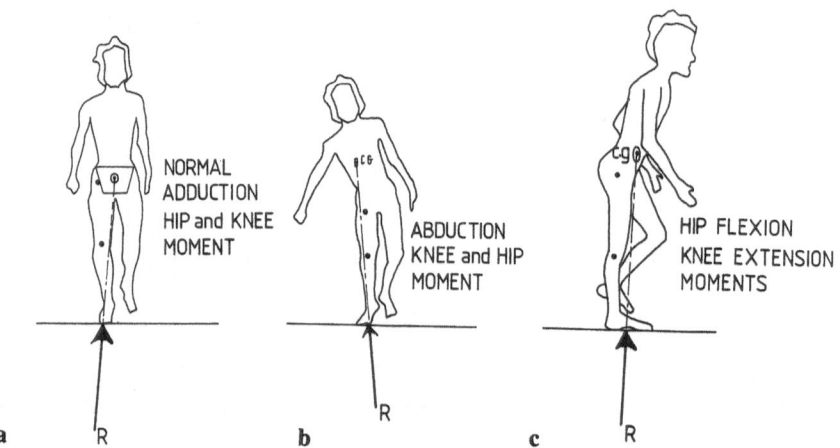

Fig. 14a–c. a Normal adduction hip and knee moment. **b** Centre of gravity shift. The body's centre of gravity (*C.G.*) shifts towards the affected joint producing an abduction knee moment (Trendelenburg lurch). **c** Sagittal plane moments. A shift of the body's centre of gravity (*C.G.*) anterior to the knee produces a knee extension moment in the sagittal plane — the most stable situation for a paralysed limb. *c.g.*, centre of gravity, *R*, resultant ground-to-foot force

undulating pathway of low amplitude and energy is conserved. In pathological gait, every attempt is made to preserve as low an energy consumption as possible by exaggerations of the motions at unaffected levels. From a clinical perspective, pathological gait is assymetrical and jerky. This requires an increased expenditure of energy, but without the displacement of normal body segments the individual with severe muscle paralysis for example would be unable to walk. These patients compensate for the inadequacy of their muscles by adopting a gait pattern that allows the centre of gravity to shift into a position in the frontal and sagittal planes so that minimal muscular effort is required to balance body weight over the affected joint. For patients with severe muscle weakness these compensatory mechanisms are obvious, however for others the changes may be very subtle.

Displacement patterns of parts of the body were analyzed by Murray et al. [19], in the assessment of walking patterns of patients with unilateral hip pain due to osteoarthritis and avascular necrosis. Murray noted abnormal vertical, lateral and forward trajectories of the head, arms and trunk. During each normal walking cycle, the head moves up and down through two peaks and two valleys. The peaks occur during single limb support, the valleys during double limb support. The heights of the two peaks and the depths of the two valleys were similar for the normal men. However for men with hip pain, the peak which occurred during weight bearing on the painful limb was notably lower than during weight bearing on the sound limb. Normal walking was also characterized by smooth lateral deviation of the head towards the side of the supporting limb during the alternating of single limb support. For the patients with painful hips there was a greater total lateral excursion. The patients also showed more asymmetry of their lateral head motion than did normal men. Most of the patients showed greater lateral deflection towards the painful side — some towards the sound side during weight bearing.

There is normally a slight increase and decrease in the rate of forward movement of the head during each walking cycle. The rate decreases as the trunk climbs to its highest and most lateral position. The rate of forward movement of the head was slower in patients with hip pain than in the normal men. Moreover, the patients' forward displacement did not follow a smooth and uniform pattern but was more rapid during weight bearing on the affected limb than during weight bearing on the sound leg.

Increased transverse rotation of the thorax and pelvis towards the affected limb was also noted. A consistent characteristic of the antalgic limp was the irregular movement of the head and trunk through space in contrast to the smooth sinusoidal movements of the normal subjects. The irregular movement was manifest by changes in three trajectories: Forward, lateral and vertical. Another consistent finding was an asymmetrical and excessive lateral shift of some part of the body towards the weight bearing side during the stance phase on the painful limb. Some patients showed a lateral lurch of the head and trunk towards the painful side, others abducted the upper limb on the painful side, and still others both shifted their trunk and abducted the arm on the painful side. Murray noted that movement of the upper extremities in the sagittal plane was also affected by a painful lower extremity. Motion of the upper limbs were markedly asymmetrical with the arm on the painful side tending to maintain a position of greater shoulder extension and greater elbow flexion the upper limbs of normal patients.

It is evident that what Murray described was a dynamic shift of the centre of gravity of the body mass which includes activity of the arms, head, and thorax towards the affected limb in both the sagittal and frontal planes. This shift effectively minimizes the magnitude of force actions transmitted by affected joints.

Walking Speed

Paul has shown that the magnitude of joint force for normal subjects depends on body weight, stride length, and walking speed. Elderly patients and those who experience discomfort while walking are likely to walk at speeds that are considerably slower than those of their more fit normal counterparts. Gait velocities in our subjects generally were quite low for those with deformity compared with normal individuals. Pain reflected in lower walking speeds and Trendelenberg gait patterns could result in joint unloading and probably accounts in part for the lower magnitude of joint force that was calculated for the subjects with deformity. Johnson et al. have commented on adaptive gait and the ability of some patients to compensate for localised pain by modifying their gait patterns suggesting that such compensatory mechanisms are probably automatic. Gainey et al. [20] for example, showed that patients with Paget's disease had a slower gait and decreased cadence compared with normal subjects. Stauffer et al. [21] also showed that patients with knee joint disease walked at a much slower rate than did normal subjects. Patients with unilateral knee disease had a shorter stance phase time on the affected side in an attempt to reduce the load on the painful knee. The percentage of stance phase was increased in patients with bilateral disease because they sought to spend as much time as possible in the period of double support thereby reducing the compressive load on both knees as much as possible.

Knee Flexion

Stauffer et al. [21] also noted that flexion of the loaded knee increases the quadriceps force across the joint and thus greatly increases the compressive joint force. Patients with painful knees characteristically attempt to reduce this compressive joint force and resulting pain by decreasing the range of knee flexion during stance phase of gait. An extension moment at the knee is the most stable situation for gait particularly if there is weakness of the Quadriceps muscle group. It is known, for example, that the only muscle required for walking is the hip extensor. Even with paralysis of all other muscles controlling the lower extremity an individual can walk if he has recurvatum of the knee and fixed equinus of the foot.

Gainey et al. [20] in a gait study of patients with Paget's disease and secondary arthritis has described a reduction in the flexion knee moment in these patients. Andriacchi [14] and Simon et al. [22] identified three distinct flexion-extension knee moment patterns in normal individuals and the changes that occur following knee arthroplasty. The most common pattern occurred in 80 percent of individuals and began with an extensor moment during weight acceptance, became a flexor moment during single limb stance, and then became an extensor torque toward the end of the single-limb stance and during weight release. The second pattern consisted of a flexor moment and the third an extensor moment throughout stance. The extensor moment was present in most of the patients who had total knee replacements. Only a few knee replacement subjects reverted to the more normal biphasic flexion-extension moment pattern.

Ground-to-Foot Forces

Similarly, Stauffer et al. [21] also showed that patients with diseased lower extremities (rheumatoid and osteoarthritis) have a reduced vertical component of floor reaction force which is an indirect but proportional approximation of the axial compression force across the knee joint that occurs in an attempt to smooth out the accelerations during stance phase. Gainey also demonstrated a reduced vertical ground reaction force and a decrease in the magnitude of lateral and fore and aft shear in patients with Pagets disease. He theorized that this was a protective mechanism to reduce knee joint force and to minimize the degree of pain during stance. Andriacchi et al. [23] has also shown that patients with knee pathology walk with a shorter step length and a higher cadence than a normal subject walking at the same speed.

Adduction Moments (Tibial Osteotomy)

Prodromos et al. [24] have shown that patients who have a low adduction moment at the knee during walking have a better result after high tibial osteotomy and maintain their postoperative alignment better than patients who have a high adduction moment. Andriachi [24] theorized that a low adduction moment during preoperative gait indicated a beneficial compensatory mechanism that tended to be maintained after tibial osteotomy. Some patients in this group exhibited abduction moments at the knee preoperatively, i.e., similar to patients with paralytic limbs such as polio patients. Andriacchi theorized that the low adduction moment was a compensatory mechanism to unload the knee. This was true even when patients were standardized for age, weight, walking speed and presence or absence of pain.

More recently Wang et al. [25] has demonstrated that patients with a low adduction moment tend to do better clinically over an extended period of time following tibial osteotomy. He has also observed that patients who *toe-in* when they walk maintain a high adduction moment whereas those individuals who *toe-out* have a low moment. Toeing-out is felt to be a compensatory mechanism that lowers the adduction moment at the knee and therefore lowers joint force transmission.

Gait Patterns After Total Knee Replacement

Even after total knee replacement, when deformity has been corrected and pain relieved, most published reports indicate that normal gait patterns and force transmission are rarely achieved. Simon et al. [22], for example, showed that stresses borne by artificial knees at least two years following surgery were less than normal. As well, the flexion-extension external moment at the knee for most patients did not approximate a normal pattern. It was concluded that these findings might be due to neural impairment of the joint or merely retention of a gait pattern that existed prior to arthroplasty.

Andriacchi et al. [14] have also shown that after total knee replacement, even asymptomatic patients with excellent clinical results have an abnormality of gait. This consists of a shorter than normal stride length, reduced mid-stance knee flexion, and abnormal patterns of external flexion-extension moments of the knee. Andriacchi also felt that these abnormal characteristics of gait after total knee replacement might be due to patients continuing to walk with a pattern they had learned prior to treatment. Another possible cause was theorized to be abnormal muscle functioning which could be caused by a partial loss of proprioceptive control or by reduction or imbalance of muscle capacity after joint replacement. A finding of practical importance was patients who had posterior cruciate retaining total knees were better able to ascend and descend stairs. He concluded that it was not possible to explain the abnormalities in gait observed in his study.

Certainly in the few total knee replacement patients tested in our study, (three surface replacements and 4 hinged joints) similar findings were observed. Even though these patients walked with what appeared to be a normal clinical gait pattern, calculation of force actions across the knee were similar to those patients with diseased knees awaiting surgery. Both types of knee replacements showed flexion-extension moments of low magnitude throughout the stance phase of gait (Fig. 15a,b). For most knee replacement patients, the knee was subjected primarily to an extension moment. This is likely a compensatory mechanism since the knee is most stable during this particular type of loading pattern. An extension moment is perhaps to be expected since hinged total knee replacements are specifically designed to lock in extension for stability. This requires less force in the quadriceps muscle which is likely weakened from surgery and/or disease. An extension knee moment of low magnitude is also likely to occur during weight bearing with surface replacement arthroplasties particularly in those designs which have little intrinsic stability such as the cruciate ligament sacrificing models. Calculation of bearing force transmitted by the implants showed an overall reduction in load compared to normals with blunting of the "a", "b," and "c" peaks, and a general flattening of the force profile curves (Fig. 16). The medial, lateral, fore, and aft shear forces at the knee also tended to be small. The torques transmitted to the tibia and femur

Fig. 15a,b. a M$_z$k knee moments (sagittal plane). **a** surface replacement arthroplasties of the knee. **b** M$_z$k knee moments (sagittal plane). Hinged total knee replacements

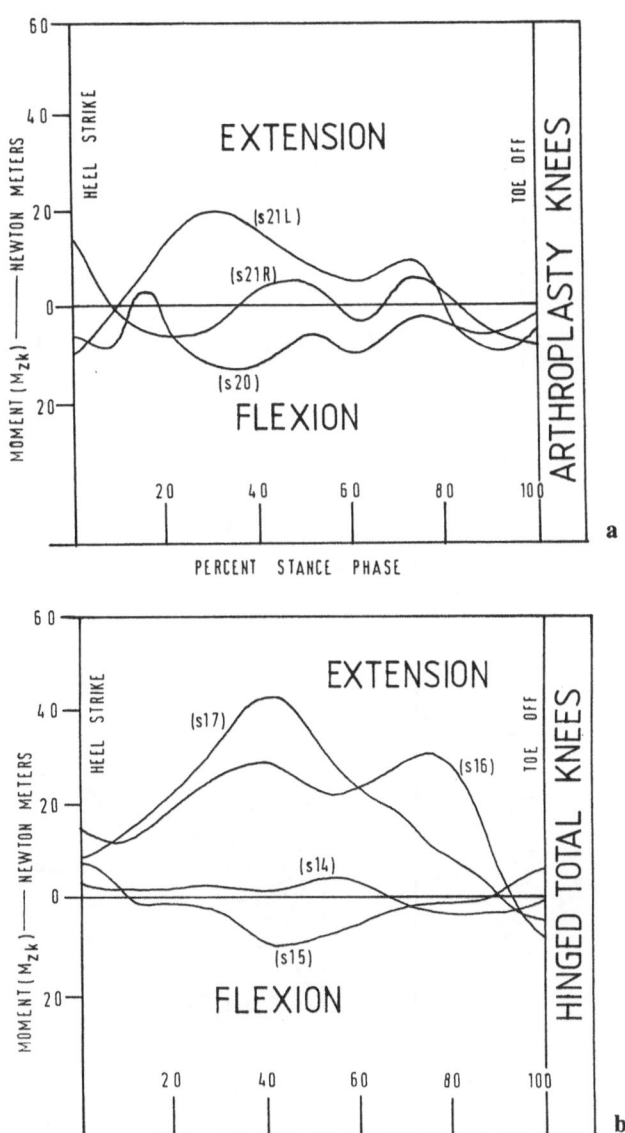

are of the same magnitude as for normal knees and generally follow the same pattern. A few arthroplasty subjects, however, exibited torques that subjected the knee to either a constant internal or external torque during the entire stance phase (Fig. 17).

From a clinical perspective, the overall effect of these abnormal gait patterns is to lower rather than increase force transmission across weight bearing joints. In the long term, this may actually be a beneficial feature rather than a

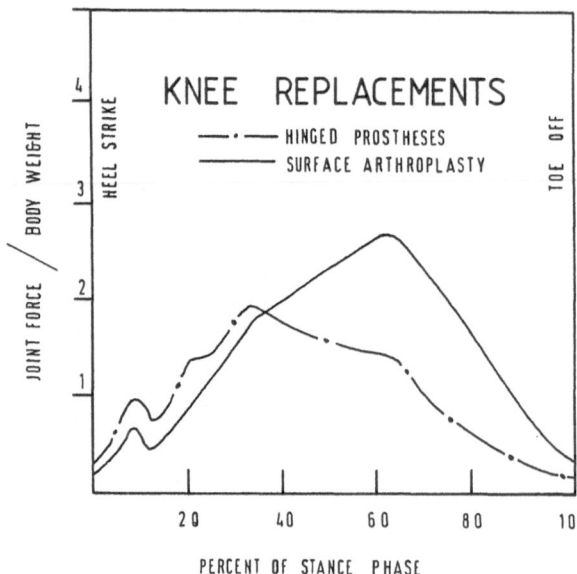

Fig. 16. Knee joint force in total knee replacements. The *a*, *b*, and *c* loading peaks are absent. The curves are similar to preoperative joint loading profiles

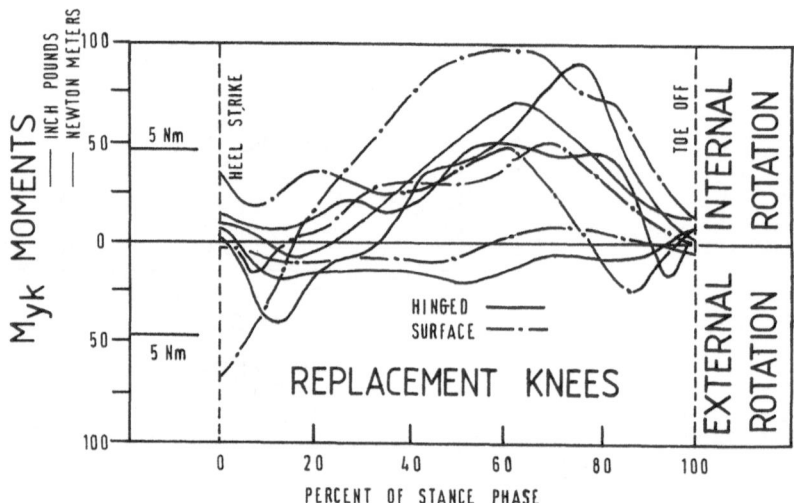

Fig. 17. Tibial shaft torque in total knee replacements. The tibial shaft torque profiles are similar to normal limbs. Some replacement knees show a constant torque during the stance phase of gait

detrimental one in terms of survival of total joint prostheses because an abnormal gait pattern that results in a lower-than-normal load transmission across the bearing surfaces may be considerably more beneficial to the lifespan of the prosthesis in terms of wear and mechanical failure than would a normal gait with a high transmission of force.

In Summary

1. Joint force transmission in normal knees is characterized by three distinct peaks corresponding to force actions generated in the major muscle groups spanning the knee — Hamstrings, Quadriceps and Gastrocnemii. The Hamstring peak is accentuated in children and joint force load tends to be greater than for adults.
2. Joint force profiles for pathological knees do not have distinct peaks. Maximum loads tend to be transmitted at the mid stance phase of gait. In general loading is reduced in less fit individuals compared to normals.
3. Knee joint bearing force tends to increase with angular deformity, particularly in subjects with flexion contracture but the magnitude of force even with severe angulation does not exceed normal values.
4. The location of the resultant force transmitted at the knee (center of joint pressure) is more dependant on the position of the center of gravity of the body's mass above the joint than the angular relationship of the tibia and femur.
5. The magnitude of knee joint force transmission in all groups regardless of deformity increases with increased walking speed.
6. Force transmission in patients with artificial knees is not normal. Compensatory mechanisms seem to exist which protect the implants from excessive loads. Attempting to design knee prostheses which produce normal loads may in the long term be disadvantageous.

References

1. Weber W, Weber E (1836) Mechanik der menschlichen Gehwerkzenge. Göttingen, Germany
2. Marey EJ, Demeny G (1887) Etudes experimentales de la locomotion humaine C R Acad Sci [III] 105:544–552
3. Braune W, Fischer O (1872) Uber den Schwerpunkt des menschlichen Körpers mit Rücksicht auf die Ausrüstung des deutchen infanteristen. Abhandlungen der Koenigl. Sächs. Gesellschaft der Wissenschaft vol. 15
4. Bernstein NA (1935) Biodynamics of locomotion (in Russian). VIEM, Moscow
5. Elftman H (1939) The function of the muscles in locomotion. Am J Physiol 125:357
6. Pauwels F (1935) Der Schenkelhalsbruch ein mechanisches problem. Ferdinand Enke, Stuttgart
7. Bresler B, Frankel JP (1950) The forces and moments in the leg during level walking. Trans ASME 72:27–36
8. Cunningham DM, Brown GW (1952) The devices for measuring the forces acting on the human body during walking. Proc Soc Exp Stress Anal IX 2:75
9. Rydell N (1965) Forces in the hip joint (II); intra-vital studies. In: Kennedi RM (ed) Biomechanics and related bio-engineering topics. Permagon Press, Oxford
10. Paul JP (1965) Bio-engineering studies of the forces transmitted by joints (II): Engineering Analysis. In: Kennedi RM (ed) Biomechanics and related bio-engineering topics. Permagon Press, Oxford

11. Morrison JB (1967) The forces transmitted by the human knee joint during activities. PhD thesis, University of Strathclyde
12. Harrington IJ (1973) Knee joint force in normal and pathological gait. MSc thesis, University of Strathclyde
13. Nissan M (1979) Review of some basic assumptions in knee bio-mechanics. J Biomech 13:375–381
14. Andriacchi TP, Galante JO, Fermier RW (1982) The influence of total knee replacement design on walking and stair climbing. J Bone Joint Surg [Am] 64:1328–1335
15. Drillis RJ (1958) Objective recordings and biomechanics of pathological gait. Ann NY Acad Sci 74:86–109
16. Kettelkamp DB, Chao EY (1972) A method for quantitative analysis of medial and lateral compression forces at the knee during standing. Clin Orthop 83:202–213
17. Johnson F, Leitl S, Waugh W (1980) The distribution of load across the knee — a comparison of static and dynamic measurements. J Bone Joint Surg [Br] 62:346–349
18. Maquet PGJ (1976) Bio-mechanics of the knee. Springer, Berlin
19. Murray MP, Gore DR, Clarkson BH (1971) Walking patterns of patients with uni lateral hip pain due to osteoarthritis and avascular necrosis. J Bone Joint Surg [Am] 53(2):259–273
20. Gainey JC, Kadaba MP, Wootten ME, Ramakrishnan MS, Siris ES, Lindsay R, Canfield R, Cochrane GVB (1989) Gait analysis of patients who have Paget's disease. J Bone Joint Surg [Am] 71(4):568–479
21. Stauffer RN, Chao EYS, Gyory AN (1977) Biomechanical gait analysis of the diseased knee joint. Clin Orthop 126:246–255
22. Simon SR, Trieschmann HW, Burdett RG, Ewald FC, Sledge CB (1983) Quantitative gait analysis after total knee arthroplasty for monarticular degenerative arthritis. J Bone Joint Surg [Am] 65(5):605–613
23. Andriacchi TP, Ogle JA, Galante JO (1977) Walking speed as a basis for normal and abnormal gait measurements. J Biomech 10:261–268
24. Prodromos CC, Andriacchi TP, Galante JO (1985) A relationship between gait and clinical changes following high tibial osteotomy. J Bone Joint Surg [Am] 67(8):1188–1194
25. Wang JW, Kuo KN, Andriacchi TP, Galante GO (1990) The influence of walking mechanics and time on the results of proximal tibia osteotomy. J Bone Joint Surg [Am] 72:905–909

Biomaterials

Materials for Bone Repair, Augmentation, and Implant Coatings

RACQUEL Z. LEGEROS[1]

Summary. This paper briefly describes some commercial and experimental materials used in maxillofacial, orthopedic and dental applications for bone augmentation, substitution and repair. These materials include:

1. Autogeneous, freeze-dried or banked bones and bovine bone-derived materials
2. Demineralized bone matrix (DBM)
3. Coral ($CaCO_3$)
4. Polymers
5. Inert ceramics (alumina, zirconia)
6. Bioactive glass ceramics
7. Calcium phosphate materials such as hydroxyapatite (HA), tricalcium phosphate (TCP), coralline HA, and non-sintered apatite
9. Composites
10. Coated and uncoated metal implants

The strength of the material/bone interface, the acceleration of bone formation and the extent of bone attachment depend on the composition and properties of the materials. Autogeneous and allogenic grafts and DBM are osteogenic or osteoinductive, inducing bone formation; other materials are osteoconductive, providing a scaffold for new bone growth. Bioactive materials such as calcium phosphate materials and glass ceramics become chemically bonded to the bone, thus providing a very strong material/bone interface. Bioinert materials such as ceramic oxides, metals and alloys, and polymers do not become directly attached to the bone, and consequently, the material/bone interface is much weaker.

HA coating on metal implants were shown to provide accelerated bone attachment and greater skeletal fixation and, in some cases, appeared to have

[1] New York University, College of Dentistry 345 East 24th Street, New York, New York 10010, USA

minimized leakage of some metal ions. However, failure between the coating and the metal substrate sometimes occurs causing the loosening and failure of the implant.

Characterization of some HA-coated implants demonstrated that the composition of the coating was different from that of pure HA ceramic depending on the method of coating. Analyses of some HA coatings showed the presence of a mixture of different calcium phosphate phases was actually present: HA, α- and β-TCP, and amorphous calcium phosphates (ACP). The concentration of these phases varied from the inner to the outer layers of the coating. Variations in the composition of HA coatings from different manufacturers were also observed. The effect of the coating composition on the in vivo performance and long-term stability of the coated implant need further study.

Table 1. Materials for Bone Repair, Substitution and Augmentation

Materials	Applications	References
1. *Natural Biomaterials*		
a) autologous and allogenic bones	maxillofacial, cranial bone defects; periodontal defects; alveolar ridge augmentation; orthopedic surgery	[6, 39]
b) demineralized bone matrix, DBM	cranial, mandibular, periodontal bone defects; other osseous reconstruction	[1, 2, 3, 4]
c) coral (Biocoral[a])	bone defects	[21]
d) coralline HA (Interpore[a])	dental, orthopedics (see Table 2)	[27, 44, 45]
e) bovine bone (Kielbone[a], Pyrost[a], Surgibone[a]; TBC[b] BonAP[b]	dental, orthopedics	[23, 41, 42]
2. *Bioactive Glass* Ceramics (Bioglass[a], Ceravital[a], A.W-GC[b])	ear implant; maxillofacial reconstruction; tooth root implants; bony defects coating for metal implants	[9, 12, 13, 24, 46, 80]
3. *Calcium Phosphate Materials*	(see Table 2); fillers for composites	(See Table 2)
4. *Ceramic Oxides* (Alumina, Zirconia)	orthopedic prostheses; tooth root implants; coating for metal implants	[63, 64, 78]
5. *Polymers* non-resorbable and resorbable	component of orthopedic prosthesis; maxillofacial reconstruction; ear implant; bone cement; dental application bone plates	[35, 65, 67, 68, 69]
6. *Metals and Alloys*		
a) Co-Cr-Mo-Ni (Vitallium[a])	orthopedic prostheses	[29, 62, 79]
b) stainless steel (Multiphase (MP)3N)	orthopedic prostheses	
c) Ti and Ti-6Al-4V	orthopedic prostheses; dental implants	[15–17, 19, 22, 25, 62, 71, 79]

[a] commercial; [b] experimental materials

Table 2. Uses of Calcium Phosphate Materials[a]

Dental applications:
 periodontal defects
 tooth root implants
 mastoidectomies
 alveolar ridge augmentation
 pulp-capping materials
 coating for dental implants
Reconstructive surgery: .
 facial bone deformities
 chin augmentation
 correction of underdeveloped mandibles
Middle ear surgery:
 middle ear implant
 middle ear ventilating tube
 reconstruction of ossicular chain
Spine surgery:
 spine fusions
 lumbar implants
Orthopedic applications:
 bone cement
 coating for orthopedic implants

[a] HA, β-TCP, BCP, coralline HA [7, 8, 10–12, 15, 17–19, 22, 34, 38, 39, 42, 43, 45, 51, 54, 56, 58, 59, 75, 80–82]

Key words. Hydroxyapatite — Tricalcium phosphate — Bone graft materials — Coatings

Introduction

From the days of the Egyptian pharaohs to President George Washington, records show that materials such as elephant's tusks, walrus teeth, and wood have been used to replace bone or missing teeth. Presently, commercial and experimental materials differing in composition, and mechanical and physicochemical properties have been proposed for bone augmentation, substitution, and repair in maxillofacial, mandibular, ear, cranial, reconstructive, orthopedic, and oral surgery (Tables 1 and 2). These materials are available as powders, particulates, or blocks of varying shapes and sizes, and with varying degrees of porosity: Solid or dense, microporous, and macroporous (Fig. 1). Bone graft materials fall into three general categories:

1. *Autografts* which are obtained from another bone source in the same subject
2. *Allografts* which are obtained from a different subject of the same species. These can be either freeze-dried or banked human bones
3. *Alloplastic* grafts which are synthetic materials
   a) Treated and sintered bovine bones
   b) Demineralized bone matrix

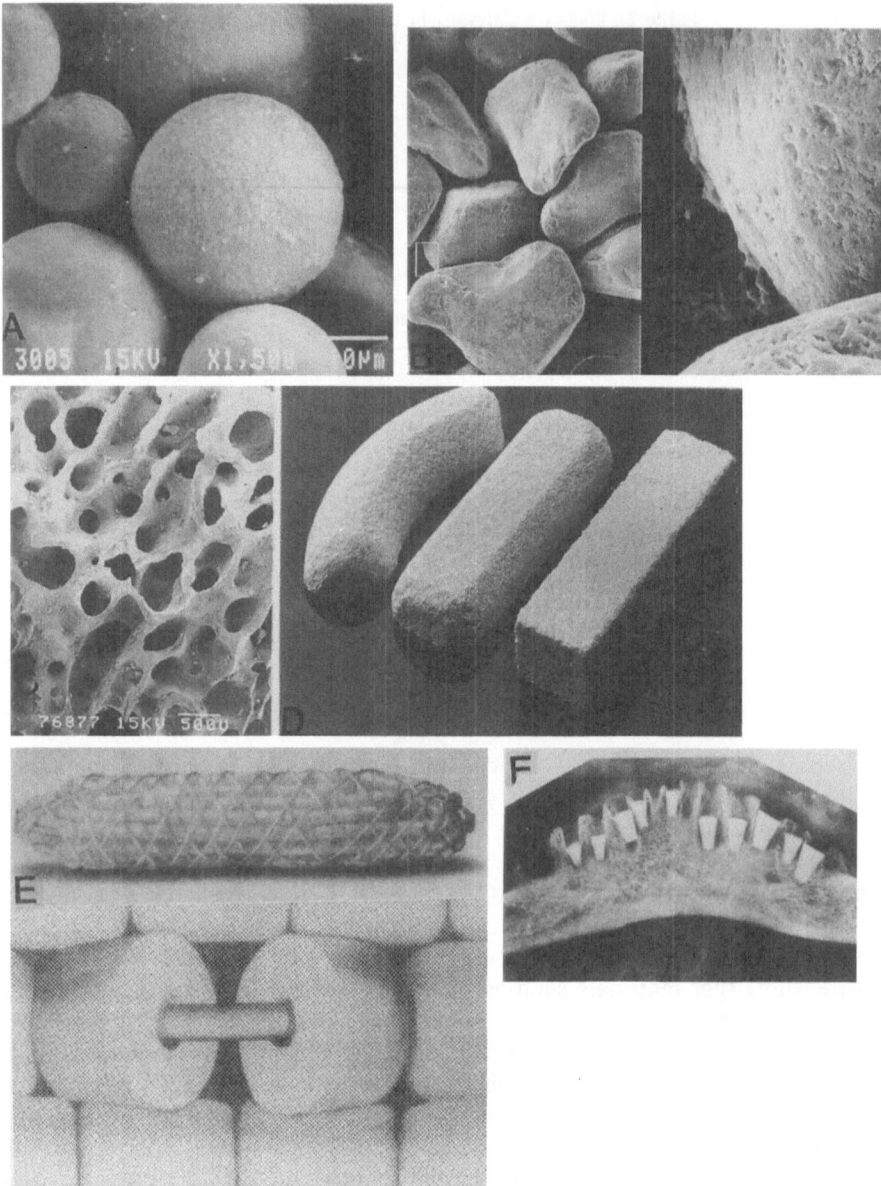

Fig. 1A–F. Different forms and shapes of materials for bone repair, substitution or augmentation. **A** SEM of HA beads (courtesy, Dr. A. Tofe, Ceramed). **B** SEM of particulates of a dense HA, Calcitite [34] **C** SEM of BonAP, an experimental material from bovine bone (LeGeros/Daculsi, 1989, unpublished). **D** Different shaped blocks from coralline HA (courtesy, Dr. E. Shors, Interpore). **E** HA beads strung together to avoid migration, used for alveolar ridge augmentation (Osteograf). **F** dense HA cones for immediate tooth root replacement (courtesy, Dr. H. Denissen). *SEM*, scanning electron microscope; *HA*, hydroxyapetite

c) Coral, $CaCO_3$
d) Resorbable and non-resorbable polymers
e) Ceramic oxides or inert ceramics
f) Bioactive glass ceramics
g) Calcium phosphate materials
h) Metals and alloys
i) Composites of materials mentioned above

Some of the alloplastic materials have also been proposed as coatings for dental and orthopedic metal implants to accelerate bone attachment, improve their skeletal fixation, and minimize loosening or failure.

In terms of promoting or inducing bone formation, bone graft materials may be described as either osteoinductive (osteogenic) or osteoconductive. Osteoinductive materials allow the formation of bone in non-osseous sites due to the presence of non-collagenous bone morphogenetic proteins which trigger non-osseous mesenchymal cells in the vicinity of the bone graft material to differentiate to chondroblasts and osteoblasts and produce new bone [1–4]. Autogenous and allogenic bones, demineralized bone matrix (DBM) are osteoinductive [1–6]. Osteoconductive materials act as templates or scaffolds for the formation of new bone [7–13]; these materials do not induce bone formation in non-osseous sites [5]. All the alloplastic materials cited above are biocompatible or biotolerant.

In terms of bone attachment, materials are described as either bioactive or bioinert. Bioactive materials allow the formation of normal tissues (new bone) on their surfaces and form a contiguous and strong bond at the bone/material interface [14]. Calcium phosphate materials, materials derived from bovine bone, and glass ceramics of specific composition are bioactive [7–9, 13–26] but are not osteoinductive [5, 8, 10, 28]. Some materials such as metals and alloys, ceramic oxides, and polymers are described as bioinert, and are characterized by the presence of fibrous layer interposed between the material and host bone [15, 16, 26]. The interfacial strength between the bone and the synthetic materials depend on the bonding mechanisms at these interfaces, which in turn, depend on the physicochemical properties of the materials that are both

Table 3. Tensile Strength of Bone and Bone Graft Materials

Material	Tensile Strength (MPa)
Cortical Bone	69–110
Co-Cr alloy	900–1540
Stainless steel	540 to 1000
Ti-6Al-4V alloy	900
Polyethylene	3
Bioglass	42
Hydroxyapatite, HA	
Dense	79–196
Porous	42

[8, 35]

different from each other and from the bone (Table 3) which they are meant to replace, augment or repair.

This paper presents a brief description of some of the commercial and experimental materials in terms of their composition, physicochemical properties and in vivo performance.

Bone

Bone, due to its architecture and composition (Figs. 2–4), has unique mechanical properties [29]. Macroscopically, there are two types of bone: Compact and cancellous bone. Compact bone is microporous while cancellous bone is macroporous. Microscopically, two types of bone are observed: Woven and lamellar bone. Bone is an intimately integrated composite of inorganic and organic phases (Fig. 3a) in the approximate inorganic/organic ratio of 75/25 by weight or 65/35 by volume [30, 31]. The organic phase consists principally of type I collagen and smaller amounts of non-collagenous proteins, acidic glycoproteins, serum proteins, and small proteoglycans [32]. The inorganic phase has been idealized as a calcium hydroxyapatite, $Ca_{10}(PO_4)_6(OH)_2$ (HA) [10, 17, 30]. Actually, the bone mineral is a carbonate-apatite (Fig. 3b), associated with minor constituents such as sodium, magnesium, chloride, HPO_4^{2-}, and fluoride, and trace constituents such as strontium, lead, zinc, copper, and iron with a calcium/phosphate (Ca/P) molar ratio below or above the stoichiometric value of 1.67 for pure HA, depending on the age and species of the bone donor [30, 31, 33, 34]. The apatite microcrystals of bone are usually rod-

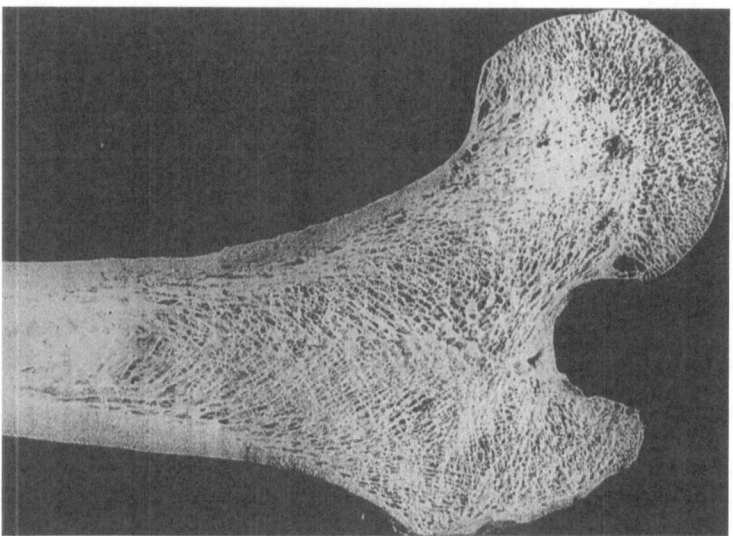

Fig. 2. Cross-section of a femur, showing the dense cortex and the highly porous trabecular bones

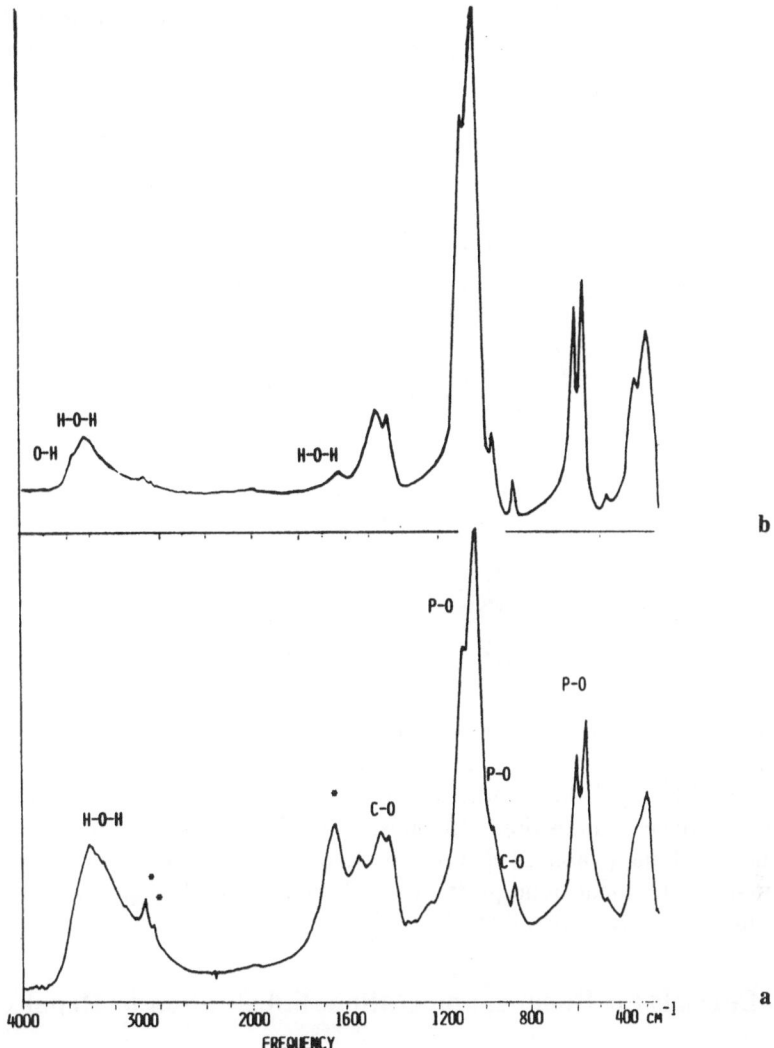

Fig. 3a,b. IR absorption spectra of powdered bone with (**a**) and without (**b**) the organic matrix. The absorption bands attributed to the organic matrix are indicated (*asterisk*), those attributed to the inorganic phase (CO$_3$-apatite) are indicated as *C-O, P-O, O-H* for the CO$_3$, PO$_4$ and OH groups, respectively, of the apatite. *H-O-H* indicate absorption bands of adsorbed H$_2$O. *IR,* infrared

like or plate-like with average dimensions of 250 × 30 Å. The crystallinity (reflecting crystal size and/or strain) is similar to that of dentin and much smaller than that of enamel (Fig. 4).

Physicochemical characterization of synthetic calcium phosphate materials and bone mineral showed that they are similar in composition in terms of having calcium and phosphate ions as the principal components. However,

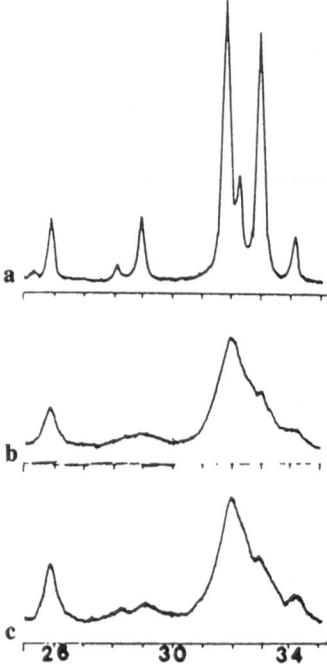

Fig. 4a–c. X-ray diffraction pattern of biological apatites: (**a**) enamel, (**b**) dentin, and (**c**) bone. The difference in the resolution of diffraction peaks reflect the difference in crystal sizes. Enamel has the largest apatite crystals. All biological apatites are CO_3-apatite [30]

differences in the Ca/P molar ratio, composition, crystalline size and shape, dissolution, and mechanical properties (Figs. 5, 6, and Table 3) are evident [8, 30, 34–37]. Other synthetic materials such as glass ceramics, polymers, and metals have composition and physicochemical properties quite different from those of bone (Table 3). Numerous research efforts continue towards development of the ideal bone graft material with properties approximating those of bone as closely as possible.

Materials for Bone Augmentation, Substitution or Repair

Bone-Derived Materials

Autogenous Bone

The use of autogenous bone for bone augmentation, substitution, and repair is an established procedure [6]. However, this requires a second surgery from donor sites (e.g., the ilium, ribs, tibia) thereby increasing risk, morbidity, discomfort, healing time, and expense. Furthermore, successful bone regeneration is not always attained due to variable resorption and unpredictable capacity to regenerate new bone [6, 13, 38].

Freeze-Dried or Banked Human Bones

Freeze-dried or banked human bones are sometimes used to eliminate the second surgical procedure required for autogenous bone grafts. Successful

Fig. 5a–c. X-ray diffraction patterns
of HA ceramic (**a**), coralline HA (**b**),
and bone mineral (**c**)

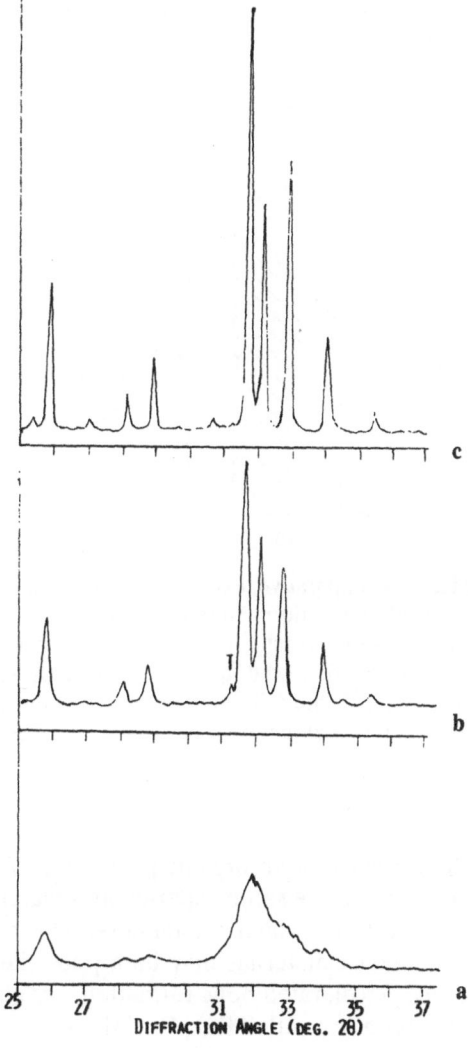

results are reported, although less so than those obtained with autogenous
bones [13, 38, 39]. However, allografts or banked bones are expensive, not
always available, and there is always a risk of acquiring some types of com-
municable diseases (e.g., AIDS) from them.

Materials Derived from Bovine Bone

Some commercial (e.g., Surgibone, Pyrost, Kielbone) and experimental (e.g.,
True-bone-ceramic [TBC], BonAP) materials originating from bovine bone are
recommended for use as bone graft materials [23, 40, 41, 42]. Infrared (IR)
analyses of some of the commercial materials (e.g., Surgibone, Pyrost) show

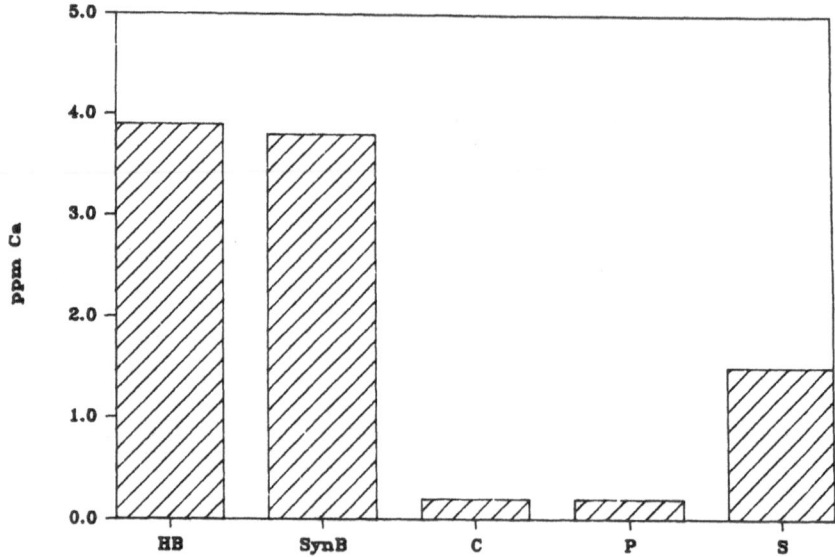

Fig. 6. Comparative extent of dissolution in acid buffer (0.1 M KAc, pH 5, 37°C; solid/solution ratio = 100 mg/25 ml; suspension time, 60 min). The extent of dissolution was measured from the calcium concentration (ppm Ca) of the acid buffer. Human bone (*HB*) dissolved to a much greater extent than β-TCP, Synthograf (*S*). HA ceramic, Calcitite (*C*), and Periograf (*P*) dissolved the least. Unsintered apatite, Syn B, an experimental material, had a similar extent of dissolution to HB [33]

the presence of an organic phase (Fig. 7a) similar to the IR spectrum of bone (Fig. 3b); while some experimental materials (e.g., BonAP) show the presence of only HA of high crystallinity. The presence of an organic phase in some commercial materials may be a potential source of inflammation or complications. Accelerated bone formation was observed with BonAP (Fig. 7a) and was also reported with TBC [41, 43].

Demineralized Bone Matrix

DBM is obtained by demineralizing cleaned human bones in 0.6 N HCl [1, 2, 4] or by other more complicated procedures [3]. Powdered DBM was shown to be effective in the correction of craniofacial defects, mandibular defects, and other osseous reconstruction [1, 4]. This natural biomaterial has advantageous osteoinductive properties, rapid resorption, and incorporation into the new bone [1, 2, 3, 4]. However, DBM has the disadvantages of being very expensive, not always available, and has obviously very weak mechanical properties limiting its use when utilized by itself. Its use in composites with HA ceramics is being investigated.

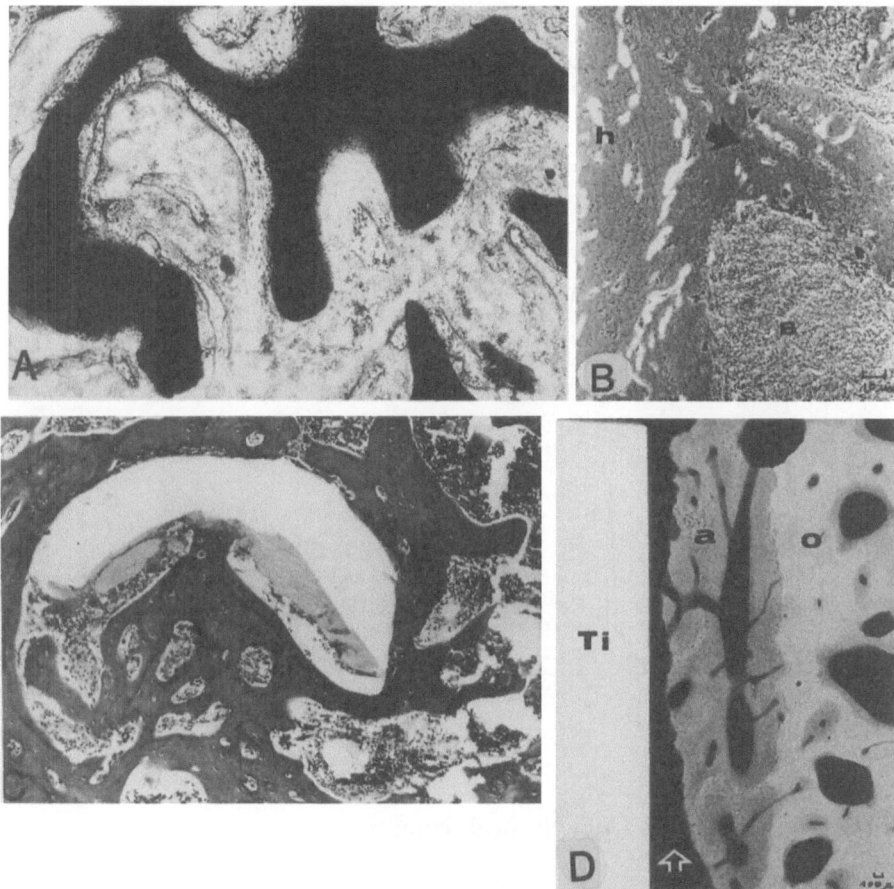

Fig. 7A–D. New bone formation associated with some alloplastic materials. **A** Light microscopy of undecalcified section of BonAP recovered from canine cortical femoral bone after 15 days showing lamellar bone directly deposited on the surface of BonAP. Bone marrow cells were also observed inside the macropores (LeGeros/Daculsi, 1989, unpublished results); **B** haversian bone (*h*) is observed in close contact with the BCP surface (*B*) and invading the macropore [16]. **C** HTR consisting of PMMA/HEMA impregnated with Ca(OH)$_2$ after 3-week implantation in rat femur bone, showing integration of the HTR bead with the bony tissue, with some capsule formation and presence of macrophages [66]. **D** X-ray microradiographs of titanium implant (*Ti*) recovered from dog femur after 3 months, a fibrous interposition (*arrow*) is observed between the titanium surface and the newly forming bone (*a*) [16]. *BCP*, biphasic calcium phosphate; *HTR*, hard tissue replacement; *PMMA*, polymethylmethacrylate; *HEMA*, hydroxymethylmethacrylate

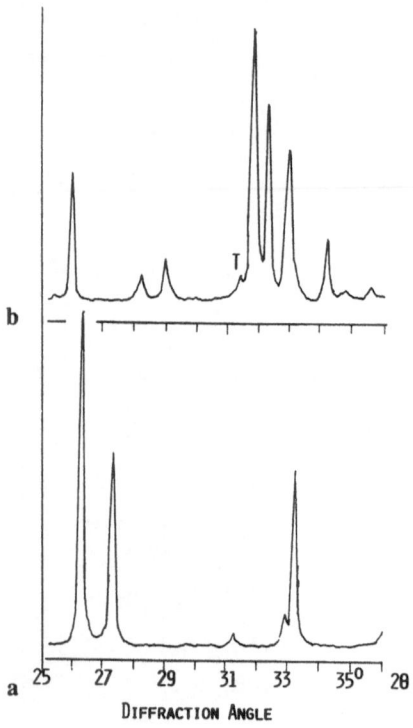

Fig. 8a,b. X-ray diffractioon pattern of coral (Biocoral) (**a**), and coralline HA, coral transformed to apatite, Interpore (**b**)

Bone Graft Materials from Natural Sources

Coral

Madreporian corals, *Porites* and *Acropora*, consist principally of a mineral phase, calcium carbonate ($CaCO_3$), in aragonite form (Fig. 8a) associated with the organic matrix. Commercially available coral such as Biocoral are reported to be free of organic phases except for some amino acids. When implanted in bone tissue, Biocoral was reported to gradually resorb and progressively become replaced by new bone [21]. The biological fate of these materials has not been sufficiently investigated.

Coralline HA

Coralline HA (Interpore) is derived from coral and is available commercially. The coral is treated to remove the organic phase and then hydrothermally converted to apatite according to the reaction ideally represented as follows:

$$10CaCO_3 + 6(NH_4)_2HPO_4 \rightarrow Ca_{10}(PO_4)_6(OH)_2$$

This method [44] conserves the desirable original interconnecting porosity of the coral which is similar to that of bone (Fig. 1c). Coralline HA has been extensively studied and successfully used in both reconstructive surgery and several dental applications (Table 2), and has been recommended for orthopedic applications [27, 45, 46]. X-ray diffraction and IR analyses showed that coralline HA is not pure HA but a CO_3-apatite mixed with a small amount of Mg-substituted beta tricalcium phosphate (β-TCP) (Fig. 8b); however the mechanism of CO_3 substitution in coralline HA is different from that in bone apatite [30, 31, 37]. Coralline HA differs from ceramic HA (Calcitite) in lattice parameters due to CO_3-for-OH substitution, in crystallinity (Fig. 5) due to the method of preparation, and in composition (CO_3-AP + Mg-substituted β-TCP vs HA) [37]. These differences may explain the observed greater reactivity of coralline HA compared to ceramic HA in vitro (in serum and in cell culture) [34, 37] and possibly in vivo.

Calcium Phosphate Materials

Unsintered Calcium Phosphate Compounds

As early as 1920, a reagent simply described as "triple calcium phosphate" was reported to promote new bone formation when applied to a bony defect [47]. Recent X-ray diffraction analyses of commercial "calcium phosphate" reagents demonstrated that identifying labels may sometimes be misleading [31, 48]. For example, reagents labeled as "Tricalcium Phosphate", "hydroxylapatite", "approximately $Ca_{10}(OH)_2(PO_4)_6$", or "approximately $Ca_3(PO_4)_2$" are actually calcium phosphates with apatitic structure and low crystallinity, sometimes mixed with another calcium phosphate phase such as monetite ($CaHPO_4$) (Figs. 9a–c). Some materials labeled "hydroxylapatite" are actually beta tricalcium phosphate, $Ca_3(PO_4)_2$ (β-TCP) (Fig. 9d). Thus, it is not possible to determine what the actual composition was of the "triple calcium phosphate" reagent used in the successful early study [48].

Experimental materials proposed for bone mineral substitution [33] or bone cement [49, 50] are characterized by X-ray diffraction as apatitic calcium phosphates. Unsintered precipitated apatites can be prepared with a wide range of crystallinity and dissolution properties similar to that of either the enamel or bone apatite [33], as shown in Fig. 6.

Another example of unsintered calcium phosphate is a commercial material, OsteoGen, described as "Resorbable HA". X-ray diffraction analyses of OsteoGen from different lot numbers demonstrated that they were not HA, but apatite of low crystallinity (Fig. 10a) or apatite mixed with monetite [41]. OsteoGen (composition not reported) was reported to be successful when used for the repair of osseous defects prior to endosseous implant surgery [51]. Properties and in vivo performance of appropriately characterized unsintered calcium phosphates need further investigation.

Calcium Phosphate Ceramics

Calcium phosphate ceramics are usually prepared by precipitation under very alkaline conditions and subsequent sintering at 900 to 1300°C [10, 31, 34,

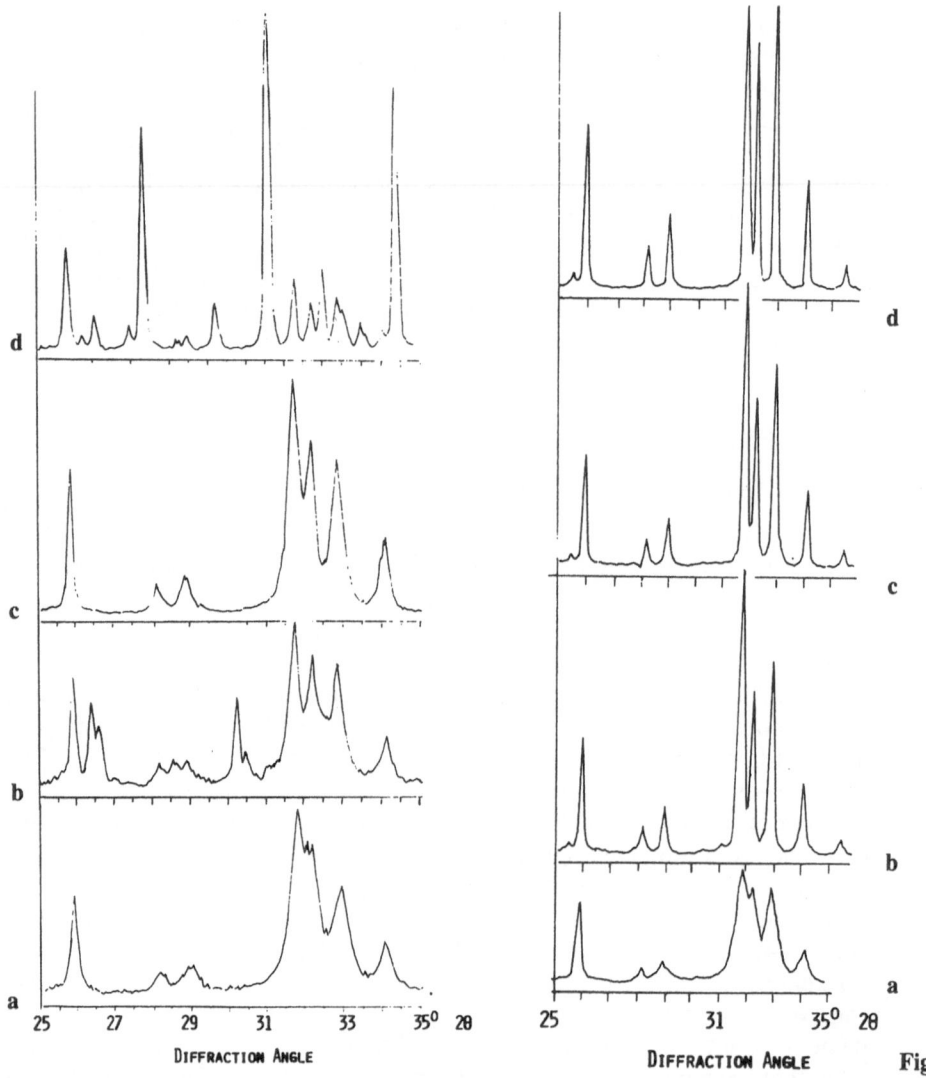

Fig. 9   DIFFRACTION ANGLE   Fig. 1

Fig. 9a–d. X-ray diffraction patterns of calcium phosphate reagents labeled as calcium phosphate, tribasic, $Ca_{10}(OH)_2(PO_4)_6$ (**a, b, c**) and "spheroidal hydroxyapatite" (**d**). **b** A mixture of apatite and monetite, $CaHPO_4$. **d** A mixture of β-TCP and HA [48]. *HA*, hydroxyapetite

Fig. 10. X-ray diffraction patterns of some commercial HA products. (**a**) resorbable HA, OsteoGen (**a**); HA ceramic from Japan, Mitsubishi (**b**); from France, Bioapatite (**c**); and from USA, Calcitite (**d**). Note that (**a**) is not a ceramic HA. Note also that (**b**), (**c**) and (**d**) differ in intensity of the diffraction peaks, reflecting differences in sintering temperature. *HA*, hydroxyapatite

48, 52]. Other methods of preparation include sintering commercial calcium phosphate reagents or precipitated apatites at temperatures above 900°C [8, 31, 34, 48]. Depending on the method of preparation of calcium phosphate reagents or apatite preparations, subsequent sintering at or above 900°C will result in the formation of HA, β-TCP, or biphasic calcium phosphate (BCP), a mixture of β-TCP and HA in varying proportions [8, 18, 31, 48, 52]. Sintering of β-TCP above 1300°C results in the formation of α-TCP and tetracalcium phosphate. The commercial calcium phosphate ceramics are available in dense (microporous) or macroporous forms, and in particulate or block forms (Fig. 1) depending on whether the application is as a filler in bony defects, in composites, alveolar ridge augmentation, or as a coating for metal implants. The differences in composition and in physical forms affect their dissolution or degradation properties in vitro [33, 34, 53] and in vivo [54, 55]. For example, macroporous HA will dissolve to a greater extent than dense HA of the same composition, and β-TCP is more soluble than HA if they are of equivalent density or porosity [26, 34, 53]. Some of the applications of calcium phosphate materials are summarized in Table 2.

Hydroxyapatite. The use of ceramic HA as a possible bone graft material and its subsequent commercial development were independently pioneered by Jarcho in the USA [10], Aoki in Japan [56], and deGroot in Europe [8]. Commercial HA (Calcitite and Osteograf in the USA; Bioapatite in France, and HA in Japan) gave similar but not identical X-ray diffraction patterns (Fig. 10), indicating differences in crystallinity. This difference in crystallinity most probably reflects the differences in the temperature of sintering. Such differences can influence their comparative reactivity in vivo [57]. Another commercial HA ceramic, Apoceram, has been shown to be a mixture of HA and β-TCP. Coralline HA (Interpore) is technically not ceramic HA since it is prepared by hydrothermal conversion of coral at much lower temperatures.

Tricalcium Phosphate. Commercial β-TCP materials (Synthograf and Peri-Oss) compared to ceramic HA (Bioapatite, Calcitite, Durapatite, Osteograf, and Orthomatrix) have been described as resorbable [11] based on its greater solubility in vitro [33] and in vivo [11, 26, 55].

Biphasic Calcium Phosphate. BCP consists of a mixture of HA and β-TCP. Commercial BCP (Triosit) consists of 40% β-TCP and 60% HA by weight. This material has been recommended for dental [58] and medical [59] applications, as it may have greater bioreactivity than ceramic HA. Studies with experimental BCP of varying β-TCP/HA ratios showed that reactivity was proportional to the β-TCP/HA ratio. The higher the ratio, the greater the reactivity [40].

In Vitro and In Vivo Reactions of Calcium Phosphate Materials

In cell culture media, several calcium phosphate materials were shown to influence the activities of specific cell types [60, 61], making the cell culture system useful in demonstrating biocompatibility and possibly predicting comparative bioreactivity. In vitro studies in serum demonstrated that coralline

Fig. 11. Microcrystals associated with large HA ceramic crystals. TEM of BCP after 3 month implantation [40]. *HA*, hydroxyapetite; *TEM*, transmission electron microscope; *BCP*, biphasic calcium phosphate

HA (Interpore) partially transformed to bone mineral like CO_3-apatite more readily than ceramic HA [37].

Whether calcium phosphate biomaterials (HA, β-TCP, BCP, and coralline HA) were implanted in osseous or non-osseous sites, these materials were found to become associated with microcrystals (Fig. 11), identified as CO_3-apatite, intimately associated with an organic matrix [16, 28, 31, 37, 40, 41, 59] shown in Fig. 12. The mechanism of this association appears to be the partial dissolution of the calcium phosphate biomaterial, resulting in an increase of the supersaturation with respect to Ca^{2+} and PO_4^{3-} ions in the microenvironment. This, in turn, causes the precipitation of biological apatites which incorporate other ions, principally CO_3 and Mg, from the biological fluid and become intimately associated with the organic matrix (Fig. 13). These biological apatite crystals become incoporated in the formation of new bone.

Ceramic Oxides (Inert Ceramics)

Alumina

Alumina, Al_2O_3, has been proposed for dental [7, 62] and orthopedic [63] implants. It is biocompatible but chemically inert [26, 63]. Boutin et al., in their review [63] reported that alumina as a component of total hip protheses outperformed other orthopedic materials such as polyethylene. They stated that the long-term success in orthopedic applications depends on its micro-structure (small grain size with uniform distribution, minimum porosity, and absence of inclusions), implant geometry, and health status of the host bone.

Fig. 12a–d. IR spectra of BCP before implantation (**a**); core of the implant (**b**); interface between implant and host bone (**c**); and bone farthest from the ceramic implant (**d**). Absorption bands due to organic matrix (*N-H*) and carbonate in apatite (*C-O*) increase from the core of the implant to the interface; and highest for the host bone [40]. *IR* infrared; *BCP*, biphasic calcium phosphate

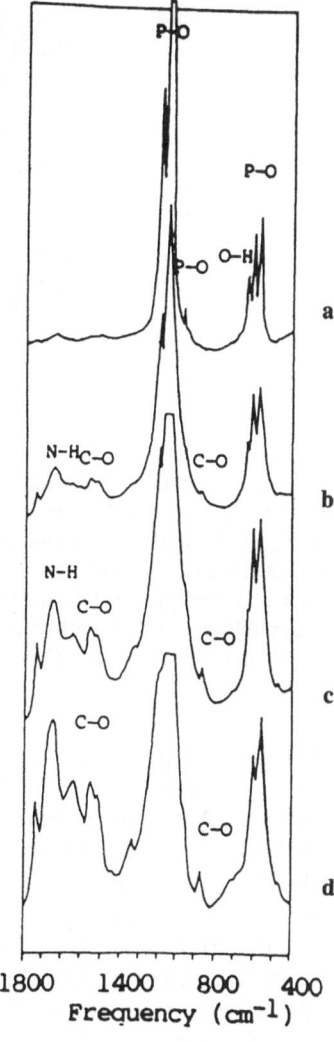

Zirconia

Zirconia (Zr_2O_3), or partially stabilized zirconia, has also been recommended for dental and medical applications (principally orthopedic) [64].

Ceramic oxides, although biocompatible, do not become directly attached to bone [20, 26].

Bioactive Glass Ceramics

Hench pioneered the development of a glass ceramic described as bioactive because it allowed the formation of new bone on its surface and provided a strong interface with the host bone [14, 45]. This bioactive glass ceramic

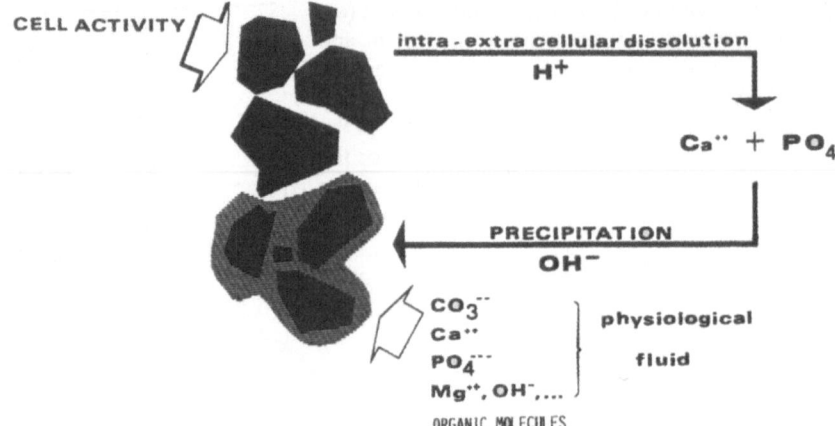

Fig. 13. Schematic presentation of biological fate of calcium phosphate materials. Cellular activity causes a decrease in pH, causing partial dissolution of the Ca-P (HA, β-TCP, BCP, AP) material and subsequent elevation of the calcium and phosphate ions in the microenvironment. This increase in concentration causes precipitation of biological apatite microcrystals (*B*) incorporating other ions (principally CO_3^{2-}) and becoming intimately associated with organic molecules [58]. *HA*, hydroxyapetite; *β-TCP*, beta tricalcium phosphate; *BCP*, biphasic calcium phosphate; *AP*, apatite

(Bioglass) is prepared by melting SiO_2 (network former), $Na_2O/K_2O + CaO$ (network modifiers), and P_2O_5 (internal nucleant for surface apatite formation) together in specific proportions (45% SiO_2, 24.5% CaO, 24.5% Na_2O and 6% P_2O_5). In vitro and in vivo, the formation of a silicon (Si) rich layer next to the Bioglass, followed by a calcium phosphate (Ca-P) rich layer (Fig. 14) was observed [14, 45, 62].

Other bioactive glass formulations include the commercially available Ceravital (glass ceramic containing apatite) and the experimental AW-GC (glass ceramic containing apatite and wollastonite). These bioactive glass ceramics, especially Bioglass, have been extensively investigated and have been recommended for maxillofacial reconstruction, ear implants, and various dental and orthopedic applications [9, 13, 24, 25]. It is also recommended as a material for coating metal implants [46, 62].

Polymers

Non-Resorbable Polymers

Ultra-high-molecular-weight polyethylene (e.g., Medpor) is used in maxillofacial reconstruction [13] and has long been used as a component of hip replacement prostheses [35, 65]. Silicone (e.g., Silastic), polyamide fibers (e.g., Nylamid), and polytetrafluoroethylene (Proplast) are used in maxillofacial reconstruction [13]. Polymethylmethacrylate (PMMA) is used as bone cement

Fig. 14a,b. a Bioglass implant (*I*) in intimate contact with bone (*B*), 3 months after implantation (hematoxylin-eosin, decalcified section). **b** bonded interface between rat tibial bone and surface active glass, Bioglass, 30 days after implantation. *BG*, bulk Bioglass; *S*, SiO$_2$-rich layer; *Ca/P*, calcium and phosphate-rich layer; *B*, bone; *O*, osteocyte. Magnification ×200. (courtesy of Drs. L. Hench and J. Wilson-Hench)

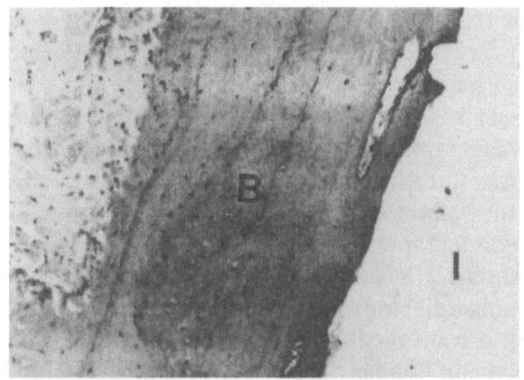

a

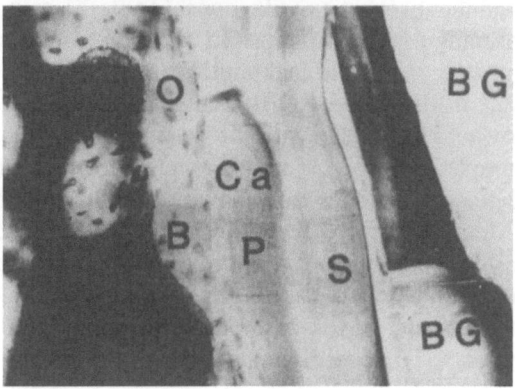

b

[35]. PMMA coated with hydroxymethylmethacrylate (HEMA) and impregnated with calcium hydroxide is commercially available (HTR), and is recommended for several dental applications [66, 67]. When implanted in bony sites, HTR beads became incorporated in the new bone [66] as shown in Fig. 7b. All of these non-resorbable polymers are available commercially.

Resorbable Polymers

Polylactide, polyglycolide, or a mixture of both have been suggested as bone plates for internal fracture fixation [68] or fracture reduction [69].

Metals and Alloys

Metals and alloys used for orthopedic implants may be classified as (a) cobalt-chromium, (b) stainless steel, and (c) titanium [70]. The first group consists of cobalt (Co), chromium (Cr), molybdenum (Mo), and nickel (Ni). For example, one commercially available alloy, Vitallium, is made up of 62% Co, 27% Cr, 5% Mo, and 3% Ni. Another commercial alloy, Multiphase (MP) 3N, consists of 35% Co, 35% Ni, 20% Cr, and 10% Mo. The stainless steel group contains 11 to 30% Cr, 11 to 14% Ni, 2 to 6% manganese (Mn), 2 to 3% Mo, and less than 1% of other elements. The titanium group consists of either pure titanium (Ti), or Ti-6Al-4V, an alloy consisting of a maximum of 6% aluminum (Al), 4% vanadium (V), with the balance made up of Ti. The titanium group, especially the Ti alloy, is used for dental implants [15, 71].

Metal implants are shown not to bond directly to bone, but appear to have a fibrous interposition [15, 16] which after a long period of time may contribute to the failure of the implant. Implants made of pure Ti are claimed to be osseointegrated, or in direct contact with the host bone [71]. Although all materials appear to be in direct contact with bone when viewed through a scanning electron microscope (SEM), at higher magnification or when viewed with a transmission electron microscope (TEM), a fibrous layer between the metal and bone is observed [15, 16]. Metal implants sometimes fail due to their dissimilar properties with bone (Table 3) causing resorption of the bone and loosening of the implants. In addition, metal alloys, especially stainless steel, tend to corrode and leak metal ions which may cause hypersensitivity reactions (e.g., in the case of Ni) or malignancy [70, 72]. Other metal ions have been shown to interfere with the crystal growth of hydroxyapatite in vitro. For example, Al, Co, Ni, and Zn were shown to suppress crystallization of apatite [73], and V and Ti were shown to inhibit apatite formation [74].

Metal alloys are the only materials which exceed the strength needed for load-bearing areas. To improve skeletal fixation of alloys, bioactive ceramics and bone cements have been used around the implants [54, 75]. Coating with porous metal beads or bioactive ceramics has also been used to stabilize the implants. The coating materials will be discussed in a later section.

Composites

Composite materials are designed to combine the desirable properties of each component, hopefully to approximate the properties of bone. The preparation and properties of some of these composite materials have been recently reviewed by Doyle [35]. HA and Bioglass have been extensively used as the inorganic filler in composites. Some examples of composites are (a) high-density polyethylene and HA [65], (b) polyhydroxybutyrate and HA [35], (c) bis-GMA and HA [36], and (d) HA and resorbable [76] or non-resorbable [77] polymers. HA/calcium sulfate has been developed as a partially-resorbable composite cement for bone repair [38, 43]. Metal fiber-reinforced Bioglass is recommended for reconstruction and repair of long bones [62].

Materials for Coating Implants

Coating orthopedic implants with porous metal beads stabilized the implants by allowing tissue ingrowth into the pores. Alumina [78], calcium phosphate ceramics (HA and TCP) [17, 19, 75], and bioactive glass ceramics [46] have been proposed as coating materials to accelerate bone attachment and improve

Fig. 15. Dental implants (Spectra system, Core-vent), uncoated (**a, b**) and HA-coated (**c, d**)

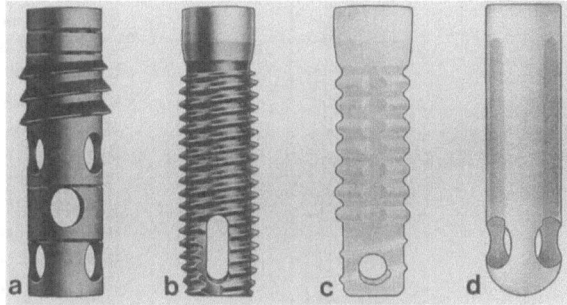

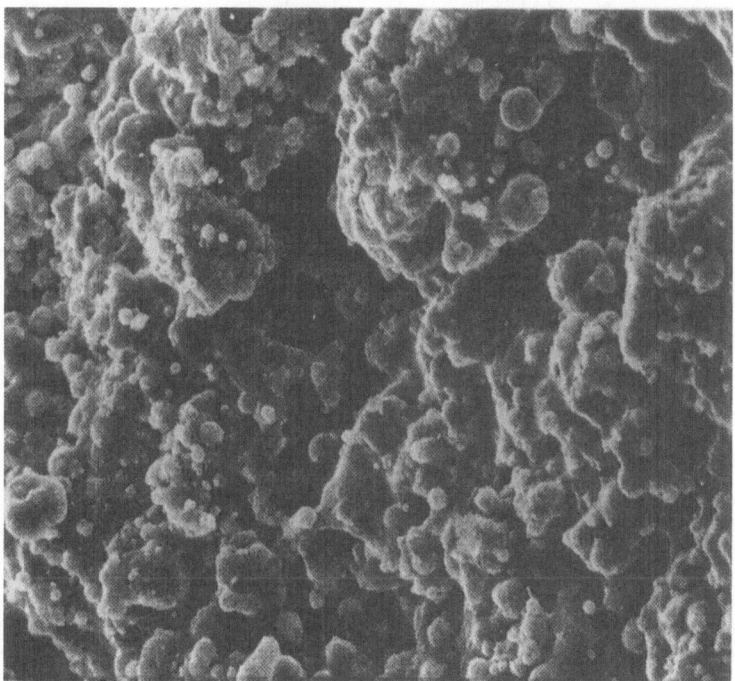

Fig. 16. SEM of HA-coating on a Ti alloy dental implant

skeletal fixation. In addition, the HA coating also appeared to minimize leakage of potentially harmful metal ions [79].

Metal implants coated with calcium phosphate materials combine the desirable bioactivity of the calcium phosphate materials and the strength of the metal. Several HA-coated dental implants are commercially available (Fig. 15). Characterization of the coatings, using SEM (Fig. 16), X-ray diffraction (Figs. 17b, 18), and IR analyses demonstrated that the composition of the coating is quite different from that of the starting material (HA) depending on the method of coating [41]. For example, coating deposited by ion-sputtering was

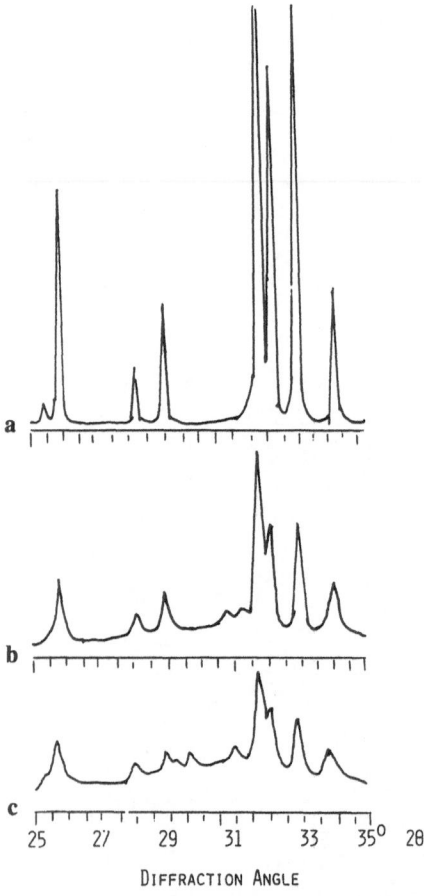

Fig. 17a–c. X-ray diffraction patterns (XRD) of **a** HA ceramic used for coating by plasma-spray method. **b** outer and **c** inner layers of the coating. The HA component of the coating has lower crystallinity than the starting material (**b, c,** compared to **a**). There is a composition gradient between the outer and inner layers of the coating. The coating has other calcium phosphate phases (α- and β-TCP, ACP) in addition to HA [63]. *HA*, hydroxyapetite; *TCP*, tricalcium phosphate; *ACP*, amorphous calcium phosphate

amorphous calcium phosphate (ACP) (Fig. 17c), while coating deposited by plasma spraying was a mixture of calcium phosphate phases: HA, α- and β-TCP, and ACP (Figs. 17b, 18). Furthermore, a composition gradient was observed where by the layer closest to the metal substrate contained more ACP than the outermost layer of the coating, and the HA component in both layers had lower crystallinity than the starting material (Fig. 17a). Coatings from different commercial HA-coated dental implants were found to differ in the relative concentrations of the different calcium phosphate phases: HA (with lower crystallinity than HA ceramic), α- and β-TCP, tetracalcium phosphate (TTCP) and ACP (Fig. 18).

The long-term stability of the implants will depend on the composition and properties of the coating. Compared to alumina, HA coating was shown to have greater bone contact [22]. HA-coated implants were shown to have greater interfacial strength than uncoated implants [15]. However, some HA-coated implants were reported to fail after a long period of time [15, 19]. The coatings were not characterized, so while the failures may be due to the

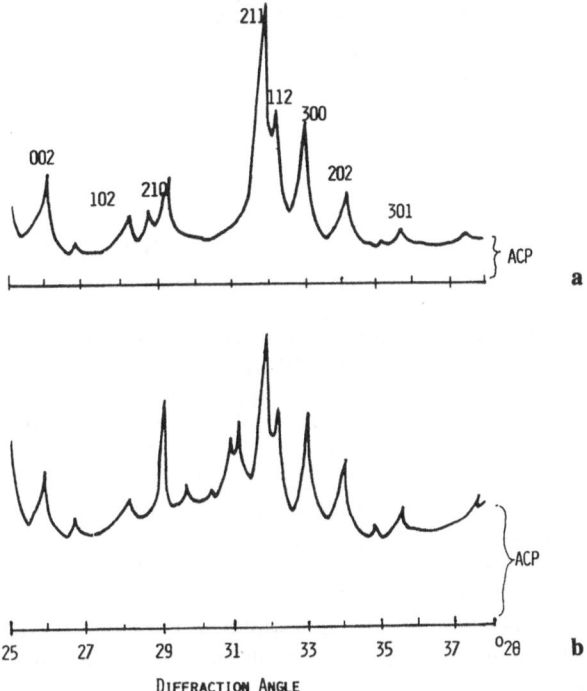

Fig. 18a,b. X-ray diffraction patterns of coatings from two commercial HA-coated dental implants showing difference in relative concentrations of ACP, α- and β-TCP

presence of high amounts of highly reactive ACP, the cause is not entirely clear. Post-treatment of the coating appears to improve the crystallinity of the HA component and decrease the amount of ACP [19].

It is evident that HA or other calcium phosphate coatings on implants should be appropriately and accurately characterized in order to better understand their effect on the stability of the implants. It also appears that the failure of HA-coated implants occurs at the interface of the metal and the coating. This is due to the much stronger bond between the HA, or calcium phosphate, coating and bone compared to that between the coating and the metal. It is apparent that a coating with improved adhesion to the metal substrate needs to be developed to minimize or totally prevent implant failure.

Acknowledgements. The author gratefully acknowledges the professional collaboration of Drs. G. Daculsi, M. Gregoire, J.P. LeGeros, I. Orly, M. Amler, and S. Niwa; and the technical assistance of J. Wong, M. Retino, and R. Zheng for the work cited in this paper. G. Merlino and D. Corn are acknowledged for their assistance in photography and V. Fronjian for word processing.

The work of the author cited in this paper was partly supported by NIH/ NIDR Research Grant Nos DE 04123 and DE 07223 and by special Calcium Phosphate Research Funds.

References

1. Glowacki J, Murray JE, Kaban L, Folkman J, Mulliken JB (1981) Application of the Biological Principle of Induced Osteogenesis for Craniofacial Defects. Lancet 1:959–963
2. Reddi AH (1985) Implant-stimulated Interface Reactions During Collagenous Bone Matrix-induced Bone Formation. J Biomed Mater Res 19:233–239
3. Urist MR, Strates BS (1971) Bone Morphogenetic Protein. J Dent Res 50:45–49
4. Vandersteehoven JJ, Spector M (1983) Histological Investigation of bone Induction by Demineralized Allogeneic Bone Matrix: A Natural Biomaterial for Osseous Reconstruction. J Biomed Mater Res 17:1003–1014
5. Amler MH (1988) Osteogenic Potential of Non-vital Tissues and Synthetic Implant Materials. J Periodontol 58:758–761
6. Enneking WF, Morris JL (1972) Human Autologous Bone Transplants. Clin Orthop 87:28–39
7. Bajpai PK (1983) Biodegradable Scaffolds in Orthopedic, Oral and Maxillofacial Surgery. In: Rubin LR (ed) Biomaterials in reconstructive surgery. CV Mosby, St. Louis, pp 312–328
8. De Groot K (1983) Ceramic of Calcium Phosphates: Preparation and Properties. In: de Groot K (ed) Bioceramics of Calcium Phosphate, CRC Press, Boca Raton, Florida, pp 100–114
9. Hench LL, Wilson J (1984) Surface-active Materials. Biomater Science 226:630–636
10. Jarcho M (1981) Calcium Phosphate Ceramics as Hart Tissue Prosthetics. Clin Orthop 157:259–278
11. Metsger DS, Driskell TD, Paulsrud JR (1982) Tricalcium Phosphate Ceramic — A Resorbable Bone Implant: Review and Current Status. J Am Dent Assoc 105:1035–1038
12. Van Blitterswijk CA (1985) Calcium Phosphate Middle-Ear Implants. PhD thesis, Rijksuniversitiet te Leiden
13. Wilson J, Merwin GE (1988) Biomaterials for Facial Bone Augmentation: Comparative Studies. J Biomed Mater Res: Appl Biomat 22:159–177
14. Hench LL, Splinter RJ, Allen WC, Greenlee TK (1971) Bonding Mechanisms at Interface of Ceramic Prosthetic Materials. J Biomed Mater Res 2:117–141
15. Cook SD, Kay JF, Thomas KA, Jarcho M (1987) Interface Mechanics and Histology of Titanium and Hydroxylapatite Coated Metal Implants. J Biomed Mater Res 23:183–199
16. Daculsi G, LeGeros RZ, Deudon C (1990) Scanning and Transmission Electron Microscopy, and Electron Probe Analysis of the Interface Between Implants and Host Bone. Scan Electron Microsc 4:309–314
17. De Groot K (1987) Hydroxyapatite Coatings for Implants in Surgery. In Vincenzini P (ed) High Tech Ceramics, Elsevier Science Publishers BV, Amsterdam, pp 381–386

18. Denissen H (1979) Dental Root Implants of Apatite Ceramics: Experimental Investigations and Clinical Use of Dental Root Implants Made of Apatite Ceramics. PhD thesis, Vrije Universiteit te Amsterdam
19. Denissen HW, Kalk W, de Nieuport HM, Maltha JC, van de Hooff A (1990) Mandibular Bone Response to Plasma-sprayed Coatings of Hydroxyapatite. Int J Prosthodont 3:53–58.
20. Gross U, Schmitz H-J, Strunz V (1988) Surface Activities of Bioactive Glass, Aluminum Oxide, and Titanium in a Living Environment. In: Ducheyne P, Lemons JE (eds) Bioceramics: Materials Chjaracteristics Versus In Vivo Behavior. Ann NY Acad Sci 523:211–226
21. Guilemin G, Patat JL, Fournie J, Chetail M (1987) The Use of Coral as a Bone Graft Substitute. J Biomed Mater Res 21:557–567
22. Hayashi K, Matsuguchi N, Uenoyama K, Kanemaru T, Sugioka Y (1989) Evaluation of Metal Implants Coated with Several Types of Ceramics as Biomaterials. J Biomed Mater Res 23:1247–1259
23. Okumura M, Ohgushi H, Yoshikawa T, Tamai S (1990) Heterotropic Ossification in Sintered Bone Ceramics and Kiel Bone by Using Bone Marrow Cells. In: Heimke G (ed) Bioceramics, vol 2. German Ceramic Society, Cologne, pp 86–93
24. Ono K, Yamamuro T, Nakamura T, Kakutani Y, Kitsugi T, Hyakuna K, Kokubo T, Oka M, Kotoura Y (1988) Apatite-wollastonite Containing Glass Ceramic-Fibrin Mixture as a Bone Defect Filler. J Biomed Mater Res 22:869–885
25. Oonishi H, Tsuji E, Ishimaru H, Yamamoto M, Delecrin J (1990) Clinical Significance of Chemical Bonds Between Bioactive Ceramics and Bone in Orthopaedic Surgery. In: Heimke G (ed) Bioceramics, vol 2. German Ceramic Society, Cologne, pp 286–293
26. Osborn JF, Newesely H (1980) The Material Science of Calcium Phosphate Ceramic. Biomaterials 1:108–111
27. Piecuch JF (1986) Augmentation of the Atrophic Edentulous Ridge with Porous Replaniform Hydroxyapatite (Interpore-200). Dent Clin North Am 30:291–305
28. Heughebaert M, LeGeros RZ, Gineste M, Guilhem, Bonel G (1988) Physico-chemical Characterization of Deposits Associated with HA Ceramics Implanted in Non-osseous sites. J Biomed Mater Res: Appl Biomat 22:257–268
29. Cowin SC (1981) Mechanical Properties of Bone. American Society of Mechanical Engineering, New York
30. LeGeros RZ (1981) Apatites in Biological Systems. Prog Crystal Growth Charact 4:1–45
31. LeGeros RZ (1991) Calcium Phosphates in Oral Biology and Medicine. In: Myers H (ed) Monographs in Oral Sciences Series, vol 15. S. Karger, Basel
32. Boskey AL, Posner AS (1984) Structure and Formation of Bone Mineral. In: Hastings GW, Ducheyne P (eds) Natural and Living Biomaterials. CRC Press, Boca Raton, Florida, pp 27–41
33. LeGeros RZ (1983) Properties of Commercial Bone Grafts Compared to Human Bone and New Synthetic Bone Biomaterials. Ninth Annual meeting of the Society for Biomaterials, Birmingham, Alabama, April Abstr. No. 86
34. LeGeros RZ (1988) Calcium Phosphate Materials in Restorative Dentistry: A Review. Adv Dent Res 2:164–183
35. Doyle C (1990) Bioactive Composites in Orthopedics. In: Yamamuro T, Hench L, Wilson-Hench J (eds) Handbook of Bioactive Ceramics Vol II, CRC Press, Florida, pp 195–207

36. LeGeros RZ, Penugonda B (1984) Potential Use of Calcium Phosphate as Fillers in Composite Restorative Biomaterials. Second World Congress on Biomaterials, Washington, DC, April

37. LeGeros RZ, Orly I, Gregoire M, Abergas T, Kazimiroff J, Tarpley T (1987) Physicochemical Properties of Calcium Phosphate Biomaterials Used as Bone Substitutes. The 13th Ann Meeting Soc of Biomaterials, Abstr. No. 84

38. Alexander H, Parsons JR, Ricci JL, Bajpai PK, Weiss AB (1987) Calcium Phosphate Ceramic Based Composite as Bone Graft Substitutes. In: William C (ed) Critical Reviews in Biocompatibility, CRC Press, Boca Raton, FL, pp 43–77

39. West TL, Burstein DD (1985) Freeze-dried Bone and Coralline Implants Compared in the Dog. J Periodontol 56:348–351

40. LeGeros RZ, Daculsi G (1990) In Vivo Transformation of Biphasic Calcium Phosphate Ceramics: Ultrastructural and Physicochemical Characterization. In: Yamamuro T, Hench L, Wilson-Hench J (eds) CRC Handbook of Bioactive Ceramics Vol 2, CRC Press, Boca Raton, Florida, pp 17–28

41. LeGeros RZ, Daculsi G, Orly I, LeGeros JP (1990) Bone Augmentation Implant Coating Materials. In: Proc. First International Conference on Implant Dentistry, Tokyo, February

42. Ueno Y, Shima Y, Akiyama T (1987) Development of a New Biomaterial as a Bone Substitute: True Bone Ceramics. In: Vincenzini P (ed) High Tech Ceramics, Elsevier Science Publishers BV, Amsterdam, pp 369–378

43. Ricci J, Alexander H, Parsons JR, Salsbury RL, Bajpai PK (1986) Partially Resorbable Hydroxyapatite-based Cement for Repair of Bone Defects. In: Saha S (ed) Biomedical Engineering V: Recent Developments. Pergamon, New York, pp 469–474

44. Roy DM, Linnehan SA (1974) Hydroxyapatite Formed from Coral Skeleton Carbonate by Hydrothermic Exchange. Nature 247:220–227

45. Holmes RE (1979) Bone Regeneration within a Coralline Hydroxyapatite Implant. Plast Reconstr Surg 63:626–636

46. Ravaglioli A, Krajewski A, Ponti P, Valmori R (1986) Bioglaze Adhesion to Metals. J Physiol 47:769–774

47. Albee FH (1920) Studies in Bone Growth. Triple Calcium Phosphate as a Stimulus to Osteogenesis. Ann Surg 71:32–36

48. LeGeros RZ (1986) Variability of b-TCP/HAP Ratios in Sintered Apatites. J Dent Res 65:292, Abstr. No. 110

49. Brown WE, Chow LC (1985) Dental Restorative Cement Pastes. US patent No. 4, 518, 430, May 21

50. LeGeros RZ, Chohayeb A, Schulman A (1982) Apatitic Calcium Phosphates: Possible Restorative materials. J Dent Res 61:343, Abstr. No. 1482

51. Wagner JR (1989) A Clinical and Histological Case Study Using Resorbable Hydroxyapatite for the Repair of Osseous Defects Prior to Endosseous Implant Surgery. J Oral Implantol 15:186–192

52. Bonel G, Heughebaert J-C, Heughebaert M, Lacout JL, Lebugle A (1988) Apatitic Calcium Orthophosphates and Related coumpounds for Biomaterials Preparation. In: Ducheyne P, Lemons JE (eds) Bioceramics: Materials Characteristics Versus In Vivo Behavior. Ann NY Acad Sci 523:115–130

53. LeGeros RZ, Parsons R, Daculsi G, Driessens F, Lee D, Metsger S (1988) Biodegradation/Bioresorption of Calcium Phosphate Ceramics: Task Group Report. In: Ducheyne P, Lemons J (eds) Bioceramics: Material Characterization vs. In Vivo Behavior. Ann NY Acad Sci 253:268–271

54. Korth MW, Meffert RM, Cassingham RJ (in press) A histological Comparison of Alloplastic Grafts Placed in Infrabony Defects Around Hydroxylapatite-coated Implants. Int J Perio

55. Osborn JR, Donath K (1984) Die Enosale Implantation von Hydroxylapatitkeramik und Tricalciumphosphatkeramik: Integration vs. Substitution. Dtsch Zahnarztl Z 39:970–976

56. Aoki H, Kato K, Ogiso M, Tabata T (1977) Studies on the Application of Apatite to Dental Materials. J Dent Eng 18:86–89

57. Niwa S, Sawai K, Takahashi S, Tagai H, Ono M, Fukuda Y (1980) Experimental Studies on the Implantation of Hydroxylapatite in the Medullary Canal of Rabbits. Biomaterials 1:65–71

58. Ellinger RF, Nery EB, Lynch KL (1986) Histological Assessment of Periodontal Osseous Defects Following Implantation of Hydroxyapatite and Biphasic Calcium Phosphate Ceramics: A Case Report. Int J Perio Restor Dent 3:223–233

59. Daculsi G, Passuti N, Martin S, Deudon C, LeGeros R, Raher S (1990) Macroporous Calcium Phosphate Ceramic for Long Bone Surgery in Humans and Dogs: Clinical and Histological Study. J Biomed Mater Res 24:379–396

60. Davies JE (1990) The Use of Cell and Tissue Culture to Investigate Bone Cell Reactions to Bioactive materials. In: Yamamuro I, Hench T, Wilson J (eds) Handbook of Bioactive Ceramics, vol I. CRC Press, Florida, pp 195–225

61. Gregoire M, Orly I, Menanteau J (1990) The Influence of Calcium Phosphate Biomaterials on Human Bone Cell Activities: An In vitro Approach. J Biomed Mater Res 24:163–177

62. Ducheyne P (1987) Bioceramics: Material Characteristics versus In Vivo Behavior. J Biomed Mater Res 21:219–236

63. Boutin P, Christel P, Dorlot J-M, Meunier A, Roquancourt A, Blanquaert D, Herman S, Sedel 1, Witvoet J (1988) The Use of Dense Alumina-Alumina Ceramic Combination in Total Hip Replacement. J Biomed Mater Res: Appl Biomat 22: 1203–1232

64. Nagai N, Takeshita N (1982) Mechanical and Biological Properties of Partially-stabilized Zirconia Ceramic Implants (in Japanese). J Dent Med 25:581–596

65. Bonfield W (1988) Hydroxyapatite-Reinforced Polyethylene as an Analogous Material for Bone Replacement. In: Ducheyne P, Lemons JE (eds) Bioceramics: Materials Characteristics Versus In Vivo Behavior. Ann NY Acad Sci 523:173–177

66. Amler MH, LeGeros RZ (1990) Hard Tissue Replacement (HTR) Polymer as an Implant Material. J Biomed Mater Res 24:1079–1089

67. Ashman A, Bruins P (1985) Prevention of Alveolar Bone Loss Postextraction with Grafting Materials. Oral Surg Oral Med Oral Pathol 60:146–153

68. Kelley BS, Casper RA, Dunn RL (1986) Totally Resorbable High Strength Bone Plate for Internal Fracture Fixation. Biomedical Engineering IV, Recent Developments, pp 26–31

69. Cutright DE, Hunsuck EE, Beasley JD (1971) Fracture Reduction Using a Biodegradable Material, Polylactic Acid. J Oral Surg 29:393–397

70. Cohen J (1983) Metal Implants: Historical Background and Biological Response to Implantation. In: Rubin LR (ed) Biomaterial in Reconstructive Surgery, CV Mosby, pp 46–61

71. Branemark PI, Zarb GA, Albrektsson T (1985) Tissue-integrated Prostheses, Osseo-integration in Clinical Dentistry. Quintessence, Chicago

72. Memon VA, Urban RM, Alroy J, Galantes JO (1986) Malignant Neoplasms Associated with Orthopedic Implant Materials in Rats. J Orthop Res 4:346–355

73. LeGeros RZ, Taheri MH, Quirolgico GM, LeGeros JP (1980) Formation and Stability of Apatites: Effects of Some Cationic Substituents. In: Proc 2nd Int Cong on Phosphorus Compounds, Boston, April, 21–25, pp 89–103

74. Blumenthal NC, Cosma V (1989) Inhibition of Apatite Formation by Titanium and Vanadium Ions. J Biomed Mater Res 23:13–22

75. Eachenroeder HC Jr, McLaughlin RE, Reger SI (1987) Enhanced Stabilization of Porous-coated Implants with Tricalcium Phosphate Granules. Clin Orthop 180: 234–247

76. LeGeros RZ, Niwa S (1989) Unpublished Studies on Apatite/Resorbable Polymer Composite

77. LeGeros RZ, Orly I (1989) Unpublished Studies on Reactive Apatite Coatings

78. Carrerot H, Rieu J, Bousquet G, Rambert A (1989) Alumina Plasma Spray coatings on Stainless Steel and Titanium Alloys for Prosthesis Anchorage. In: Heimke G (ed) Bioceramcs, vol 2, (Proc 2nd International Symposium on Ceramics in Medicine, Heidelberg, 1989

79. Ducheyne P, Healy KE (1988) The Effect of Plasma-sprayed Calcium Phosphate Ceramic Coatings on the Metal Ion Release from Porous Titanium and Cobalt-chromium Alloys. J Biomed Mater Res 22:1137–1163

80. Reck R (1984) Bioactive Glass Ceramics in Ear Surgery: Animal Studies and Clinical Results. Laryngoscope 94:1–54

81. Benque EP, Gineste M, Dufforts JF, Louise F, Borchetti A, Heughebaert M (1986) A Propos des Bioapatites dans la Chirurgie de Comblement. J Periodontol 5:89–101

82. Cranin AN, Tobin GP, Gelbman J (1987) Applications of Hydroxyapatite in Oral and Maxillofacial Surgery, Part II: Ridge Augmentation and Repair of Major Oral Defects. Compend Contin Educ Dent 8:334–345

Modern Trends of Orthopedic Biomechanics in Japan

Fracture

Visco-Elastic Behavior of Callus in Rabbit Femurs

Yoshinori Miyasaka,[1] Minoru Sakurai,[1] Toshimitsu A. Yokobori[2], and Shigeru Sasaki[2]

Summary. Experimental callotasis was induced in rabbit femurs and mechanical tests were performed for the callus in callotasis in order to investigate its mechanical behaviors when tensile loads were applied.

Whole-length femurs were removed after sacrificing eight rabbits in the course of callotasis. Femurs were osteotomized at the level of the midshaft with a bone saw and fixed with a mini-model external fixator. Callotasis was started at the rate of about 0.35 mm twice a day one week after the osteotomy. Twelve osteotomized femurs with applied external fixation were left without callotasis and served as controls.

The callus showed the characteristics of hysteresis in the load-unload curves and also of stress relaxation in stress relaxation tests. Therefore, the elongated callus in callotasis had visco-elasticity characteristics, which were very similar to those of callus in the control group one week after the osteotomy. The immature zone of elongated callus is thought to be responsible for visco-elasticity.

Key words. Callotasis — Callus — Visco-elasticity — Mechanical property

Introduction

Callus distraction has long been considered harmful to fracture healing or bone formation [1, 2], but a new method called callotasis, in which callus is distracted gradually, achieved many successful clinical results and gave impetus to research in the field of long bone lengthening. However, the basic mechanism of callotasis remains to be elucidated. We made an experimental callotasis in the rabbit femurs and carried out mechanical tests for elon-

[1] Department of Orthopedic Surgery, Tohoku University, Seiryo-machi 1-1, Aoba-ku, Sendai, 980 Japan
[2] Department of Mechatronics and Precision Engineering, Tohoku University

gated callus of the femur removed in the course of callotasis. We investigated mechanical behaviors of callus with and without callotasis, the latter being utilized as the control.

Materials and Methods

Twenty Japanese white rabbits weighing from 2.5 to 3.0 kg were used. Under intravenous pentobarbital anesthesia, femoral shafts of the rabbits were osteotomized with a bone saw and fixed with a mini-model external fixator with 4 screws, and their incised skin sutured. After about one week, femoral lengthening was started at the rate of 0.35 mm twice a day. The mini-model external fixator used and an X-ray taken after the operation are shown in Fig. 1. In the process of lengthening, the rabbits were sacrificed and their femurs were removed and frozen for later testing. The control femurs without lengthening were removed, one, two, and three weeks respectively after the osteotomy.

   Mechanical tests were performed on these specimens with a Multi-Capable-Micro-Strength test machine (YSGS type) [3] which was developed by one of the authors. This machine system was mainly developed for the evaluation of small loads and displacements, and the system made it possible to perform mechanical tests on the whole length of a rabbit's long bone [4]. The load-unload hysteresis test and stress relaxation test were the mechanical tests used in this study. The rate of deformation was 11 mm per minute [5].

Results

Load-Unload Hysteresis Test

Load-deformation curves (= load-deflection curves) in two representative control femurs removed one week after osteotomy are represented in Fig. 2. In both cases, the unload curve did not show the same trace as that of the load

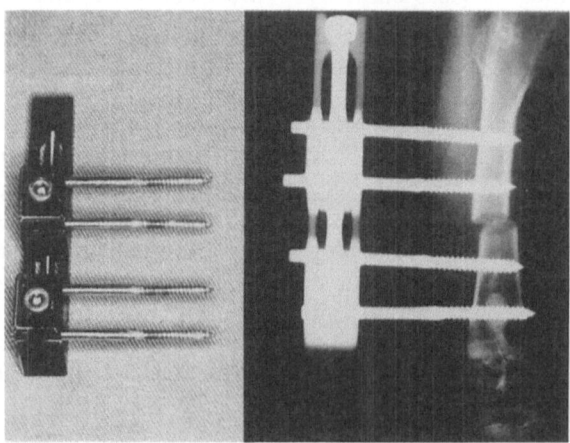

Fig. 1. The mini-external-fixator used and a radiograph after osteotomy of the femur

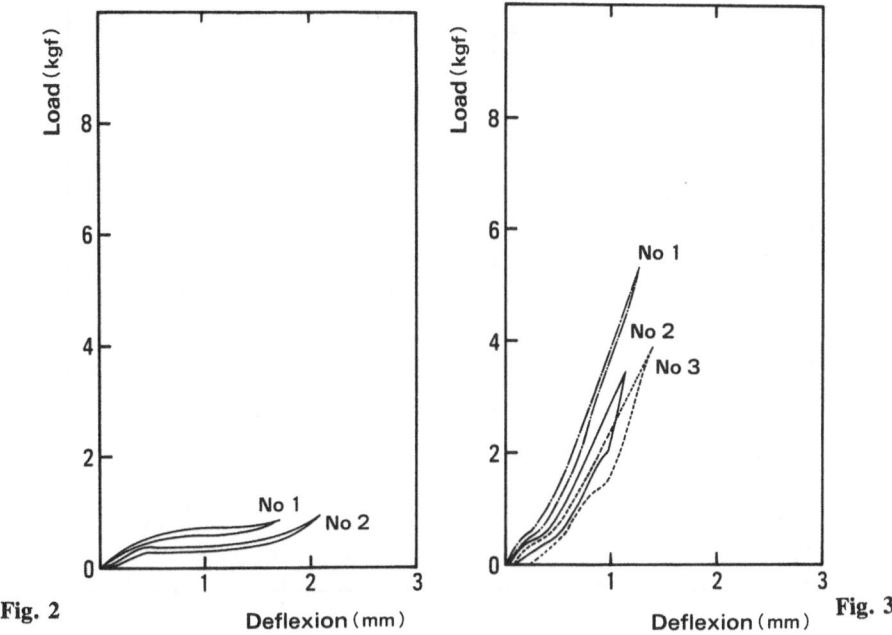

Fig. 2. Two typical examples of the load-deformation curve obtained from callus after 1 week

Fig. 3. Three typical examples of the load-deformation curve obtained from callus after 2 weeks

curve; that is to say, they showed evidence of hysteresis. The compliance of these curves was very large. Three control femurs after two weeks also revealed a characteristic of hysteresis, but had smaller values of compliance than those after one week (Fig. 3). Two control femurs after three weeks also exhibited hysteresis, but further loading was needed to induce displacement as much as that after two weeks (Fig. 4).

The broken line in Fig. 5 represents a typical example of the load-unload curve in elongated callus produced by experimental callotasis. Figure 6 shows the X-ray of a femur removed after 16 days of lengthening at the rate of 0.35 mm two times per day with a final elongation of 11 mm. As to the mechanical behavior of elongated callus to tensile loads, the compliance was very large in a range of small deformation, but it decreased abruptly when deformation became larger than 1.5 mm. However, even when a load of as much as 5 or 6 kg was applied, callus was not disrupted and hysteresis was shown by gradual removal of the applied load (Fig. 5).

Stress Relaxation Test

Figure 7 shows two examples of stress relaxation curves of the callus in the experimental callotasis. The curves of both femurs revealed characteristics of stress relaxation which showed a time-dependent mechanism. Although the

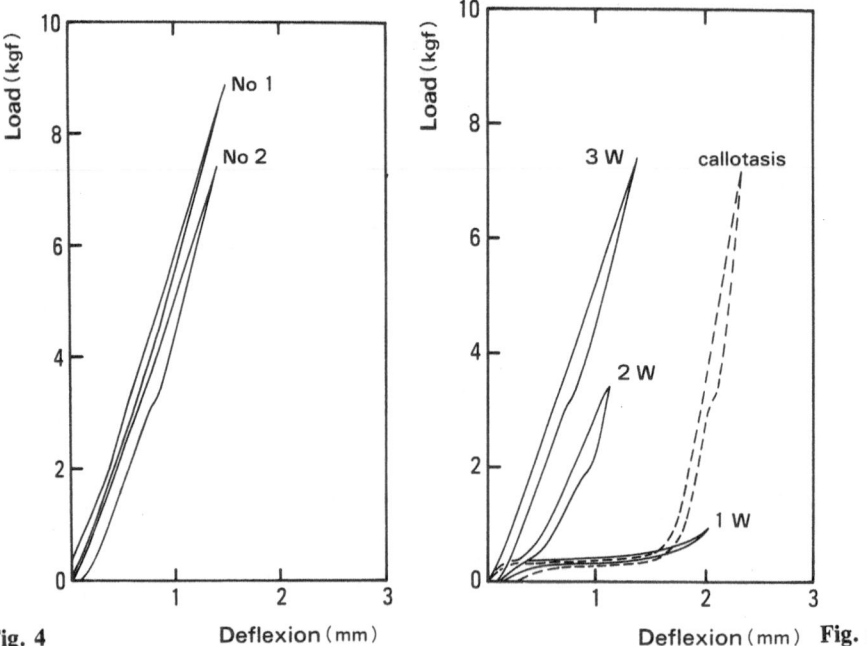

Fig. 4 Deflexion (mm)

Deflexion (mm) **Fig. 5**

Fig. 4. Two typical examples of the load-deformation curve obtained from callus after 3 weeks

Fig. 5. The comparison between four typical examples of the load-deformation curve: 1, 2, and 3 weeks after osteotomy (controls), and during callotasis

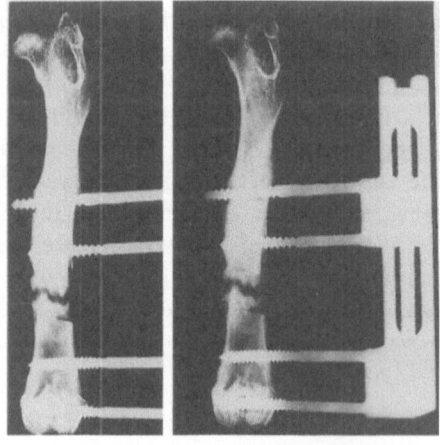

Fig. 6. Two different conditions of the radiograph of experimental callotasis in the femur removed after 16 days of lengthening. The elongated callus of the femur reveals three radiographic zones in structure

Fig. 7. Two typical examples of stress relaxation curves obtained from callus during callotasis

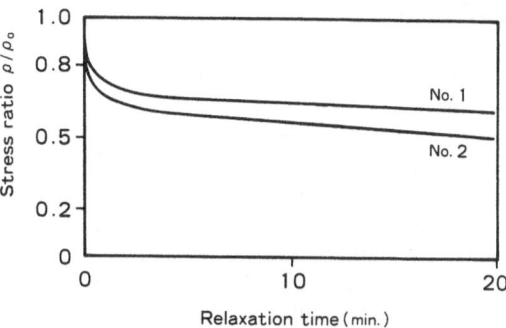

degree of relaxation in the other cases differed, they also demonstrated stress relaxation.

In summary, elongated callus had the property of hysteresis and stress relaxation; in short, it had visco-elasticity.

Discussion

Many clinical cases of callotasis have successfully demonstrated that even callus under tension was not disturbed in its mineralization and finally developed to be mature bone. Callotasis has achieved very excellent clinical results, but its basic mechanism has not yet been clarified. To clarify its mechanism, approaches from two aspects should be essential: Mechanical behavior of callus under tensile loads, and the biological reaction of callus and its surrounding tissues. The present study addressed the former aspect and the results led to the conclusion that callus has visco-elasticity according to its hysteresis loop and stress relaxation curve.

The comparison of load-unload curves was made in Fig. 5, between the callotasis femur (*broken line*) and representative controls, one, two, and three weeks after osteotomy (*solid lines*). The curve pattern in callotasis was very similar to that for the one week control in the deformation range less than 1.5 mm. In such a small range, for example, at the value of 1.0 mm per day as in clinical cases, elongated callus is thought to maintain the state which is easy to be stretched, as the one week controls do. All of the load-unload curves in the three groups of controls showed hysteresis behavior. The two week control curve had a steeper inclination than the one week control, and the three week curve was the steepest of the three. Put another way, the more immature the callus is, the larger its compliance is and the easier it is to stretch.

Our experience reveals that the callus has three radiographic zones in structure during the process of clinical callotasis. The central zone is radio-transparent and represents immature callus. The surrounding two zones are radiopaque due to mineralization and comprise more mature callus [6]. The results of this study lead to the consideration that the central immature zone is the portion which becomes elongated and is responsible for the visco-elastic behavior.

References

1. Albright JA, Brand RA (1987) Scientific Basis of Orthopaedics. Appleton & Lange, Norwalk, Connecticut
2. White AA, Panjabi MM, Southwick WO (1977) The four biomechanical stages of fracture repair. J Bone Joint Surg [Am] 59:188–193
3. Yokobori AT Jr, Ohkuma T, Yoshinari H, Maeyama T, Yokobori T (1990) Mechanical test method of micro blood vessel. The first biomechanical symposium. Japanese Society for Mech Engineers, pp 21–22
4. Miyasaka Y, Sakurai M, Yokobori AT Jr, Sasaki S (1990) Mechanical test method for immature callus formed after the experimental osteotomy of long bones. (in Japanese) J Jpn Orthop Assoc 64(8):1152
5. Miyasaka Y, Sakurai M, Yokobori AT Jr, Sasaki S (1990) Visco-elasticity of callus in a long bone after experimental transverse osteotomy. (in Japanese) Nihon Kikai Gakkai, vol E, pp 119–121
6. Miyasaka Y, Kashimoto O, Nobuta S, Sakurai M, et al. (1991) A radiographical requisite for successful callotasis (in Japanese). J Jpn Soc External Fix 2:129–131

A New Methodology with an Application of Robotics to Control the Mechanical Environment Around Experimentally Fractured Bone

KIYOSHI MABUCHI,[1] HIROMICHI FUJIE,[2] YOSHIRO YAMATOKU,
MAKOTO YAMAMOTO[3], and TADASHI SASADA[4]

Summary. An experimental study was conducted to evaluate the effect of the mechanical environment on the fracture healing of bone. Robotics were utilized to effect various modes of stimulation which included cyclic bending and cyclic torsion. A fractured leg of a rabbit received intermittent stimulation of 1 hour per day during a 60 day period. Differences in healing rate were clearly distinguished between the various modes of stimulation applied. It is evident that robotics are particularly suited to control of the mechanical environment surrounding an experimentally prepared fracture site in bone.

Key words. Bone fracture — Healing of bone — Mechanical environment — Robotics

Introduction

A number of clinical and experimental studies have focused on identifying the optimum conditions for fracture healing [1, 2, 3, 4, 5, 6]. Following these studies, the prevailing opinion is that some motion or mechanical stimulation is required to promote the fracture healing of bone [7, 8]. The problem remains, however, of specifying the nature and quantity of this motion [9] in terms of (a) what directions of motion are required, (b) what level of loading should be

[1] Department of Biomedical Engineering, School of Medicine, Kitasato University, 1-15-1, Kitasato, Sagamihara, Kanagawa, 228 Japan
[2] Musculoskeletal Research Laboratories, School of Medicine, University of Pittsburgh, M272 Scaife Hall, Pittsburgh, PA 15261, USA
[3] Department of Orthopaedic Surgery, School of Medicine, Kitasato University, 1-15-1, Kitasato, Sagamihara, Kanagawa, 228 Japan
[4] Tokyo Institute of Technology, 2-12-1, Ookayama, Meguro-ku, Tokyo, 152 Japan

Table 1. Summary of studies in which control over the mechanical environment surrounding a fractured bone has been attempted

Studies	Device	Species (part)	Mode	Maximum amount	Cycle	Period per day
Eggers, et al. (1949)	Rubber bands	Rat (skull)	Compression, bending or tension	Unknown	Constant	Full time
Friedenberg, et al. (1952)	Spring clamp	Dog (ulna)	Compression	130 N	Constant	Full time
Yamagishi, et al. (1955)	Spring and plate	Rabbit (tibia)	Compression, tension or bending with shear	Unknown	Constant	Full time
Panjabi, et al. (1977)	Spring clamp Pneumatic cylinder	Rabbit (tibia)	Compression	20 N	i) Constant ii) 0.9 Hz	Full time 4 hours
Panjabi, et al. (1979)	Spring clamp Pneumatic cylinder	Rabbit (tibia)	Compression	i) 8 or 80 N ii) 40 N	Constant 0.9 Hz	Full time 4 hours
O'Connor, et al. (1982)	Pneumatic cylinder	Sheep (ulna)	Bending with compression	±4% strain	0.5 Hz	1 hour
Radin, et al. (1984)	Motor and cam	Rabbit (tibia)	Compression	80 N	1.0 Hz	40 minutes
Goodship, et al. (1985)	Pneumatic cylinder	Sheep (tibia)	Compression	1 mm or 360 N	0.5 Hz	17 minutes
Kenwright, et al. (1989)	Servohydraulic actuator	i) Sheep (tibia) ii) Patient (tibia)	Compression	0.5, 2 mm, 200 or 1000 N 1.0 mm	0.5 Hz	17 minutes
Present study	Robot-arm	Rabbit (tibia)	Compression, bending or torsion	5.5 N, 1 or ±5 degrees	0.5 Hz 0.33 Hz	30 minutes 1 hour

used, (c) what frequency of loading should be used, and (d) over what period of time should this treatment be conducted.

To solve these problems, strict control of the mechanical environment surrounding the fracture site is necessary. Although this has been attempted by a number of investigators [4, 7, 8, 10–14] the variety of conditions tested has been limited, particularly with respect to the direction of motion. The modes investigated to date include constant and cyclic compression and constant bending; these studies are summarized in Table 1. No one has rigorously investigated the effect of torsion, or shear or cyclic bending on the fracture healing rate. Such modes of mechanical stimulation are inevitably, to some extent, present during fracture healing in the animal body.

In addition to the problem of environmental control, quantitative assessment of fracture healing has also proved to be an area of concern. In the majority of studies to date, each quantitative assessment has required the euthanasia of an individual animal [4, 8, 9, 13, 15, 16]. Thus, a large number of animals were necessary to study the effect of even a single variable.

In the present study, we will show how robotics can be utilized to totally control the mechanical environment around a fracture site. This methodology basically involves the application of a robot-arm to manipulate the distal end of a fractured tibia of a rabbit, and the specific mode of stimulation is controlled by a computer program. Moreover, it will be shown how robotics can be used to quantitatively assess the fracture healing of bone via the micro-torsion test. The torsional rigidity of a rabbit tibia was estimated with this test both prior to and following (on an intermittent basis) the introduction of an experimentally prepared fracture site. From this data, a healing ratio was calculated which permitted the changing pattern of stiffness to be displayed for the fracture site whilst healing took place under the application of loads applied in cyclic compression, cyclic bending, and cyclic torsion. The effect of zero load or non-stimulation facilitated via rigid fixation was studied in the same manner.

Materials and Methods

The specifications of the robot-arm (Movemaster II, RM-501, Mitsubishi Electronic Corp. Nagoya, Japan) (Fig. 1) used in the present study are as follows:

1. Freedom of motion   = 5 degrees
2. Maximum lifting capacity = 12 N
3. Position accuracy   = ±0.5 mm
4. Control language   = BASIC

The robot arm can be controlled by any computer and either sequential or real time can be selected.

A healthy adult rabbit was chosen for this experiment as its size was suitable to the specification of this robot-arm. The forelegs and right hind leg were set in place on a fixation table using gypsum, such that the upper half of the rabbit

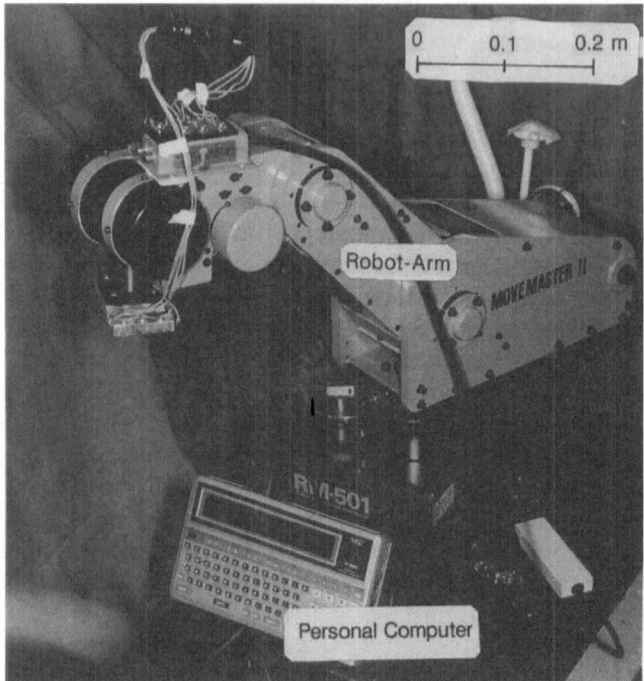

Fig. 1. Robot-arm controlled by a personal computer

body was in an upright position and the lower half was laid on its side (Fig. 2). Following anesthesia, 2 mm Kirschner wires were drilled in a crosswise direction both in the proximal and distal ends of the left tibia so that percutaneous external fixation could be established. Utilization of this method of fixation enabled the rabbit to be fed throughout the two month duration of the experiment. Following fixation, a mid-diaphyseal tibial osteotomy was performed to introduce an experimental fracture site.

The technique used in the present study has been newly developed and we have termed it the "pin-hole osteotomy." This procedure involves driving a pin hole through the rabbit leg percutaneously. Subsequently, a fretsaw blade encircled by two sheaths was inserted into the hole and the tibia was cut. The two sheaths acted to protect the soft tissue and thus minimized bleeding.

Mechanical stimulation was applied by the robot-arm through the distal span of resin following the introduction of the experimental fracture site. All stimulation were cyclic (0.33 Hz) and intermittent (60 min. per day), yielding a total of 1200 cycle per day. Previous studies have selected the cycle of stimulation within the range 0.5–1.0 s and the duration per day between 17–240 min. (Table 1).

In the present study, we analysed five loading modes: Compression, torsion, two different amounts of bending, and non-stimulation. Non-stimulation was

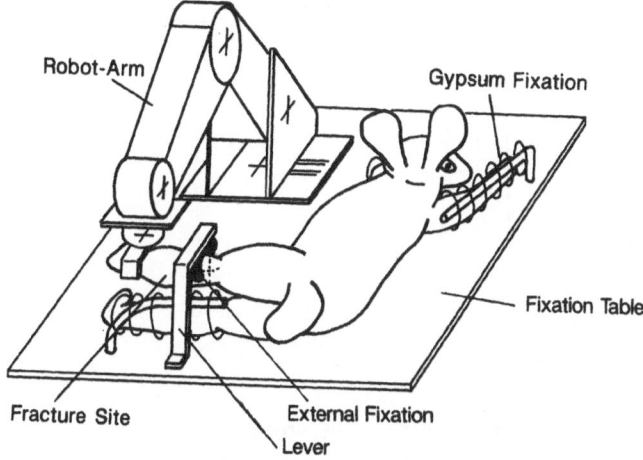

Robot-Arm

Gypsum Fixation

Fixation Table

Fracture Site

External Fixation

Lever

Fig. 2. Schema of the arrangement of devices for the application of mechanical stimulation on the rabbit leg

Table 2. Mode and amount of stimulation applied related to robot-arm motion

Mode of stimulation	Amount	Motion angle of each joint of robot-arm (degree)				
		Waist	Shoulder	Elbow	Wrist-pitch	Wrist-roll
Compression	0–5.5 N	1.82	0.58	−0.11	−0.06	1.82
Torsion	0–1 degree	0.00	0.81	−1.60	−0.21	−1.79
Bending	0–1 degree	0.00	0.50	−0.98	1.48	0.52
Bending	±5 degrees	0.00	−0.51	1.00	−1.49	−0.50

achieved by fixing the rabbit leg as rigidly as possible. In total, six cases were studied; two of non-stimulation and one case each of compression, torsion, smaller bending and larger bending. The modes of stimulation applied by the robot-arm were controlled by interchanging the software; parameters set on the robot-arm for each mode of stimulation are shown in Table 2.

The force or torque applied to the fracture site was the result of a displacement made by the robot-arm. Under compressive stimulation, this displacement was relatively small due to the fact that displacement started from a point at which the two opposing fracture surfaces were in direct contact. The level of compression was varied from 0–5.5 N, as shown in Fig. 3. This square wave pattern was common for all stimulations. The other active stimulations involved cyclic torsion at 0–1 degree, and cyclic bending of 0–1 degree and ±5 degrees. It should be noted, however, that these values diminished with the healing of bone due to the fact that the robot-arm could not move beyond the point where the resisting force exceeded the load capacity. These limits of moment were 0.54 Nm under torsion and 0.94 Nm under bending; these values were obtained by the product of the load capacity 12 N and the moment of the arm, 0.045 m and 0.078 m respectively.

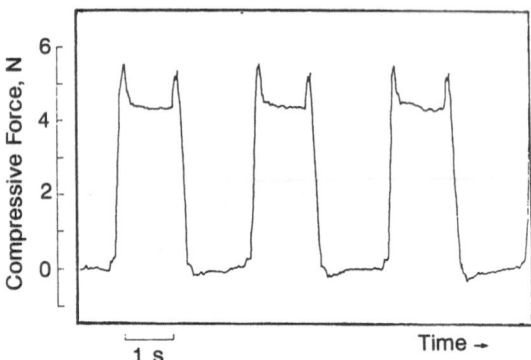

Fig. 3. Change in the compressive force with time

As the healing bone increased in strength, the robot-arm automatically stopped any further motion once the set limits were attained. Consequently, this resulted in a reduction in the displacement. If this limitation on the motion-angle had not been incorporated in the experiments involving torsion and bending, the tibia of the rabbit would have ruptured each time the stimulation was applied. Such a limitation was not, however, necessary under the mode of compressive stimulation as this motion did not act to separate the opposing surfaces forming the interface of the fracture site.

Robotics were also used to measure the stiffness of the healing tibia following fracture; this represented an *in vivo* mechanical test. This method of fracture healing assessment was termed the "micro-torsion test".

The technique involved the detachment of the distal fixation and replacement with a ball bearing having a central steel rod attached to an aluminum plate, as shown in Fig. 4. The force transducer (U-gauge T1, Toyo Baldwin Co., Tokyo) was attached at one end to the fingers of the main robot and at the other end to the aluminum plate. The transducer indicated the torsional force induced by the robot-arm during lifting. The torque was calculated from the product of the force and the moment of the arm. The torsional angle was measured using a linear variable displacement transducer (LVDT, Gap sensor AEC-2525, Applied Electronics Corp. Tokyo) that was gripped in place by the assistant robot-arm. The LVDT indicated the distance separating the probe head and the aluminum metal surface. The curve of the torsional angle vs. torque was plotted by an X-Y plotter.

The amount of torsion was limited to less than $0.005\,\mathrm{Nm}$ and the angle to less than $2 \times 10^{-3}\,\mathrm{rad}$. These parameters are much smaller than the level of torsional stimulation, 2×10^{-2} rad, applied to study the effect on fracture healing. The torsional rigidity was obtained by dividing the torque at the maximum torque angle by the maximum torque angle. Prior to the introduction of the experimental fracture, the micro-torsion test was performed on the intact tibia to obtain the natural torsional rigidity. The healing ratio was calculated by dividing the torsional rigidity of the healing tibia by that of the intact tibia. In addition to this, X-rays of the healing tibia were taken once every three days.

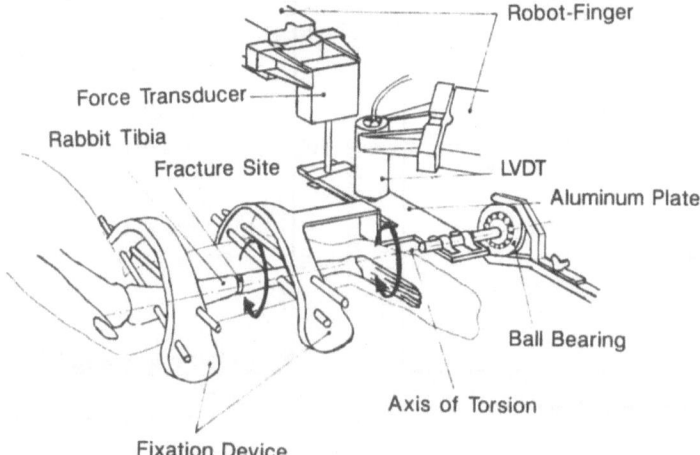

Fig. 4. Schema of the arrangement of devices for the micro-torsion test. *LVDT*, linear variable displacement transducer

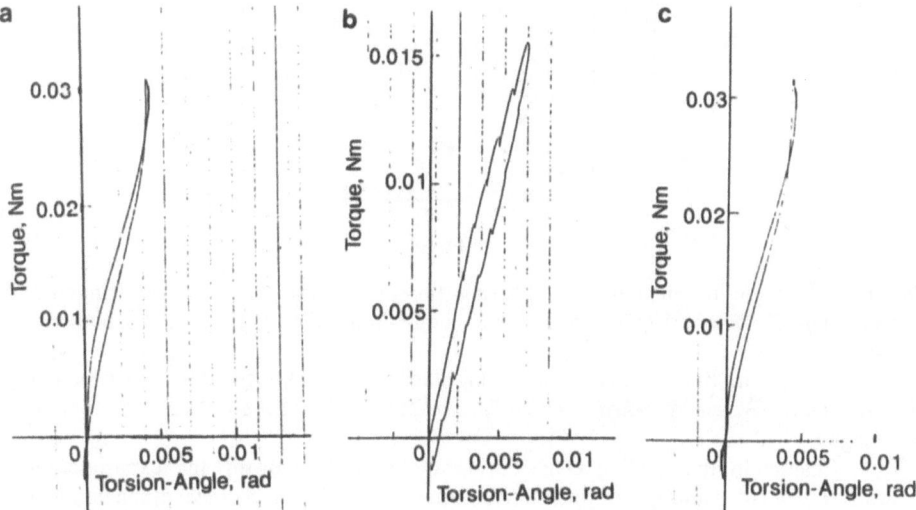

Fig. 5a–c. Torque vs. torsion angle recorded with an X-Y recorder before the osteotomy of the tibiae for **a** Rabbit 1, before the experiment of non-stimulation, **b** Rabbit 2, before the experiment of cyclic compression, **c** Rabbit 3, before the experiment of cyclic torsion

Results

The torsional rigidity of the intact tibiae of the rabbits studied was about 12 Nm/rad (Fig. 5). Following fracture, this value fell to less than 0.1 Nm/rad, resulting in a healing ratio close to zero (Figs. 6–8).

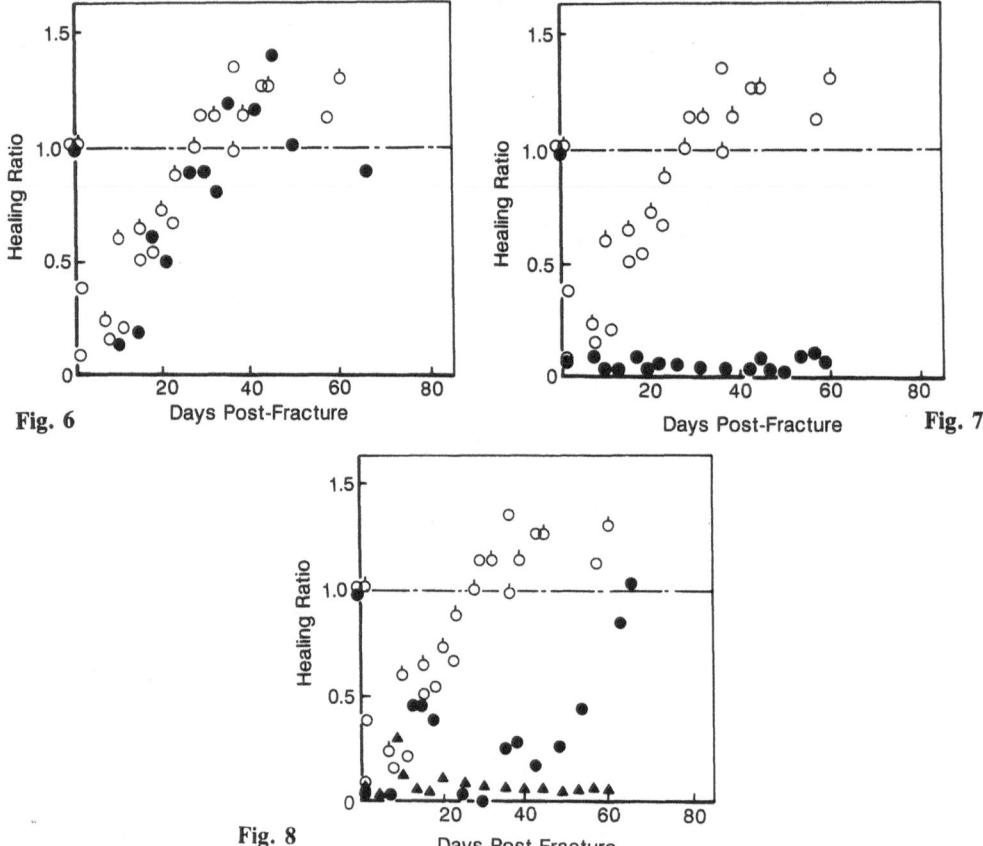

Fig. 6. Change in the healing ratio during the period following the experimental fracture. *Open circle* and *spiked open circle*, non-stimulation; *solid circle*, cyclic compression 0–5.5 N

Fig. 7. Change in the healing ratio during the period following the experimental fracture. *Open circle* and *spiked open circle*, non-stimulation; *solid circle*, cyclic torsion 0 1 degree

Fig. 8. Change in the healing ratio during the period following the experimental fracture. *Open circle* and *spiked open circle*, non-stimulation; *solid circle*, cyclic bending 0–1 degree; *solid triangle*, cyclic bending ±5 degree

With regard to both cases involving the application of non-stimulation, the healing ratio increased rapidly over 30 days and exceeded a value of 1.0 as shown in Fig. 6. The experimentally prepared fracture sites thus appeared to heal over this time period; the fracture lines evident in X-rays disappeared completely after 60 days. Callus formation around the fracture site was only slight as shown in Fig. 9a. This healing mode appeared to be typical of that associated with primary fracture healing.

Under compressive stimulation, the healing ratio increased as rapidly as that under non-stimulation over 30 days (Fig. 6). Again, healing of the fracture appeared to be complete over this time period; the fracture line, similarly disappeared completely after 60 days. Callus formation was evident, as shown in Fig. 9b.

Under cyclic torsion, 0–1 degree, the healing ratio remained at zero throughout a 60 day period as shown in Fig. 7. Although the amount of callus produced was greatest of the five cases studied, the fracture line was clearly evident after 60 days (Fig. 9c). The experimental fracture did not heal under this condition.

Under cyclic bending, 0–1 degree, the healing ratio increased over the first 15 days as shown in Fig. 8. Following this period the healing ratio returned to zero, only to increase again after 40 days. Ultimately, the ratio attained a value of 1.0, and the fracture line completely disappeared after a 60 day period (Fig. 9d). A small amount of callus was evident around the fracture site. Fracture healing was delayed under this condition.

Under cyclic bending, ±5 degrees, the healing ratio remained at zero throughout a 60 day period as shown in Fig. 8. Although a significant amount

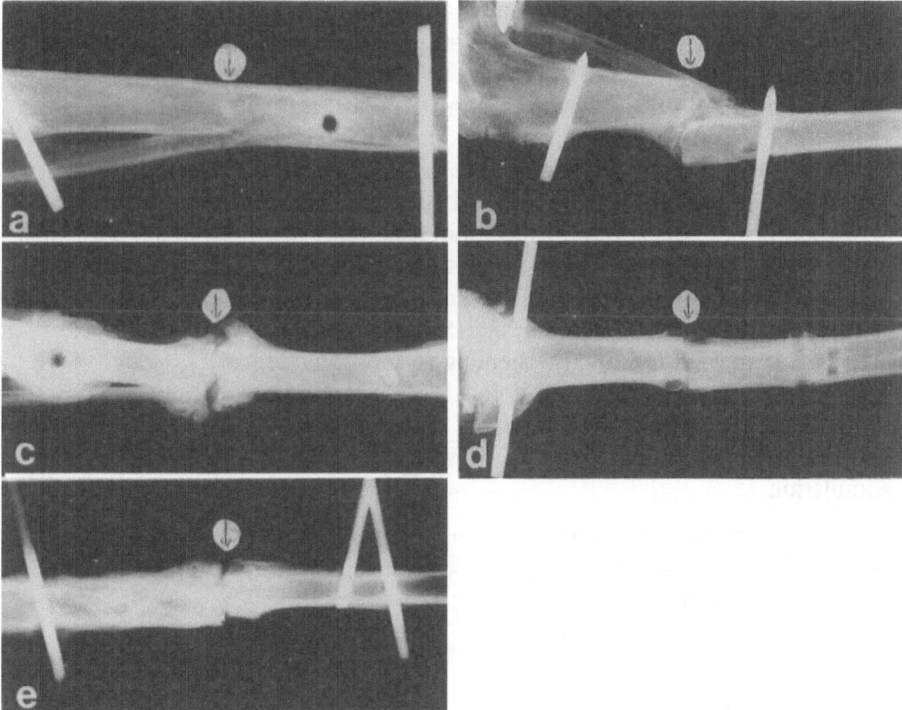

Fig. 9a–e. X-ray photographs of fracture sites 60 days after the experimental fractures. These were taken as soon as the experiments had been finished. *Small arrows* show the fracture sites for **a** non-stimulation, **b** cyclic compression, **c** cyclic torsion, **d** cyclic bending 0–1 degree, **e** cyclic bending ±5 degrees

of callus formation was present, the fracture line was clearly evident after the 60 day period (Fig. 9e). The fracture did not heal under this condition.

Discussion

In the present study, we have shown how robotics can be utilized to totally control the mechanical environment around an experimentally prepared fracture site in the tibial bone of a rabbit. This method permits the application of the torsion and bending modes of stimulation, which have not previously been studied. Differences in healing rate were clearly distinguished for the various modes of stimulation applied.

The maximum load applied in the present study was 5.5 N under the compressive stimulation mode. This value represented about 15% of the body weight of the rabbit. Although this value is somewhat lower than that used by other investigators (Table 1), the load was sufficient to cause callus formation (Fig. 9b) which was not apparent in the case of non-stimulation (Fig. 9a).

The amount of torsion, 0–1 degree, was lower than the values of displacement observed clinically with healing bones [16]. The tibial fracture site in the rabbit leg did not heal under this small amount of torsion despite the formation of a large callus. This result is obviously distinct from that found under compressive and non-stimulation modes.

A maximum of one degree bending is realistically found in the clinical situation. Fracture healing under this small degree of bending was delayed. Bending at the level of ±5 degrees prevented fracture healing from occurring. These results show that the degree of bending is important.

In the present study, due to the limited number of experiments performed, it is difficult to draw any firm conclusion, however, it can be surmised that torsional stimulation is particularly detrimental to fracture healing. Consequently, the concept proposed by Sarmiento [1, 2] that any direction of motion is beneficial for fracture healing is questionable.

More studies are required to determine the optimum treatment to promote the healing of a fractured bone. It is anticipated that the use of the present method will contribute towards identification of the optimum modes of stimulation.

Acknowledgement. This study was supported in part by funds provided by the Ministry of Education, Japan, Science and Culture, Grant-in-Aid, No. 61570727, 63570708.

References

1. Sarmiento A (1967) A functional below-the-knee cast for tibial fracture. J Bone Joint Surg [Am] 49:855–875
2. Sarmiento A (1974) Functional bracing of tibial fracture. Clin Orthop 105:203–219

3. Bradley GW, Mckenna GB, Dunn, HK, Daniels AU, Statton WO (1979) Effects of flexural rigidity of plates on bone healing. J Bone Joint Surg [Am] 61:866–872
4. Eggers GWN, Sindler TO, Pomerat CM (1949) The influence of the contact-compression factor on osteogenesis in surgical fractures. J Bone Joint Surg [Am] 31:693–716
5. Woo SLY, Lothringer KS, Akeson WH, Coutts RD, Woo YK, Simon BR, Gomez MA (1984) Less rigid internal fixation plates: Historical perspectives and new concepts. J Orthop Res 1:431–449
6. Kenwright J, Goodship AE (1989) Controlled mechanical stimulation in the treatment of tibial fractures. Clin Orthop 241:36–47
7. Panjabi MM, White III AA, Southwick WO (1977) Temporal changes in the physical properties of healing fractures in rabbits. J Biomech 10:689–699
8. Panjabi MM, White III AA, Wolf JW Jr (1979) A biomechanical comparison of the effects of constant and cyclic compression of fracture healing in rabbit long bones. Acta Orthop Scand 50:653–661
9. White III AA (1975) Fracture treatment — The still unsolved problem. Clin Orthop 106:279–284
10. Friedenberg ZB, French G (1952) The effect of known compression forces on fracture healing. Surg Gynecol Obstet 94:743–748
11. Yamagishi M, Yoshimura Y (1955) The biomechanics of fracture healing. J Bone Joint Surg [Am] 37:1035–1086
12. O'Connor JA, Lanyon LE (1982) The influence of strain rate on adaptive bone remodeling. J Biomech 15:767–781
13. Radin EL, Martin RB, Burr DB, Caterson B, Boyd RD, Goodwin C (1984) Effects of mechanical loading on the tissues of the rabbit knee. J Orthop Res 2:221–234
14. Goodship AE, Kenwright J (1985) The influence of induced micromovement upon the healing of experimental tibial fractures. J Bone Joint Surg [Br] 67:650–655
15. Brighton CT (1984) Principal of fracture healing. Instr Course Lect 33:60–82
16. Takao Y, Kobayashi A, Yamada K, Hachiya S, Suzuki K, Morioka T (1980) Functional bracing for the fractures of the humeral shaft. (in Japanese) Orthop Surg Traumatol 23:1071–1083

Spine

Three-Dimensional Kinematic Analysis of the Human Whole Lumbar Spine and Natural Lumbosacral Spondylolysis

Isao Yamamoto, Kiyoshi Kaneda[1], and Manohar M. Panjabi[2]

Summary. The purposes of this in vitro study were to analyse the three-dimensional (3D) movements of the fresh whole human lumbar spines and natural L5 spondylolysis spine to clarify the instability of this lesion. The representative values of the movements of the human whole lumbar spine were presented. In flexion, the mean values showed an increasing range of motion from L1/2 to L5/S1, and L5/S1 had the greatest motion (10.0°). In extension, similar magnitudes of motions were seen at L1/2, L2/3, L3/4, and more motion took place at L4/5 (5.6°) and L5/S1 (7.8°). In axial rotation, the maximum range of motion was under 3°. The least motion took place at L5/S1 (1.3°). In lateral bending, similar magnitudes of motions were seen on the left and right sides. The greatest motion was seen at L2/3 (7.0°) and the least motion was seen at L1/2 (4.9°). The main motions of the lumbosacral junction in the natural L5 spondylolysis specimen were 14.8° in flexion, 9.0° in extension, 7.7° in left axial rotation, 7.5° in right axial rotation, 10.7° in right lateral bending, and 7.6° in left lateral bending. The lumbosacral movements of the L5 spondylolysis specimen were bigger than those of the normal specimens in all directions. Especially the rotational instability was remarkable in the natural L5 spondylolysis specimen.

Key words. 3D Kinematics — Whole lumbar spine — Lumbosacral spondylolysis

Introduction

The lumbar spine has three important functions: The flexibility of trunk motions, the support of body weight, and the protection of nervous tissues.

[1] Department of Orthopaedic Surgery, Hokkaido University School of Medicine, Sapporo, Japan
[2] Department of Orthopaedics and Rehabilitation, Yale Medical School, New Haven, Connecticut, USA

194

The clinical pathologic conditions such as spinal trauma, degeneration, and diseases may potentially produce unstable conditions and abnormal stress distributions in the lumbar spine. These conditions may lead to low back pain and neurologic disturbances in the lumbar spine. In these pathologic conditions, abnormal motion is certainly one of the signs of typical symptoms of instability [1]. For evaluating abnormal motions in these pathologic conditions and biomechanical properties of the lumbar spine, knowledge of the physiologic lumbar spine movements are necessary. To investigate more natural movements of the lumbar spine, which include lordosis and integrity of the ligaments, it is necessary to use fresh whole lumbar and lumbosacral specimens. However, there are few experimental studies with whole lumbar spine kinematics. One purpose of this study was to determine systematically the 3D physiologic movements of the whole lumbar spine using fresh human cadaveric specimens.

Lumbosacral spondylolysis is one of the important pathologic conditions which produce instability in the lumbar spine. However, there have been few reports concerning the instability of L5 spondylolysis. Accurate estimation of the instability of this lesion is difficult in in vivo studies due to muscle spasm associated with pain [2]. In this series of biomechanical in vitro studies, we were able to study specimens with natural L5 spondylolysis. Another purpose of this study is to investigate the 3D movements of natural L5 spondylolysis and clarify the instability of this lesion by comparison with the movement of the normal lumbar spine, also by three dimensional analysis. This study was done in the Biomechanic Laboratory in the Department of Orthopaedics and Rehabiliation, Yale University School of Medicine.

Methodology

The fresh cadaveric specimens used in this study, including whole lumbar spines and the spine with natural L5 spondylolysis, were freeze-stored at $-20°C$ until tested. Muscles of the specimen were carefully dissected and all ligamentous components were preserved. The sacra were embedded in a quick-setting epoxy block to align the superior surface of L4 horizontally and to provide fixation to the test table. The top vertera was provided with another epoxy block that included a specially designed loading jig. Markers made of 1.5 mm-thick Plexiglas were affixed to each vertebra (Fig. 1). Each marker contained three 0.5 mm diameter steel balls distributed noncolinearly. Two oblique and one lateral roentgenogram of the specimen with markers were taken. The marker points were digitized to provide geometric relationships between the anatomic coordinate system and the corresponding marker points. Such relationships made it possible to transform the motions of the markers, as measured by the stereophotogrammetry, into the relative motions of the vertebra. Six pure moments were applied to the specimen at the top vertebra: Flexion, extension, bilateral axial rotations, and bilateral lateral bendings. A whole lumbar specimen was a long specimen. Therefore, pure moment loading was chosen since it produces uniform loading to each segment of a long

Fig. 1. Experimental apparatus showing the specimen with markers

specimen. Such is not the case with eccentric force application. The pure moments were produced using a specially designed loading jig and loading system. This system was made up of paired vacuum-operated low friction glass cylinders that moved freely in accordance with the motion of the top vertebra of the specimen in order to produce pure moments. The maximum moment applied to the normal specimens was 10 N-m reached in five equal steps, which, from preliminary experiments, was judged to be sufficient to produce physiologic motions but small enough not to damage the specimen. The maximum moment applied to the L5 natural spondylolysis was 7.5 N-m reached in four equal steps. Each moment was applied in three load-unload cycles, allowing 30 seconds of creep to occur at each load step to precondition the specimen and minimize the viscoelastic effects of the specimens. At each load-step of the third load cycle, stereophotographs were taken of the markers attached to the specimen by two 35 mm cameras. The stereophotographs of the marker points (three per vertebra) were digitized on an Altek digitizer (62 × 63 cm^2, accuracy ±0.075 mm), and the 3D coordinates of each point were calculated by applying the Direct Linear Transformation method. The 6 df motion was determined for each moving vertebra. This technique allows the spine to move freely in all of its natural motion patterns and to provide noncontacting motion measurement of the specimen. The standard error of the mean (SEM) of the angular movement was 0.2°.

Fig. 2. Natural L5 spondylolysis

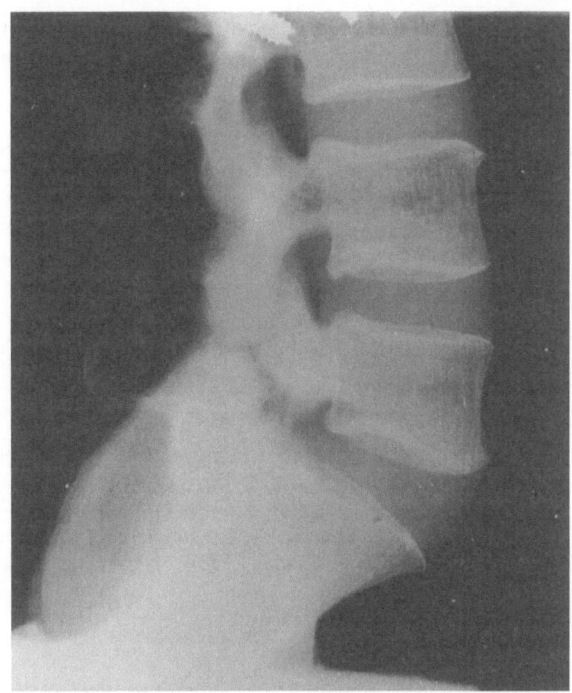

The 3D Kinematics of the Human Whole Lumbar Spine

The materials used were ten fresh cadaveric whole lumbar spine specimens including the sacrum. The ages of the seven specimens were between 25 and 63, however those of the remaining three specimens were unknown. The average disc degeneration grades were 2.4 at L1/2 and L2/3, 2.6 at L3/4 and L4/5, and 2.8 at L5/S1. This was obtained on direct macroscopic examination of the mid-disc surface on a scale of one to four by Nachemson's criteria [3].

The 3D Movement Analysis of Lumbosacral Spondylolysis

A fresh cadaveric lumbar specimen (L2 to the sacrum) with bilateral L5 spondylolysis was used. The age of the specimen was 19 years old. There were no degenerative findings of the discs on X-ray (Fig. 2). The specimen was carefully cleaned of muscles, and all ligamentous components remained intact.

Results

The specimens were additionally tested using the same method mentioned above but with 50 N, 100 N and 150 N pre-load to align the center of the L2 and L5 vertebral bodies, and the pullies for the weight of pre-load were located

Table 1. Ranges of motion in normal whole lumbar spines. The numbers in parentheses represent the standard error of the mean

Level	Type of motion					
	+MX	−MX	+MY	−MY	+MZ	−MZ
L1/2	5.8° (0.6°)	4.3° (0.5°)	2.6° (0.5°)	2.0° (0.6°)	5.2° (0.4°)	4.7° (0.4°)
L2/3	6.5° (0.3°)	4.3° (0.3°)	2.2° (0.4°)	3.0° (0.4°)	7.0° (0.6°)	7.0° (0.6°)
L3/4	7.5° (0.8°)	3.7° (0.3°)	2.7° (0.4°)	2.5° (0.4°)	5.8° (0.5°)	5.7° (0.3°)
L4/5	8.9° (0.7°)	5.8° (0.4°)	1.7° (0.3°)	2.7° (0.5°)	5.9° (0.5°)	5.5° (0.5°)
L5/S1	10° (1.0°)	7.8° (0.7°)	1.5° (0.2°)	1.3° (0.2°)	5.7° (0.4°)	5.3° (0.4°)

+MX, flexion; −MX, extension; +MY, left axial rotation; −MY, right axial rotation; +MZ, right lateral bending; −MZ, left lateral bending

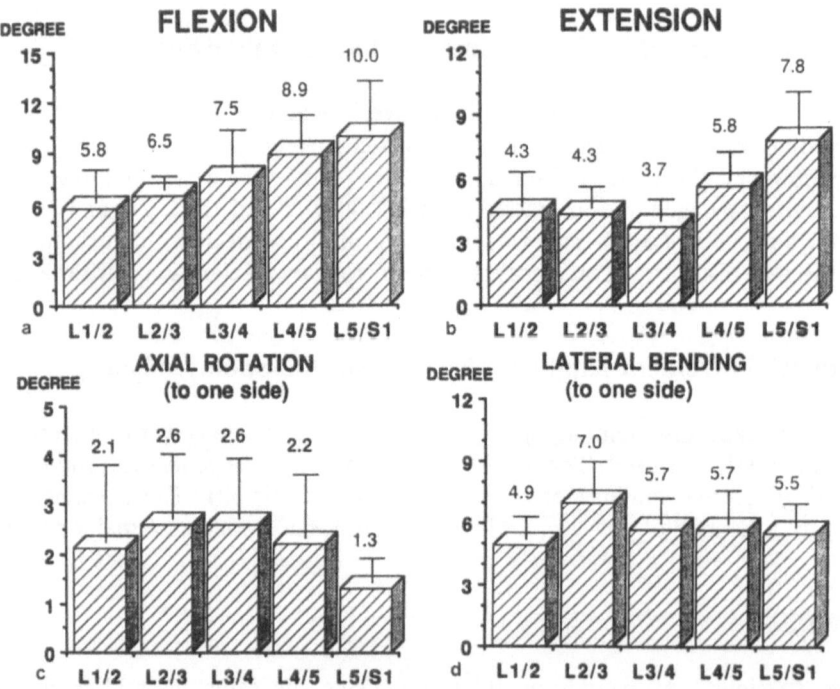

Fig. 3a–d. 3D movements of whole lumbar spine in **a** flexion, **b** extension, **c** axial rotation to one side, and **d** lateral bending to one side

bilaterally at the L5/S1 disc level. In this situation, the pre-load acted mainly as a compressive load to the specimen without causing additional moments. The results with the pre-load are not statistically different from the results without pre-load. Therefore, only the set of results without pre-load are presented. Although all the 6 *df* were measured, the main motions (rotations in the direction of the applied moments) in degrees are presented.

3D Kinematics of the Normal Whole Lumbar Spine

Ranges of motion for flexion, extension, right and left axial rotation, and right and left lateral bending are listed in Table 1. The results are illustrated in Fig. 3a–d and the right and left values of axial rotation and lateral bending were averaged. In flexion, the mean values showed an increasing range of motion progressively from L1 to L5, and L5/S1 had the greatest range of motion (Fig. 3a). In extension, similar magnitudes of motion were seen at L1/2, L2/3, L3/4, and more motion took place at L4/5 and L5/S1 (Fig. 3b). In axial rotation, the maximum range of motion was under 3°. The least motion took palce at L5/S1 (Fig. 3c). In lateral bending, similar magnitudes of motion were seen on the left and right sides. The greatest motion was seen at L2–3 and the least motion was seen at L1/2 (Fig. 3d).

3D Kinematics of the Natural L5 Spondylolysis

The range of motion (ROM) of main motions for flexion, extension, right and left axial rotation, and right and left lateral bending are shown in Table 2. The natural L5 spondylolysis specimen showed more unstable movements in all directions at the lumbosacral junction than the normal spines, especially in axial rotational movement.

Discussion

The history of the study of spinal motion analysis began from Galen who used animal spines (described in [4]). Weber et al. did the first analysis of lumbar spinal motion using cadavers [5]. Since then, there have been reported a

Table 2. Ranges of motion for spondylolysis specimens

Level	Type of motion					
	+MX	−MX	+MY	−MY	+MZ	−MZ
L4/5	5.8°	3.3°	1.6°	0.9°	5.8°	5.1°
L5/S1	14.8°	9.0°	7.7°	7.5°	10.7°	7.6°

+MX, flexion; −MX, extension; +MY, left axial rotation; −MY, right axial rotation; +MZ, right lateral bending; −MZ, left lateral bending

variety of studies on the movements of the lumbar spine—various kinds of clinical studies and experimental research using cadavers. Most of the previous in vitro studies analyzed only the movements of a functional spinal unit or a few lumbar motion segments. However, to investigate more natural movements of the lumbar spine, which include lordosis and integrity of the ligaments, it is necessary to use whole lumbar spine specimens. There are few papers with whole lumbar spine kinematics. Quinnell and Stockdale [6] and Lehmann et al. [7] tested multilevel lumbar spinal segments. However, they applied off-centered loads at T12, with resulting different and unknown moments, as well as forces applied at each lumbar level. Goel et al. [8] performed the first systematic in vitro study of the 3D kinematics of the whole lumbar spine. He applied a maximum load of 3.0 N-m. Hilton et al. [9] studied the mobility of the lumbar motor segments with 2 kg and 4 kg loads. It was found that applying a 4 kg load increased total mobility by up to 25%. To evaluate the physiologic movements of the human lumbar spine in vitro, it is necessary to load the spine with the maximum possible load that does not cause injury to the specimen. We applied a load of 10 N-m as the maximum moment that, from preliminary experiments, was judged to be large enough to produce physiologic motions but small enough not to injure the specimen. We also developed a specially designed loading system, and great efforts were taken to load the specimen with pure moments in order to avoid complex loads with unknown magnitude at each disc level. The direction and the magnitude of the applied load are thought to be important in testing long specimens.

Various kinds of clinical studies and experimental research using cadavers has been performed to investigate the movement of the lumbar spine. In clinical studies, direct measurement of living subjects are limited by the accuracy of the measurement instruments and the identification of body landmarks. The radiographic methods are considered to provide more accurate measurement results. However, conventional radiography does not demonstrate the extent of any simultaneous movements in the horizontal and coronal planes because it cannot quantify movements not in the plane being studied. In in vitro studies, it is possible to apply a uniform load to the lumbar spine and obtain accurate load-displacement characteristics of each intervertebral joint. Table 3 summarizes the results of previous studies [10–18]. On the whole, our results are in good agreement with most of the recent reports. The significant differences were seen at L5/S1 between in vivo and in vitro studies. The differences at L5/S1 might be caused by the presence of the iliolumbar ligament. Our results, with an intact iliolumbar ligament in four specimens, were very similar to those of the in vivo studies [19]. In this study, the iliolumbar ligament was shown to have the function of restricting lateral bending in the lumbosacral junction. When bilateral iliolumbar ligaments were present, flexion and extension movements of the lumbosacral junction were restricted, but this is not the case when only one ligament is present. This ligament also restricted motion to the contralateral side under torsional load.

A progressive decrease in range of motion with age has been demonstrated in in vivo and in vitro studies [17, 20]. In our experiments, the average disc degeneration grade by Nachemson was 2.5. Our results are in good agreement with those of Pearcy et al. [15, 16], who tested males aged 21 to 37 years. So we think our results accurately represent the average characteristics of adult human lumbar spine movement.

Spondylolysis is one of the important pathologic conditions which produce instability in the lumbar spine. The instability of this spinal lesion may cause low back pain and associated leg symptoms. This lesion is sometimes indicated for the surgical treatment of spinal fusion due to its instability. Kessen et al. [21] reported the significant displacement of the translation on lateral X-ray film of patients with spondylolysis in three postures. However, he did not study the movements of these patients. Accurate estimation of the instability of this lesion is difficult in in vivo studies due to muscle spasm associated with pain. Pearcy et al. [2] reported decreased motion in patients with lumbar spondylolytic spondylolisthesis using the stereoradiography method. The causes are thought to be muscle spasm and defensive action to protect patients from pain. In the present study we had the chance to study the 3D movements of the specimen with natural L5 spondylolysis. The results were compared with the average values of normal whole lumbar spine movement at a load of 7.5 N-m (Fig. 4a–c). The 3D in vitro analysis of L5 spondylolysis has not been reported yet in the literature. From our results, the lumbosacral movements of the natural L5 spondylolysis specimen were greater than those of the normal specimens in all directions. The rotational instability was especially remarkable

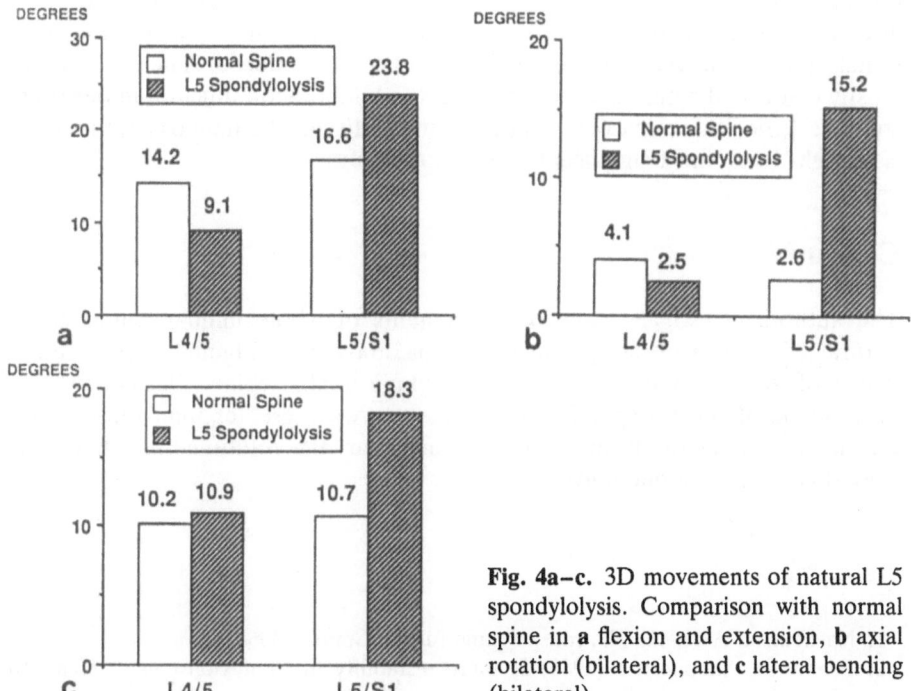

Fig. 4a–c. 3D movements of natural L5 spondylolysis. Comparison with normal spine in **a** flexion and extension, **b** axial rotation (bilateral), and **c** lateral bending (bilateral)

Table 3. Previous reports

	Year		Cases	Range of Motion (Degree)				
				L1/2	L2/3	L3/4	L4/5	L5/S1
1. Sagittal Movements (Flexion-Extension)								
Tanz	1953	in vivo	14	5.6	7.6	8.6	12.2	8.2
Allbrook	1957	in vivo	20	6.0	8.0	13.0	19.0	18.0
Clayson	1962	in vivo	26	12.6	15.8	15.9	17.7	18.7
Froning	1968	in vivo	30	9.0	11.0	13.0	16.0	17.0
White	1978	in vitro	—	12.0	14.0	15.0	17.0	20.0
Pearcy	1984	in vivo	11	13.0	14.0	13.0	16.0	14.0
Hayes	1989	in vivo	59	7.9	9.0	10.0	13.0	14.0
Authors	1989	in vitro	10	10.1	10.8	11.2	14.5	17.8
2. Lateral Bending (to one side)								
Tanz	1953	in vivo	14	5.4	7.8	8.0	8.2	2.3
White	1978	in vitro	—	6.0	6.0	8.0	6.0	3.0
Pearcy	1984	in vivo	10	5.0	5.5	5.0	3.0	1.5
McGlashen	1987	in vitro	9	—	—	—	—	4.4
Authors	1989	in vitro	10	4.9	7.0	5.7	5.7	5.5
3. Axial Rotation (to one side)								
White	1978	in vitro	—	2.0	2.0	2.0	2.0	5.0
Pearcy	1984	in vivo	10	1.0	1.0	1.5	1.5	1.0
Authors	1989	in vitro	10	2.1	2.6	2.6	2.2	1.3

in the natural L5 spondylolysis specimen. White and Panjabi [22] discussed the instability of spondylolysis, that is, when there is a defect in the pars interarticularis, and the role of the posterior elements in stabilizing the functional spinal unit was thought to be significantly reduced. The facet joints mostly controlled rotational movement in the lumbosacral junction in our other study [23]. For these reasons, rotational instability in the lumbosacral junction is thought to be most significant in L5 spondylolysis.

Conclusions

The authors investigated the 3D movements of whole lumbar spines and a natural L5 spondylolysis specimen. In this study, we obtained representative values of the movements of the human whole lumbar spine. The lumbosacral movements of the L5 spondylolysis specimen were greater than those of the normal specimens in all directions. In particular, the rotational instability was remarkable in L5 spondylolysis.

References

1. Nachemson A (1981) The role of spine fusion. Spine 6:306–307
2. Pearcy M, Shepherd J (1985) Is there instability in spondylolisthesis? Spine 10: 175–177

3. Nachemson A (1960) Lumbar intradiscal pressure. Acta Orthop Scand [Suppl]:43
4. Andersson N, Ekström T (1940) Über die Beweglichkeit der Wirbelsäule. Morph Jb 85:135–185
5. Weber EH (1827) Anatomisch-physiologische Untersuchung über einige Einrichtungen im Mechanismus der menshlichen Wirbelsäule. Arch F Anat Physiol:240–271
6. Quinnell RC, Stockdale HR (1981) Some experimental observations of the influence of a single lumbar floating fusion on the remaining lumbar spine. Spine 6:263–267
7. Lehmann TK, Wilson MA, Crowninshield KD (1982) Load response characteristics of lumbar spine following surgical destabilization. Presented at the 28th Annual Meeting of the Orthopaedic Reserach Society, New Orleans, January 16–18
8. Goel VK, Goyze S, Clark C, Nishiyama K, Nye T (1985) Kinematics of the whole lumbar spine: Effect of disectomy. Spine 10:543–554
9. Hilton RC, Ball J, Benn RT (1979) In vitro mobility of the lumbar spine. Ann Rheum Dis 38:378–383
10. Allbrook D (1957) Movements of the lumbar spinal column. J Bone Joint Surg [Br] 39:339–345
11. Clayson SJ, Newman IM, Debevec DF, Anger RW, Skowlund HV, Kottke FJ (1962) Evaluation of mobility of hip and lumbar vertebrae of normal young women. Arch Phys Med Rehabil 43:1–8
12. Froning EC, Frohman B (1968) Motion of the lumbosacral spine after laminectomy and spine fusion. J Bone Joint Surg [Am] 50:897–918
13. Hayes MA, Howard TC, Gruel CR, Kopta JA (1989) Roentgenographic evaluations of lumbar spine flexion-extension in asymptomatic individuals. Spine 14: 327–331
14. McGlashen KM, Miller JAA, Schultz AB, Andersson GBJ (1987) Load displacement behavior of the human lumbosacral joint. J Orthop Res 5:488–496
15. Pearcy MJ, Portek I, Shepherd J (1984) Three-dimensional X-ray analysis of normal movement in the lubmar spine. Spine 9:294–297
16. Pearcy MJ, Tibrewal SB (1984) Axial rotation and lateral bending in the normal lumbar spine measured by three-dimensional radiography. Spine 9:582–587
17. Tanz SS (1953) Motion of the lumbar spine: A roentgenologic study. AJR 69: 399–412
18. White AA, Panjabi MM (1978) The basic kinematics of the human spine. A review of past and current knowledge. Spine 3:12–20
19. Yamamoto I, Panjabi MM, Oxland TR, Crisco JT (1990) The role of the iliolumbar ligament in the lumbosacral junction. Spine 15, 1138–1141
20. Taylor J, Twomsey L (1980) Sagittal and horizontal plane movement of the human lumbar vertebral column in cadavers and in living. Rheumatol Rehab 19:223–232
21. Kessen W, During J, Beeker TW, Goudfrooij H, Crowe A (1984) Recordings of the movement at the intervertebral segment L5-S1: A technique for the determination of the movement in the L5-S1 spinal segment by using three specified postural positions. Spine 9, 83–90
22. White AA, Panjabi MM (1978) Clinical biomechanics of the spine. Lippincott, Philadelphia, p 38
23. Yamamoto I, Panjabi MM, Crisco JJ, Orland TR (1992) Three-dimensional evaluation of the biomechanical instability of the sequential lumbosacral injuries. Presented at International Society for the Study of the Lumbar Spine, Chicago, May 20–24

Hip Joint

Micromotions and Strains of Cementless Femoral Prostheses

Yasuhiro Okumura, Shinichi Imura, Hironori Oomori,
Kouzou Ichihashi, and Hideyuki Takedani[1]

Summary. Using cadaver femurs, strains on the femoral surface and micro-motions between stem and bone were measured.

Stress shielding in the proximal medial region after stem insertion was apt to be stronger with a collarless stem than with a collared stem.

It was considered important to obtain a close proximal fit in order to minimize micromotion in the proximal region of the stem. Distal fixation, however, did not exert any influence on micromotion in the proximal region. With a collared stem, micromotion in the proximal region was largest in proximal-loose and distal-fit. The collarless stem had generally smaller micro-motion regardless of proximal and distal fixation. Therefore, it was considered more favorable to use a collarless stem to obtain proximal fixation of the stem.

Key words. Total hip arthroplasty — Femoral stem prostheses — Strain distribution — Micromotion measurement

Introduction

Cementless femoral stem prostheses generally used have been manufactured using normal femurs of European and American people as a model. Such prostheses are not always considered compatible in Japanese patients with secondary hip osteoarthritis. Accordingly, we designed prototype stems by analysis of femoral cavity morphology using computer graphic methods with computed tomography (CT) images of the hip both in normal subjects and patients with osteoarthritic hips. It is important for cementless hip arthroplasty that implant fixation occurrs as quickly as possible after surgery, and that strain distributions on the implanted femur approximate those of the normal femur as

[1] Department of Orthopaedic Surgery, Fukui Medical School, Fukui, Japan

Fig. 1. Five strain gauges were bonded on each of the medial, lateral, anterior, and posterior surfaces of the femur

Fig. 2. Loading condition and strain measurement apparatus

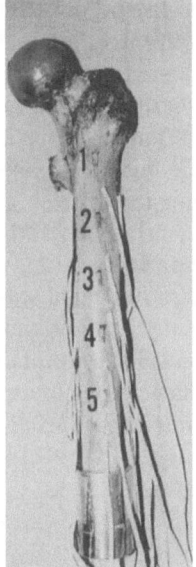

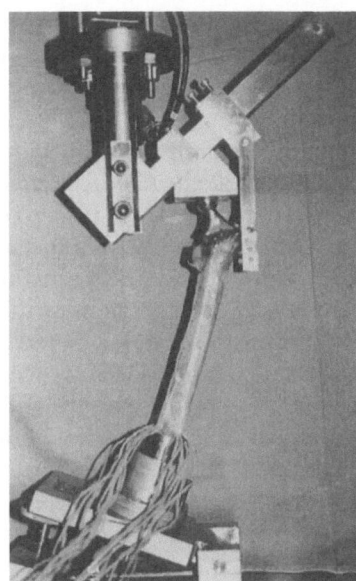

Fig. 1   **Fig. 2**

closely as possible. The purpose of this study is to understand the mechanical behavior of micromotions between bone and stem and the strain distributions on the femur immediately after insertion of the prototype stems into the preserved cadaver femurs.

Methods

The prototype stems we used were smooth-surfaced titanium alloy stems with a distal tip structure, and both with and without a collar. After removal of the soft tissues on the surface of the cadaver femurs preserved by freezing at $-86°$, radiography was at first undertaken in anteroposterior and mediolateral views, and then 20 strain gauges were bonded on the surface of the femur at 5 locations in the medial, lateral, anterior, and posterior portions. Five locations had equal intervals, and the upper location was at 1 cm distal from the femoral neck cut level and the fourth location at 1 cm proximal to the tip of the stem (Fig. 1).

These femurs were fixed on a table at an angle of 25° adduction from the vertical surface and applied with 1 Hz and 2000 N vertical cyclic loading by a cyclic loading machine (Marui, Osaka, Japan). The abductor forces were set to undergo a force 0.74-times the load at on the femoral head at an angle of 30° from the femoral axis. First, strains on the intact femoral surface were measured. Then, the femoral head was excised and a stem was inserted to measure the surface strains of the implanted femurs (Fig. 2).

Prototype stems were implanted using various surgical techniques. These techniques were divided into groups according to type of mechanical fixation.

Group 1 included stems with proximal-fit and distal-fit, group 2 included stems with proximal-fit and distal-loose, group 3 included stems with proximal-loose and distal-fit, and group 4 included stems with both proximal- and distal-loose. Proximal-fit stems are those where the stem is inserted using a press-fit technique and contact is made with the medial, anterior, and posterior cortices. Proximal-loose stems are those that are one size smaller than the press-fit stem. Distal-fit stems use a distal tip equivalent in size to the diameter of the femoral canal isthmus. Distal-loose stems use a distal tip 4 mm smaller than the diameter of the femoral canal isthmus. In addition, the prototype stems were prepared by 8% enlargement of the femoral canal with a rasp of the same size.

Micromotions were measured using a non-contact deviation meter by applying a 1 Hz and 2000 N cyclic load 30 times. They were measured in the mediolateral, anteroposterior, and superoinferior directions. Rotation at the proximal region and in mediolateral micromotions at the distal tip region were also measured.

Results

Femoral Surface Strains after Insertion of the Prototype Stems

With a collared stem, the proximal medial compression strains were 73%–75% of those of the intact femur in proximal-loose, and they were 50%–55% in proximal-fit. On the other hand, with a collarless stem, the strains resulted in 15%–20% of those of the intact femur. This means that stress shielding in the

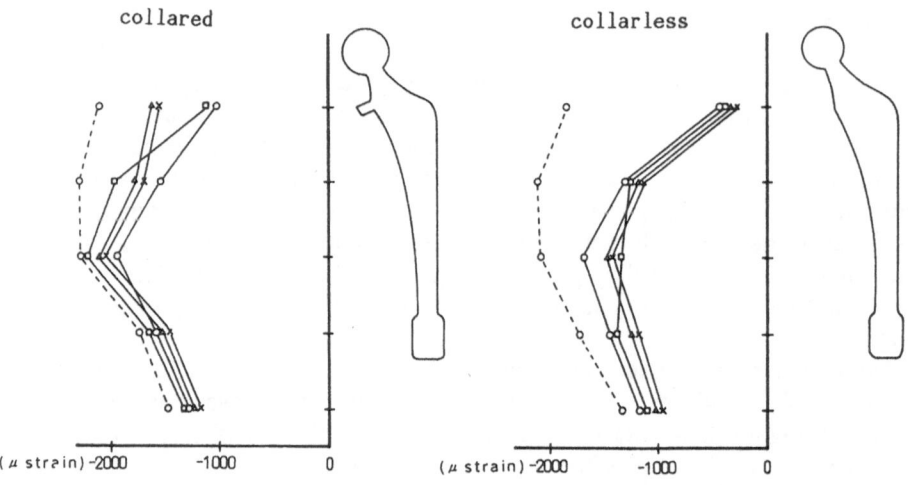

Fig. 3. Strain distributions on the medial surface of the proximal femur. The *dotted line with open circles* represents intact femur, the *solid line with open circles* represents proximal-fit and distal-fit, the *solid line with open squares* represents proximal-fit and distal-loose, the *solid line with open triangles* represents proximal-loose and distal-fit, and the *solid line with x's* represents proximal-loose and distal-loose

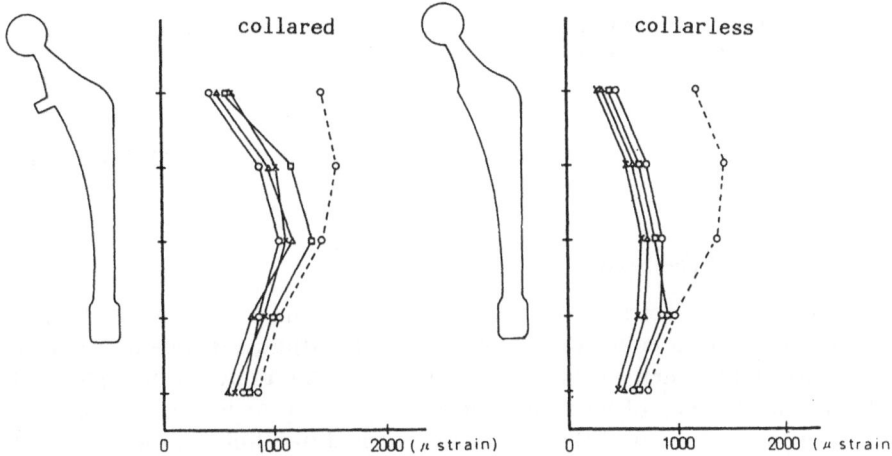

Fig. 4. Strain distributions on the lateral surface of the proximal femur. For definitions, see Fig. 3

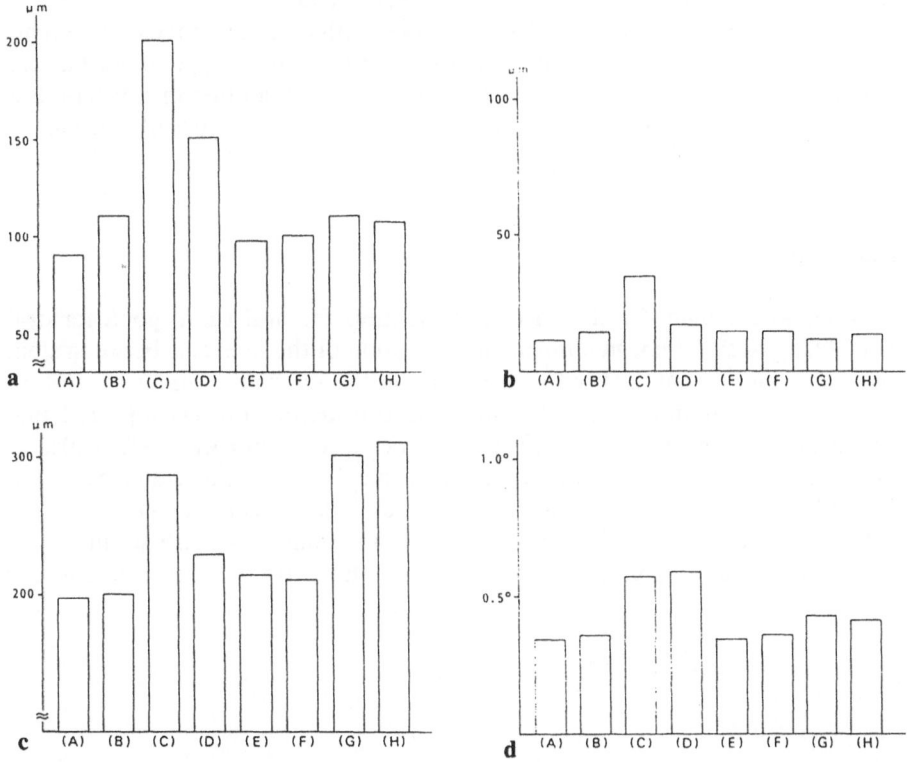

Fig. 5a–d. Proximal micromotion in **a** the mediolateral direction, **b** the anterioposterior direction, **c** the superioinferior direction, and **d** rotation. (*A*), collared, proximal-fit, distal-fit; (*B*) collared, proximal-fit, distal-loose; (*C*), collared, proximal-loose, distal-fit, (*D*), collared, proximal-loose, distal-loose, (*E*) collarless, proximal-fit, distal-fit; (*F*) collarless, proximal-fit, distal-loose; (*G*), collarless, proximal-loose, distal-fit, and (*H*), collarless, proximal-loose, distal-loose

proximal region of the femur with a collarless stem was larger than that with a collared stem. However, there was little difference in the type of mechanical fixation used with collarless stems (Fig. 3). The proximal lateral tensile strains were 20%–25% of those of intact femur with a collared stem, and 15%–25% with collarless one. There was no significant difference between both stems (Fig. 4).

Micromotions of the Stem

In proximal mediolateral directions, micromotion in a proximal-loose and distal-fit stem with a collar were a maximum of 200 μm. Micromotion ranged from 95 to 100 μm when a collarless stem was used, but there was no significant difference in the type of mechanical fixation. In the proximal anteroposterior directions, micromotion of the proximal-loose and distal-fit stem with a collar were a maximum of 30 μm, and all other types of mechanical fixation were less than 15 μm. In the proximal superoinferior directions, micromotion showed higher values in the proximal-loose type than those in proximal-fit type regardless of whether or not the prosthesis had a collar.

Rotation of the stem with a collar was larger with a proximal-loose stem than with a proximal-fit one, but there was little difference in the type of mechanical fixation (Fig. 5). In distal mediolateral directions, micromotion in a distal-loose stem with a collar showed the largest values out of the four types of mechanical fixations examined (Fig. 6).

Discussion

In order to evaluate the design of a cementless femoral stem, good femoral canal filling of the stem, minimizing micromotion of the interface between stem and bone, and having a strain distribution on the implanted femur as close to those of intact femur as possible must be considered. Oh [1] reported that strains on the femoral calcar surface using a cadaver femur were 30%–40% of those of the intact femur with complete collar-calcar contact, and were less than 10% of those of the intact femur without collar-calcar contact.

In our present study, with a collared stem, the femoral surface strains at the proximal medial region were 50%–75% of those of the intact femur and

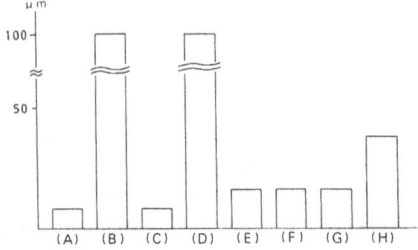

Fig. 6. Distal mediolateral micromotion. For definitions, see Fig. 5

15%–20% with a collarless stem. These results were similar to those reported by Oh et al. [1].

It is also important to obtain fixation at the proximal region of the stem in order to minimize micromotion of stems with and without collars. Walker et al. [2] similarly reported by micromotion analysis of the femoral stem in cementless total hip arthroplasty that the proximal-fit stem had evidently less micromotion than the proximal-loose stem. Although Whiteside et al. [3] stated that it was important to fix the distal region of the stem in order to minimize micromotion of the stem in the proximal region, there was no indication in our present study that distal fixation of the stem influenced micromotion in the proximal region.

Moreover, with a collared stem, micromotion showed the largest value with a proximal-loose and distal-fit prosthesis. This condition resulted in stem migration due to three point support when an undersized stem with a collar was inserted into the femur, and partial thickening or thinning of the femoral cortex was observed clinically. On the other hand, micromotion in collarless stems was generally smaller with proximal-fit prostheses. It seems to be the case that the stem will continue to settle until it reaches a position where stem fixation is possible under loading even if an undersized stem is inserted.

It is possible to decrease stress shielding in the proximal medial region by collar-calcar contact, but it is very difficult to obtain complete collar-calcar contact clinically, and micromotion in the proximal region increases when an undersized stem with a collar is inserted. Therefore, it was considered more favorable to use collarless stems in proximal femoral fixation.

References

1. Oh I, Harris WH (1978) Proximal strain distribution in the loaded femur. J Bone Joint Surg [Am] 60:75–85
2. Walker PS, Schneeweis D, Murphy S, Nelson P (1987) Strain and micromotion of press-fit femoral stem prostheses. J Biomech 20:693–702
3. Whiteside LA, Easley JC (1989) The effect of collar and distal stem fixation on micromotion of the femoral stem in uncemented total hip arthroplasty. Clin Orthop, 239:145–153

A Finite Element Analysis
for the Dysplastic Hip:
About Rotational Acetabular Osteotomy

YUKIO YOSHIDA, NOBUO MATSUI, YOICHI TANEDA,
and HIROTAKA IGUCHI[1]

Summary. Acetabular dysplasia is the major cause of osteoarthritis (OA) of
the hip in Japan. Pre- or early stages of osteoarthritic cases due to dysplastic
hip underwent rotational acetabular osteotomy (RAO) at Nagoya City
University Hospital. A two-dimensional finite element model was employed to
reveal the effects of RAO. The dynamic characteristics revealed by finite
element analysis showed that the center of gravity of the stress on the loaded
articular cartilage had shifted medially and that the average of the stress had
decreased postoperatively.

Key words. Finite element analysis — Hip joint — Rotational acetabular
osteotomy — Osteoarthritis

Introduction

In Japan about 90% of cases of osteoarthritis (OA) of the hip are caused by
congenital hip dislocation and/or congenital acetabular dysplasia. Once the
OA has progressed, there is only one way to treat this disease, that is by
THA (total hip arthroplasty). From this point of view it is very important to
carry out preventive therapy while the condition is still prearthritic. Our prior
statistics revealed that OA of the hip tends to progress in the future in patients
with a center edge (CE) angle of less than 10° [1].

Sixty-five patients were treated with rotational acetabuler osteotomy (RAO)
at our clinic. RAO is an operative method involving spherical osteotomy
around the acetabulum and rotation of the fragment. The theory behind
the technique is to horizontalize the acetabulum and decrease the stress concentration. However, the efficacy of this method has not yet been proved

---

[1]Department of Orthopaedic Surgery, Nagoya City University Medical School, 1
Kawasumi, Mizuho, Nagoya, 467 Japan

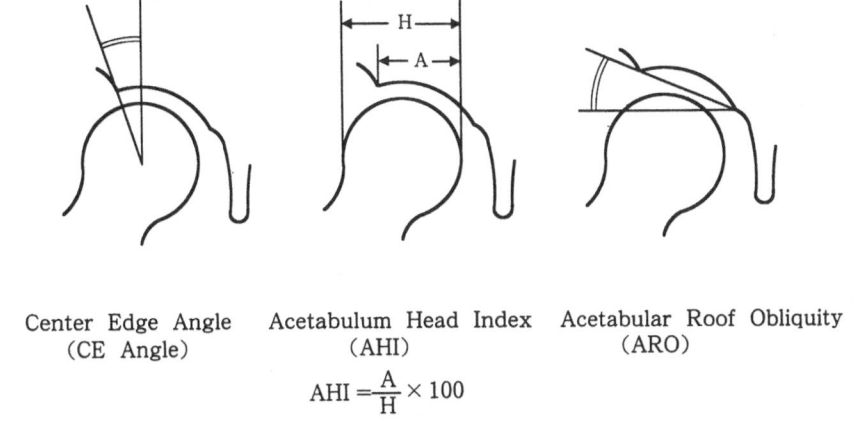

Center Edge Angle   Acetabulum Head Index   Acetabular Roof Obliquity
(CE Angle)   (AHI)   (ARO)

$$\mathrm{AHI} = \frac{A}{H} \times 100$$

Fig. 1. Roentgenographic indices

by biomechanical studies, so in the present study the preoperative and post-operative stress condition of the hips were analyzed by Finite Element Analysis (FEA) and the results were compared with the clinical results of the actual cases.

Materials and Methods

Thirty-five hips (3 of 3 male patients and 32 of 30 female patients) 25 pre-arthrotic hips, and 10 early-staged osteoarthritic hips were studied in the present report. The mean age at operation was 25.8 ± 11.1 years, and the mean follow-up period was 48.0 ± 13.5 months. For the clinical results, the following parameters were assessed:

1. JOA hip score
2. Roentgenographic indices (Fig. 1)
   a) Center edge angle (CE angle)
   b) Acetabulum head index (AHI)
   c) Acetabular roof obliquity (ARO)
3. Stress analysis using finite element analysis (FEA)

A rating scale adopted by the Japanese Orthopaedic Association (JOA score) is an evaluation system for clinical conditions which is mainly used in Japan. The scoring consists of pain (40 points), range of motion (20), activities of daily living (ADL) (20), and gait ability (20), making a total of 100 points if all conditions are normal.

As a biomechanical analysis two-dimensional FEA was performed using PEL (program knows elements' location) software developed by ourselves with specialized pre- and post-processing for orthopaedic usage. The system has got artificial intelligence to locate proper material to the proper position.

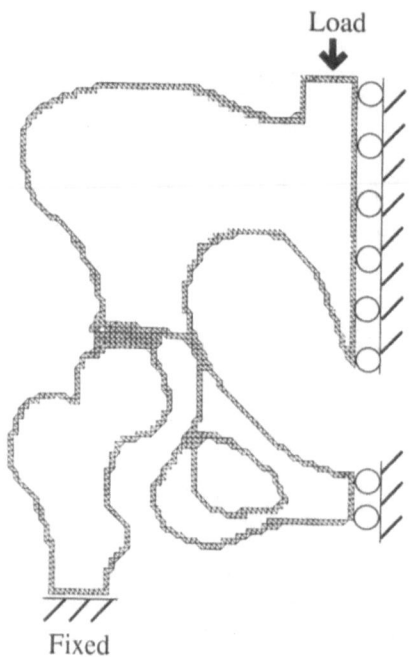

Load   **Fig. 2.** Weight-bearing condition

Fixed

Table 1. Properties of the material

	Young's modulus (N/mm²)	Poisson's ratio
Cortical bone	1500	0.3
Trabecular bone	500	0.3
Cartilage	100	0.3

Plain anterioposteior X-rays of the pelvis including the fifth lumbar vertebra and proximal femur were digitized and divided into around 9000 elements automatically and default materials were set automatically as well by the position of each element. Cortical bone, trabecular bone, and articular cartilage were located and the examiner verified the material. Half of the pelvises were analyzed using their symmetricity. Assuming a standing position, the center line of the pelvises were fixed horizontally but left free vertically to make it possible to analyze the whole pelvis (Fig. 2). The material properties are shown in Table 1.

The following indices were calculated from FEA:

1. Stress pattern (X-direction, Y-directon, destructive stress) (Fig. 3)
2. Average of the stress
3. Center of gravity index

The center of gravity index was devised to analyze the pattern of the stress of the joint cartilage (Fig. 4). The index represents the position of the center of

Fig. 3. Stress patterns

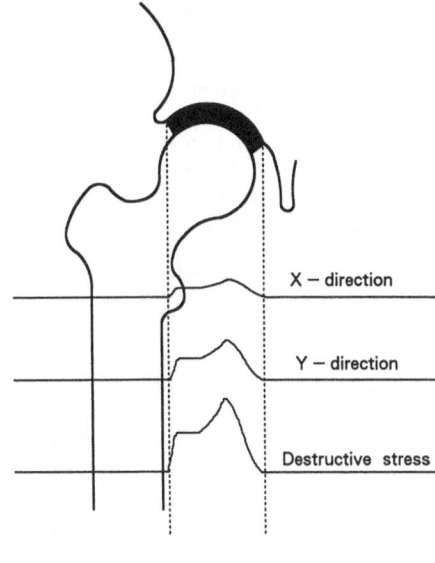

X – direction

Y – direction

Destructive stress

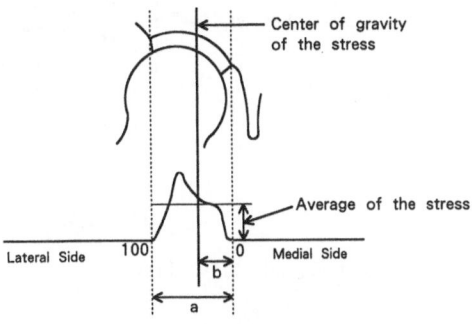

Center of gravity of the stress

Average of the stress

Lateral Side   100   0   Medial Side

b

a

Fig. 4. The indices assessed in the FEA

Center of gravity index $= \dfrac{b}{a} \times 100$

Table 2. JOA hip score

	Preoperation	Postoperation
Pain	26.1 ± 7.8*	36.2 ± 4.8*
Gait	17.4 ± 3.1	16.6 ± 3.8
ROM	18.9 ± 2.2	18.2 ± 2.2
ADL	18.5 ± 2.4	18.8 ± 2.0
Total	80.9 ± 10.7	89.8 ± 8.6

ROM, range of motion; ADL, activities of daily living
* $P < 0.01$

gravity of the stress on articular cartilage; the medial end of the cartilage is rated as 0 and the lateral end as 100. In normal cases, the average of this index was approximately 45.

Table 3. Range of motion of the hip joint

	Preoperation	Postoperation
Flexion	112.6 ± 12.7*	98.1 ± 12.0*
Extension	14.3 ± 7.9	13.7 ± 6.0
Abduction	34.0 ± 9.8	30.0 ± 8.7
Adduction	17.9 ± 3.9	17.6 ± 3.5
External rotation	31.3 ± 15.6	31.9 ± 15.7
Internal rotation	46.6 ± 17.2	39.7 ± 17.4

*$P < 0.01$

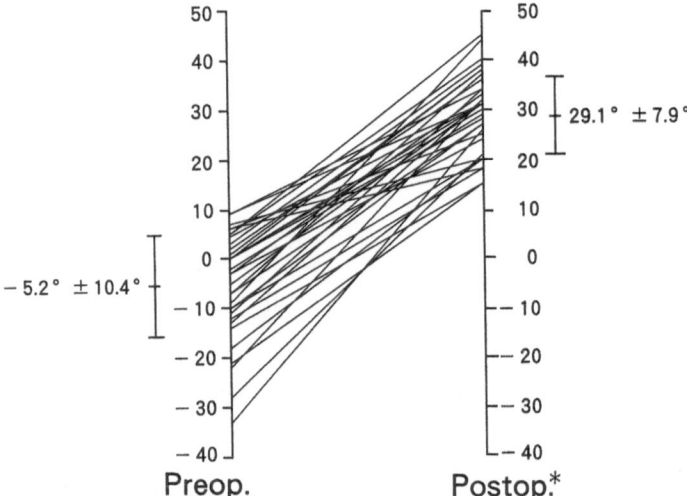

Fig. 5. CE angle.*$P < 0.05$

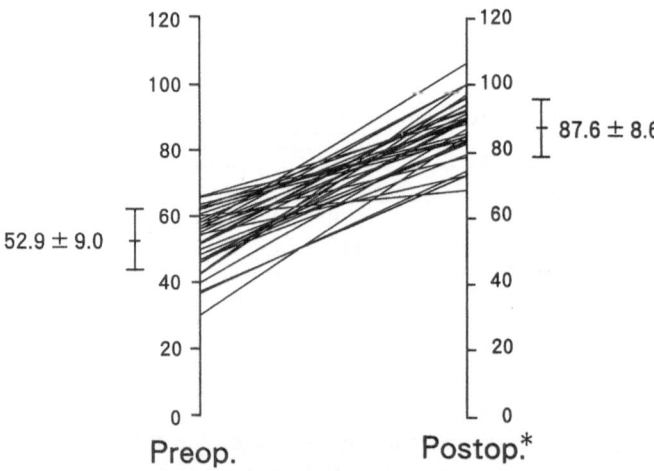

Fig. 6. Acetabulum head index (AHI).*$P < 0.05$

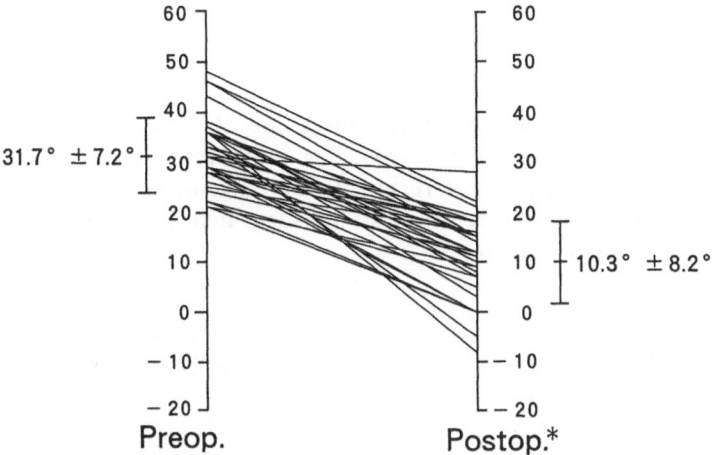

Fig. 7. Acetabular roof obliquity (ARO).*$P < 0.05$

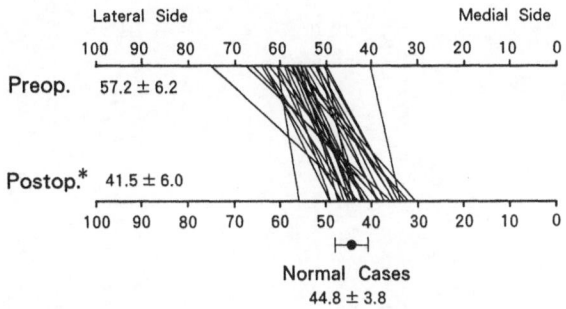

Fig. 8. Center of gravity index.*$P < 0.05$

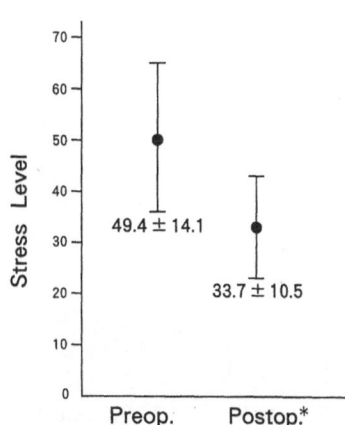

Fig. 9. Average of the stress.*$P < 0.05$

Results

JOA Score

The clinical results of 35 hips were favorable. Preoperative JOA scores averaged 80.9 and postoperative scores 89.8. Pain scores improved especially from 26.1 to 36.2 (Table 2). With regards to the range of motion (ROM) of the hip joint, flexion decreased by about 10° postoperatively, but other directions of ROM did not change (Table 3).

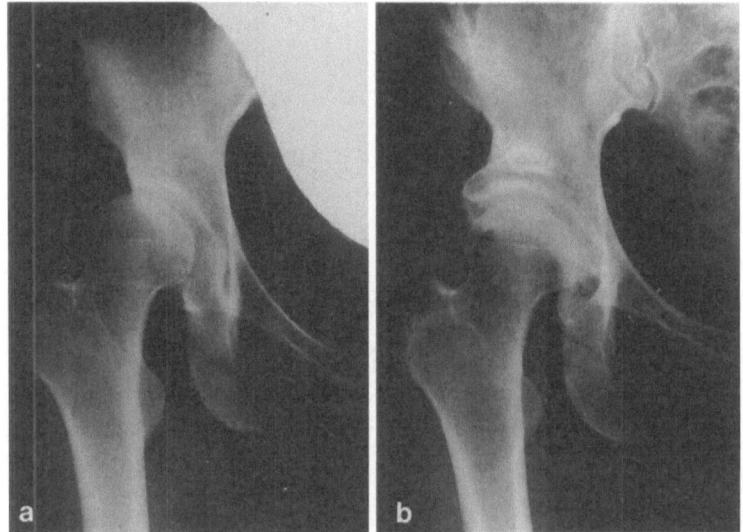

Fig. 10a,b. Case 1: A 41-year-old housewife. **a** Preoperation. **b** Postoperation, 5 years after surgery

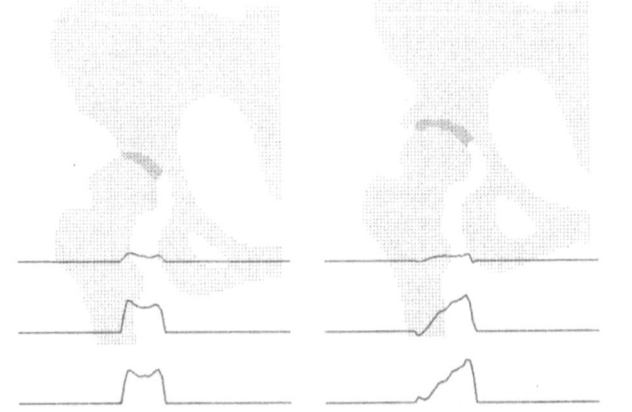

Fig. 11a,b. Case 1: Stress pattern. **a** Preoperation. **b** Postoperation

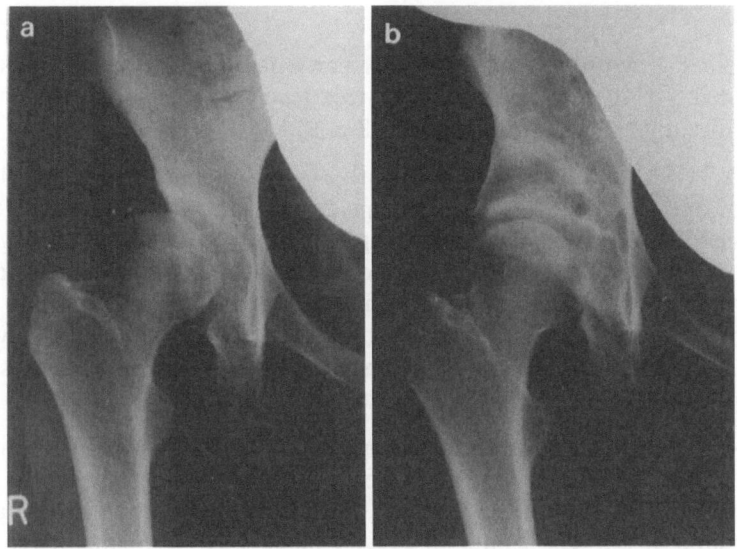

Fig. 12a,b. Case 2: A 13-year-old girl. **a** Preoperation. **b** Postoperation, 4 years after surgery

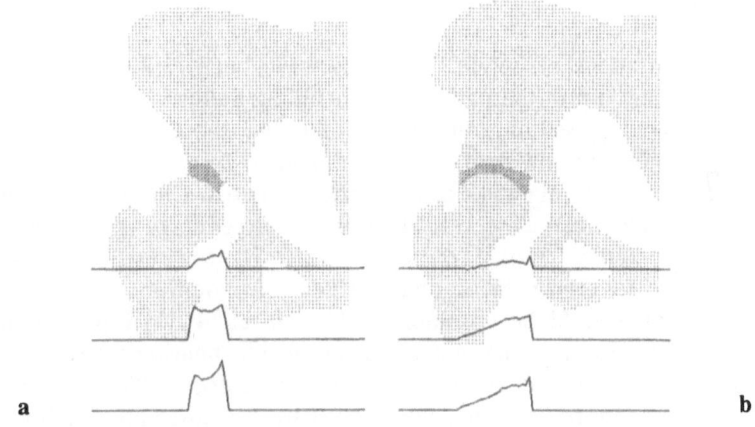

Fig. 13. Case 2: Stress pattern. **a** Preoperation. **b** Postoperation

Roentgenographic Indices

All of the indices improved postoperatively. The CE angle improved from an average of −5.2° to 29.1° postoperatively (Fig. 5), AHI improved from 52.9 to 87.6 (Fig. 6), and ARO improved from 31.7° to 10.3° (Fig. 7).

Stress Analysis

The center of gravity of the stress was located laterally but moved medially postoperatively (Fig. 8). The average stress levels on the articular cartilage were significantly reduced after surgery (Fig. 9).

Case 1: A 41-year-old housewife with early-staged osteoarthritis (Fig. 10). At the age of 38, she noted the right hip pain and limp. Her preoperative CE angle was 5°. RAO was performed, and a good acetabular covering was obtained. The JOA hip score was 67 points preoperatively, but rose to 100 points postoperatively. The patient's course has been favorable. In this case, the center of gravity index was 53 preoperatively, but shifted medially and became 37 postoperatively, and the magnitude of the stress level was also decreased (Fig. 11).

Case 2: A 13-year-old girl with prearthrosis (Fig. 12). At the age of 12, she felt right hip pain after walking long distances. RAO was done, and the CE angle changed from −3° to 33° postoperatively. Currently, she does not complain of pain nor limp at all. The center of gravity index was 52 preoperatively, but changed to 37 postoperatively (Fig. 13). The magnitude of the stress level decreased.

Discussion

A variety of methods of pelvic osteotomy have been reported in the treatment of dysplastic hip in the past [2–9], including Salter's osteotomy, Chiari's osteotomy, Shelf operation, and RAO. Among these procedures Salter's osteotomy is indicated only for the children, and so we will omit this procedure here as we are considering adult osteoarthritic patients. Chiari's osteotomy and shelf operation also increase the loading area, but coverage is not obtained by the cartilage which is created postoperatively. By contrast, with RAO the femoral head is covered with original cartilage. In addition, the possibility of bone absorption should be considered. Tagawa et al. [7] reported that the advantages of RAO are enlargement of the weight bearing area, decrease of the shearing force, and decrease of the resultant force. The biomechanical basis of this procedure has not been proved by a mathematic method. However, by the present study, it was proved that the peak stress is decreased and the stress is divided to a wider area. On the basis of our study, it seems that RAO is effective in normalizing stress in the hip joint, and is a good procedure for the treatment of the dysplastic hip.

References

1. Yoshida Y, Matsui N, Taneda Y, Wada I, Iguchi H (1989) A study of natural history of prearthrosis of the hip (in Japanese). Jpn J Rheum Joint Surg 8:213–218
2. Chiari K (1955) Ergebnisse mit der Beckenostetomie als Pfannendach-plastik. Z Orthop 87:14–26

3. Ninomiya S, Tagawa H (1984) Rotational acetabular osteotomy for the dysplastic hip. J Bone Joint Surg [Am] 66:430–436
4. Salter RB (1961) Innominate osteotomy in the treatment of congenital dislocation and subluxation of the hip. J Bone Joint Surg [Br] 43:518–539
5. Steel HH (1973) Triple osteotomy of the innominate bone. J Bone Joint Surg [Am] 55:343–350
6. Sutherland DH, Greenfield R (1977) Double innominate osteotomy. J Bone Joint Surg [Am] 59:1082–1090
7. Tagawa H, Hijikata H, Ninomiya S (1984) Indications and results of rotational acetabular osteotomy. In: Ueno R, Akamatsu N, Itami Y, Tagawa H, Yoshino S (eds) The Hip — Clinical studies and basic research. Elsevier, Amsterdam, pp 117–122
8. Tagawa H (1975) Rotational acetabular osteotomy (in Japanese). Hip Joint 1: 108–113
9. Wagner H (1976) Osteotomies for congenital hip dislocation. In: The Hip — Proceedings of the 4th Open Scientific Meeting of the Hip Society. Mosby, St. Louis, pp 45–66

Knee Joint

Function of the Ligament Augmentated Device Under Simulated Anterior Cruciate Ligament Reconstruction and Rehabilitation

HIROYUKI NAKAMURA, MASAHIRO INOUE, KONSEI SHINO, NORIMASA NAKAMURA, and KEIRO ONO[1]

Summary. In this study, functional properties of artificial ligaments (LAD and Dacron, n = 4 respectively) were examined biomechanically. The artificial ligaments underwent a testing procedure which simulated surgical and post-surgical conditions including initial tensioning, fixation, and range of motion exercise. First, the specimen was set on an Instron type testing machine and it underwent a creep test with a 49 N load for 5 minutes. Then the specimen underwent a cyclic stress relaxtion test (CSRT) where it was cyclically stretched between 0 and 5% deformation 100 times. Load during the cycling was recorded. Finally, the specimen was loaded to failure. One human tibialis posterior tendon was tested in the same way, and the relaxation properties of the artificial ligaments and the human tendon were compared with each other. Based on the experimental data, the sharing of loads during the CSRT between the artificial ligament and the human tendon was calculated when these two materials were used in combination.

The load in either of the artificial ligaments decreased from 49 N to zero after the CSRT while a load of 7 N was still existent in the human tendon. The ratio of load between the ligament augmented device (LAD) and the human tendon at 5% deformation was calculated to change from 65:35 to 74:26 during the CSRT.

This study demonstrated functional differences between the artificial ligaments and the human tendon. This difference should be considered when composite grafts (artificial ligament and human tendon) are used in ligament reconstruction.

Key words. Anterior cruciate ligament — Ligament augmentated device — Biomechanics

---

[1] Department of Orthopaedic Surgery, Osaka University Medical School 1-1-50 Fukushima, Osaka, 553 Japan

Introduction

Recently, autogenous tissues such as patella tendon, semitendinous tendon, and iliotibial tract have been used as graft materials in anterior cruciate ligament (ACL) reconstruction. However, such grafts may weaken soon after transplantation [1, 2, 3], become stretched in the early postoperative period, or be ruptured or elongated if excessive stress is applied before they have matured sufficiently [1, 2, 4]. It thus seems necessary to protect these grafts by immobilization after surgery. On the other hand, it is well-known that immobilization leads to joint constrictures, loss of ligament strength [5, 6], and adverse effects on the articular cartilage [7]. In order to overcome these para-doxical conceps of graft protection vs. early motion, the ligament augmentation device (LAD) has lately come to be used in ACL reconstruction.

The LAD was first reported by Kennedy et al. in 1980 [1]. Initially, they attempted to reinforce the feeble prepatellar tissue of the ACL graft using the LAD. Subsequently, the LAD has come to be used widely in the augmentation of various autogenous tissues. McPherson et al. [8] conducted mechanical testing of the LAD. They reported that 6 mm braid LAD maintained a failure load of over 80% after fatigue testing at a failure load of 1513 N, and also possesed the satisfactory mechanical characteristic of only 3% deformation in the cyclic creep test. They further experimented with ACL reconstruction in goats, and compared the tensile strength of graft in presence and absence of augmentation by the LAD. As a result, both immediately after transplantation and two years after transplantation, the failure load of the augmentation group was significantly larger than that of the nonaugmentation group, but there was no difference in the failure load in the intermediate period (at 3, 6 and 12 months). Jackson et al. [4, 9] also used goats to study the effect of the LAD on patellar tendon allografts in reconstruction of the ACL. They determined that, at one year after transplantation, the failure load of the LAD group was significantly greater, but there was no difference in A-P laxity. As for the clinical results of the LAD, Roth et al. [10] compared augmentation and nonaugmentation groups, and reported a significant improvement in the augmentation group with respect to Lachman's sign, the pivot shift sign, and the anterior drawer sign, but found no difference in the KT-1000 anterior laxity test.

In all these studies, the efficacy of the LAD in graft augmentation was recognized. In the reports by Jackson et al. [4, 9] and Roth et al. [10], however, the degree of A-P laxity was not different between the augmented and nonaugmented groups, but a difference was evident when compared with the contralateral side for these studies. If the composite graft with the LAD maintained the initial state after transplantation, then the difference from the sound side would be smaller. Thus, the function of the LAD seems less than optimal.

Accordingly, we performed this study to clarify the functional characteristics of the LAD in the simulated ACL reconstruction. When an artificial ligament is inserted, first the ligament was placed under the appropriate tension (initial tensioning), and then fixed in the tensioned state. Subsequently, range of motion (ROM) exercises were started after the operation. The function of the LAD was evaluated with regard to the degree of stress relaxation, creep, and changes in the initial load distribution between the LAD and the graft.

Materials and Methods

We tested LAD artificial ligaments (3M 6mm braid, 120mm, n = 4), Dacron artificial ligaments for A-C separation (Stryker 120mm, n = 4), and a human tibialis posterior tendon graft (n = 1).

In the present experiment, considering that specific strains occur repeatedly (rather than repeated identical stresses) when ROM exercises of the knee joint are performed after surgery, changes of function of the artificial ligament under repeated strain were investigated.

First, the artifial ligament was held by clamps as reported by Woo et al. [11], and was set in the testing machine. To simulate the initial tension applied when fixing an LAD at the time of operation, a creep test was conducted by applying a load of 49N to the ligament for 5 minutes, and the length after the test was designated as the initial length. At this time, the ligament would be fixed during an actual operation. Waiting for 30 minutes while maintaining the initial length, the load applied to the ligament was recorded at specific time intervals (stress relaxation test), and the load on the artificial ligament was measured 30 minutes later, the simulated end of the operation (step 1). Then, to simulate the ROM exercises, we used a cyclic stress relaxation test (CSRT) that was cycled 100 times at a crosshead speed of 40mm/min in a range causing 5%

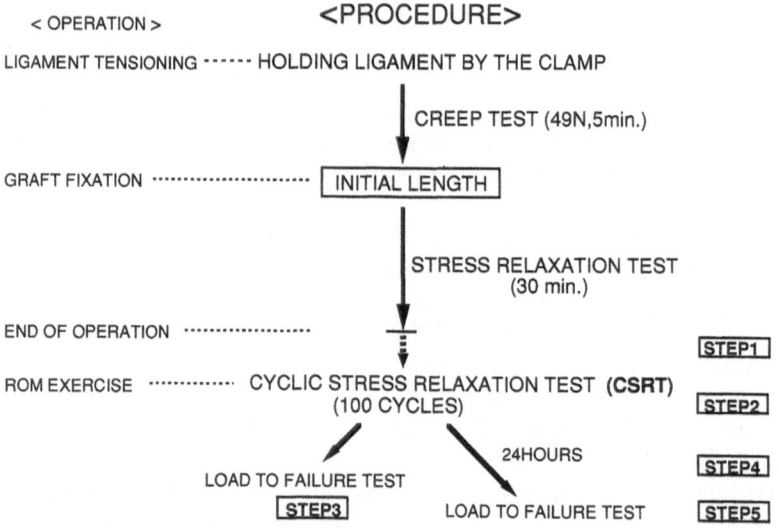

Fig. 1. Testing procedure simulating surgery and rehabilitation

deformation of the initial length. After the 100th cycle, the tension of the ligament when the clamp returned to the original state was measured (step 2).

All the ligaments were studied through step 2. Then, using two artificial ligaments and the human tibialis posterior tendon, the ligament was stretched at a crosshead speed of 20 mm/min (load to failure test), and a load-deformation curve was plotted (step 3). On the basis of this curve, the strain was calculated when the ligament was loaded with 49 N and this value was defined as the ligament laxity.

Another group of ligaments (LAD, n = 2; Dacron, n = 2) were left at room temperature for 24 hours following the CSRT while maintaining the clamped position, and then the load on each ligament was measured (step 4). Afterwards, a similar load to failure test was conducted to determine the ligament laxity (step 5). In the experiment using a human tibialis posterior tendon, the test was conducted through step 3 in a thermostatic oven filled with phisiological saline kept at 36°C (Fig. 1).

Results

In step 1, the set load of 49 N was decreased to 43.1 N for the LAD, to 46.1 N for the Dacron graft, and to 44.1 N for the human tibialis posterior tendon.

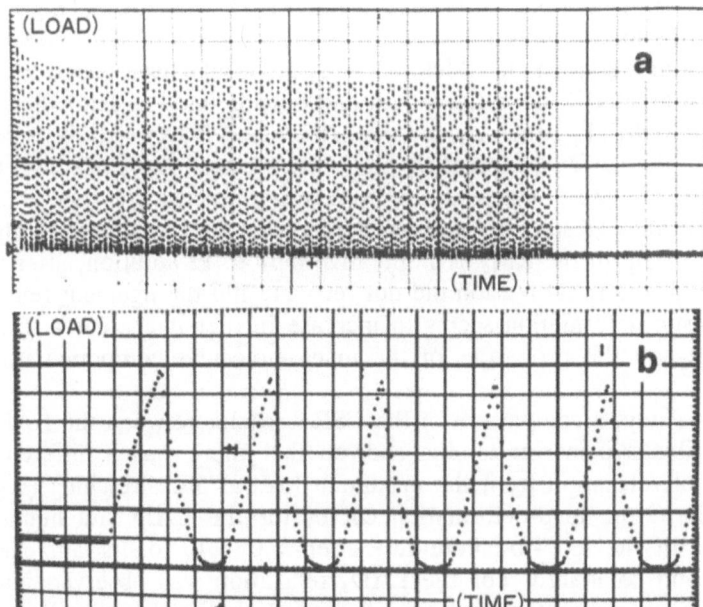

Fig. 2a,b. Cyclic stress relaxation test (CSRT). **a** Whole process. **b** Initial five cycles. Note that the load on the ligament is 0.0 N when the clamps were returned to its initial position after first cycle

The CSRT results for the LAD are shown in Fig. 2. The whole test is recorded in Fig. 2a, and the initial five cycles after the start of the test are magnified in Fig. 2b. In each waveform, the crest corresponds to the maximum load on the ligament at 5% strain, and the lowest point coincides with the load when the clamp returned to the original position. A decrease of the maximum load on the ligament is noted over the course of time. In Fig. 2b, when the clamp returned to the original position after the first tensioning, the load on the ligament was already 0.0N for both the LAD and the Dacron grafts. In step 2, the load was 0.0N for all artificial ligaments, so that relaxation was noted and plastic deformation took place. On the other hand, the human tibialis posterior tendon was loaded with 6.9N in step 2. In step 4, for both the LAD and the Dacron graft, the load was 0.0N. The ligament laxity in step 3 was 2.7% for the LAD, 3.1% for the Dacron graft, and 1.7% for human tendon. In step 5, the ligament laxity was 1.7% for the LAD and 3.1% for the Dacron graft.

Discussion

According to McPherson et al. [8], when defining the initial length as that when a load of 50N was applied after the initial 10 cycles of the cyclic creep test and when the peak load was between 50N–500N, a creep of 3% occurred after one million cycles. In our study, however, plastic deformation was induced by only one cycle which caused a 5% increase in length. Furthermore, in the load-to-failure test after CSRT, the laxity of the LAD was 2.7%. These findings suggest that McPherson et al. would have found a creep that was much greater if the initial length had been defined as that when 50N was applied initially. In other words, when the same load or strain is repeatedly applied to the LAD, the initial change is the greatest. In clinical use, the initial length is the length of a ligament when it is fixed for the first time at operation. Therefore, these results suggest that the LAD is already stretched when ROM exercises begin in the early postoperative period. In addition, after a 24-hour rest period, the LAD tension did not recover, and the ligament remained lax. Such plastic deformation seems to increase further if a greater strain occurs depending on the properties of the macromolecular compound used as the LAD.

In our study, a comparison of the CSRT results between the LAD and the human tibialis posterior tendon showed that there was a difference in the manner of relaxation from the maximum load at 5% stretching (Fig. 3). On the basis of the load-deformation curves for the LAD and human tendon obtained in the load-to-failure test after a CSRT, load-strain curves were plotted and compared. For the LAD, relaxation was observed initially and there was no load on the ligament, after which the curve climbed sharply up to around 1% lengthening. However, for the human tibialis posterior tendon, a load was present initially and the curve rose more slowly as the strain increased when compared with the LAD curve (Fig. 4). That is, the function of the LAD

Fig. 3a,b. CSRT. **a** LAD. (LOAD)
b Human tibialis posterior
tendon

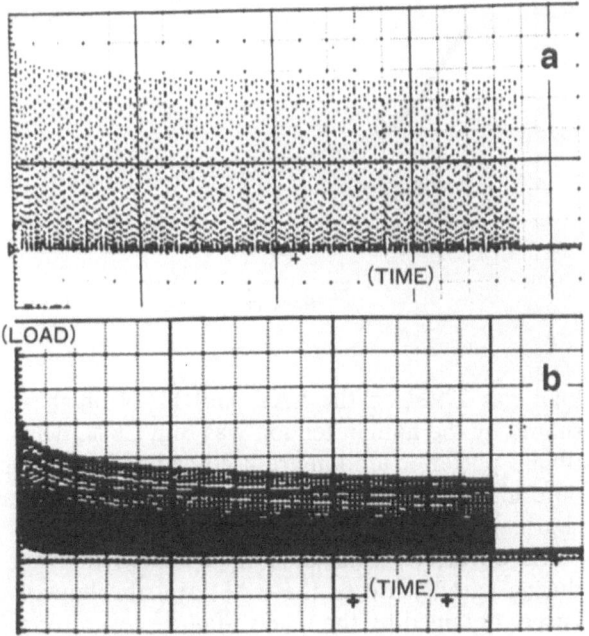

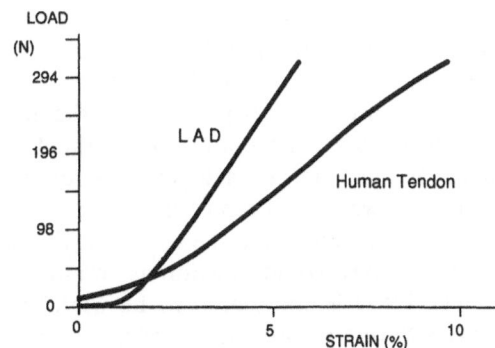

Fig. 4. Load-strain curves of LAD
and human tendon after CSRT

and the human graft differed greatly. In actual clinical use, however, human
tendon grafts and the LAD are handled alike.

Hanley et al. [12] experimented with cadaveric tendons to assess load sharing in a composite graft (LAD plus transplanted ligament) using a buckle
transducer. This study directly measured the load applied to the different
grafts, but the results were obtained by applying a single stress at various
angles of knee joint flexion, and the measurement error was large. In our
experiment, at the time of the first 5% deformation obtained from the CSRT
curve, the load on the LAD was 363 N and that on the human tibialis posterior
tendon was 196 N (35% of the total load). At the time of the 100th cycle, the

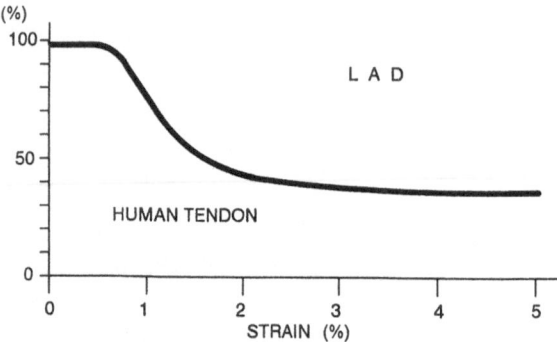

Fig. 5. Load shared by human tendon after CSRT

load was 299 N on the LAD and 114 N on the human tendon, and the load shared by the human tendon was only 26%. Thus, when the LAD and human tibialis posterior tendon are used as a composite graft, the load sharing at 5% deformation differs at the end of a CSRT when compared with the initial load sharing.

Moreover, we calculated the load on the human tibialis posterior tendon during parallel use of the LAD and the human tendon from the load-strain curves obtained by the load-to-failure test after a CSRT (Fig. 5). Initially, the load on the human tendon was 100%, and this was reduced to about 40% at 3% lengthening. In other words, when the LAD was used in graft augmentation, it showed a protective function during simulated ROM exercises, but this was completely different from the load sharing set initially. Thus, even if ideal load sharing were obtained at operation, it would not be maintained.

In our present experiment, for the ease of analysis the tension test was conducted by fixing both ends of the LAD. Mendenhall [13] used an LAD fixed at one end so as to avoid stress shielding of the autograft. Since his report, however, no report has clearly indicated the load sharing by ligaments fixed at one end. From the results of this study, however, it was at least shown that the LAD would function as an augmentatory device if fixed at both ends.

The LAD is considered to be useful to protect grafts as an augmentation device during reconstruction of the ACL. When we use the LAD, it is best to fix both ends when it is in a stretched state. However, since plastic deformation occurs in the LAD and since initial load sharing can not be maintained, care should be taken not to apply excessive stress before a graft is matured.

Since 5% lengthening could occur in the ACL in both ROM exercises and in daily activities, and since a stretch of this degree might be necessary for graft remodeling, our experiment was conducted by setting the deformation at 5%. Moreover, the reason for setting the deformation at 5% in this study was that we took into account the fact that a composite graft of the LAD and human tendon was fixed onto a bone tunnel by fibrous tissues 3–4 weeks postopetatively. If the entire deformation of the graft is defined as that which occurs only within the joint and excludes the immediate postoperative period when the graft fixes onto the bone tunnel, a lengthening of 5% with the LAD fixed at both ends may be too much. To investigate the function of the LAD in the immediate postoperative period, a CSRT under a smaller strain level should be performed. Also, since rehabilitation is a daily activity, the CSRT should be conducted on several consecutive days to assess changes in LAD function and also to determine the changes in load sharing.

In this study, the function of artificial ligaments used for augmentation was studied by simulating the conditions of an actual operation. For both the LAD and the Dacron graft, 100 cycles of 5% lengthening caused plastic deformation. This plastic deformation was the obvious functional difference between the artificial ligaments and human tendons. In particular, in a simulation of clinical use, the initial load sharing set between the LAD and the autograft was not maintained and the function of the LAD as an augmentation device was reduced. Therefore, the LAD is considered to have a protective effect on the graft when ROM exercises are done in the early postoperative phase, but if a greater stress is applied before the graft is mature the level of protection the LAD affords is reduced. Thus, even when using the LAD, a patient should not return to the activities of daily living or sporting activities soon after an ACL operation.

References

1. Kennedy JC, Roth JH, Mendenhall HV, Sanford JB (1980) Intra-articular replacement in the anteriorcruciate ligament-deficient knee. Am J Sports Med 8(1):1–8
2. Kennedy JC (1983) Application of prosthetics to anterior cruciate ligament reconstruction and repair. Clin Orthop 172:125–128
3. Noyes FR, Butler DL, Paulos LE, Grood ES (1983) Intra-articular cruciate reconstruction I. Perspectives on graft strength, vascularization, and immediate motion after replacement. Clin Orthop 172:71–77
4. Jackson DW, Grood ES, Arnoczky SP, Butler DL, Simon TM (1987) Cruciate reconstruction using freeze-dried anterior cruciate ligament allograft and a ligament augmentation device (LAD). Am J Sports Med 15(6):528–538
5. Noyes FR, Torvik PJ, Hyde WB, Delucas JL (1974) Biomechanics of ligament failure. J Bone Joint Surg [Am] 56:1406–1418
6. Noyes FR (1977) Functional properties of knee ligaments and alterations induced by immobilization. Clin Orthop 123:210–242
7. Enneking WF, Horowitz M (1972) The intra-articular effects of immobilization on the human knee. J Bone Joint Surg [Am] 54:973–985
8. McPherson GK, Mendnhall HV, Gibbon DF, Plenk H, Rottmann W, Sunford JB, Kennedy JC, Roth JH (1985) Experimental mechanical and histologic evaluation of the Kennedy ligament augmentation device. Clin Orthop 196:186–195
9. Jackson DW, Grood ES, Arnoczky SP, Butler DL, Simon TM (1987) Freeze-dried anterior cruciate ligament allografts, preliminary studies in a goat model. Am J Sports Med 15(4):295–303
10. Roth JH, Kennedy JC, Lockstadt H, Mccallum CL (1985) Polypropylene braid augmented and nonaugmentated intra-articular anterior cruciate ligament reconstruction. Am J Sports Med 13(5):321–336

11. Woo SL-Y, Gomez MA, Akeson WH (1985) Mechanical behaviors of soft tissues: Mesurement, modifications, injuries and treatment. In: Nahum and Melvin (eds) The biomechanics of trauma. Appleton-Century-Croft, Norwalk, pp 109–133
12. Hanley PH, Lew WD, Lewis JL, Hunter RE, Kirstukas S, Kowalczyk C (1989) Load sharing and graft forces in anterior cruciate ligament reconstructions with the ligament augmentation device. Am J Sports Med 17(3):414–422
13. Kampen CL, Mendnhall HV, McPherson GK (1987) Synthetic augmentation of biological anterior cruciate ligament substitutions. In: Jackson DW, Drez D Jr (eds) The anterior cruciate deficient knee. Mosby, Tront, pp 226–238

Topographic Variation of Indentation Stiffness in the Tibial Subchondral Plate of the Knee Joint

SENEKI KOBAYASHI[1] and TSUNENORI TAKEI[2]

Summary. The topographic variation of indentation stiffness of the proximal tibial subchondral plate was studied. A flat-ended cylindrical indenter 4 mm in diameter was used at a descending rate of 2.4 mm/min. The maximum load and the maximum slope of the load-deformation curve during the initial compression of 0.3 mm were defined as the indentation stiffness of the subchondral plate. Ten pig knee joints were used.

Variance analysis showed that indentation stiffness of the subchondral plate differed not only among 45 selected areas but also among six defined anatomical regions into which the 45 areas were classified. Indentation stiffness of the tibial subchondral plate listed by region in descending order of magnitude was (a) the region around the free-edge of the meniscus, (b) the meniscus-covered region, (c) the meniscal-horn region, (d) the intercondylar region (no difference in stiffness was noted among the three regions b, c and d), (e) the sloping region toward the intercondylar eminence, and finally (f) the periarticular region. The medial regions had greater indentation stiffness values than the corresponding lateral regions around the free-edge of the meniscus, the meniscus-covered regions, and the meniscal-horn regions.

Key words. Indentation test — Knee joint — Subchondral plate — Biomechanics — Tibial plateau — Topography — Stiffness

Introduction

The forces acting at the human knee joint during level walking have been demonstrated to be more than three times body weight [1]. To elucidate the biomechanics of the knee joint under such a heavy load, the changes in

---

[1] Department of Orthopaedic Surgery, Shinshu University School of Medicine, Asahi 3-1-1, Matsumoto, 390 Japan
[2] Nagano Rehabilitation Center, Shimokomazawa 618, Nagano, 380 Japan

the joint space and deformation of the articular cartilage were observed by the load-freezing technique [2]. The articular cartilage came into contact and the collagen network of the articular cartilage underwent compression at a much lower load when subchondral bone was replaced with polymethylmethacrylate [3]. Based on these observations, it was hypothesized that the subchondral bone played an important role in maintaining the joint space between the opposing articular surfaces and in protecting the articular cartilage from abnormal deformation. To examine the bony structure of the knee joint, the mechanical properties of the cancellous bone in the knee joint were studied [4]. The results of investigation of the indentation stiffness of the tibial subchondral plate, the region between the overlying articular cartilage and the underlying cancellous bone, are reported in the present study.

Materials and Methods

Knee joints were obtained from freshly slaughtered pigs which were about ten months old and weighed approximately 110 kg. In the indentation test, a flat-ended cylinder 4 mm in diameter descended at a rate of 2.4 mm/min. The maximum load and the maximum slope of the load-deformation curve during the initial compression of 0.3 mm were measured and defined as the "indentation stiffness". An initial load of 5 N was used to define the "contact position".

In a preliminary experiment, the maximum load and the maximum slope of the load-deformation curve during the initial 0.3 mm compression in the indentation tests and the maximum strength and elastic modulus in compression tests on cylindrical bone plugs were compared using cancellous bone in the pig knee joints. As described in a previous study [4], each knee joint was flexed at 30 degrees and cut parallel to the plane that included the margin of the tibial articular surface. Three bony plates of 8.5 mm thickness were obtained from the distal end of the femur and another three were taken from the proximal end of the tibia. On a cut face of each bony plate, the standardized locations for bone plug specimens [4] were marked with Indian ink. The indentation tests were performed at these locations. Cylindrical bone plugs with a cross-sectional diameter of 8.5 mm and a height of 8.5 mm were then removed from the marked and indented sites. The bony plugs were compressed at a rate of 2.4 mm/min in a loading machine, and the maximum strength and elastic modulus of the specimen were measured. Statistical correlation was investigated between two of the four measured values (the maximum load and the maximum slope of the load-deformation curve in the indentation tests, and the maximum strength and the elastic modulus in the compression tests on cylindrical bone plugs).

Topographic variation of the indentation stiffness of the tibial subchondral plate was studied using ten pig knee joints. To prepare the specimens, disarticulation was performed above both menisci, the fibula was removed, and the tibia was cut parallel to, and 5 cm distal to, the proximal tibial articular

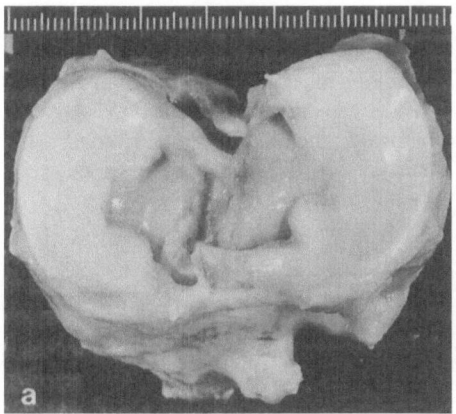
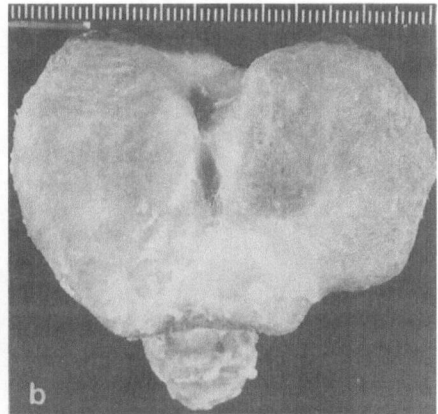
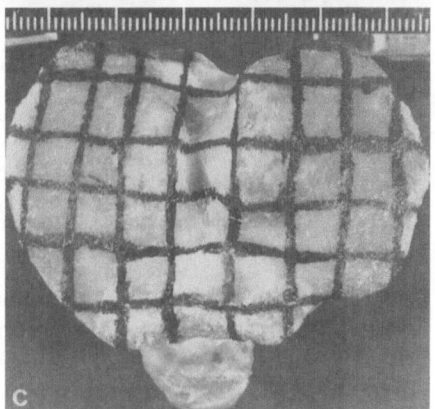

Fig. 1a–c. Superior view of the proximal end of the left tibia with the tibial tuberosity directed to the *bottom* of the photograph **a** with menisci, **b** with the subchondral plate exposed, and **c** with a 7.5 × 7.5 mm grid inked on the exposed subchondral plate

Fig. 2. Anatomical regions of tibial subchondral plate. *I*, intercondylar region; *E*, sloping region toward intercondylar eminence; *F*, region around free-edge of meniscus; *M*, meniscus-covered region; *H*, meniscal-horn region; *P*, periarticular region; (*asterix*), unclassified region

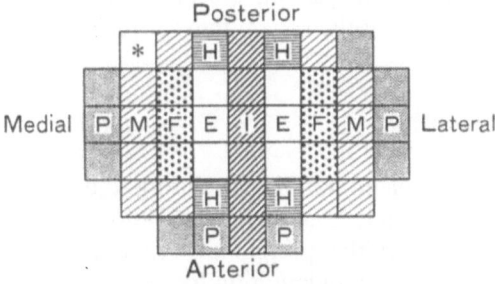

surface. After photographing the tibial articular surface (Fig. 1a), both menisci, cruciate ligaments, and the articular cartilage were removed with a scalpel and an airdrill to expose the subchondral plate (Fig. 1b). The subchondral plate was inked to produce a 7.5 × 7.5 mm grid and photographed in the same manner (Fig. 1c). The grid employed for the indentation tests, shown in Fig. 2, was divided into six anatomical regions; an intercondylar region (*I* in Fig. 2), a sloping region toward the intercondylar eminence (*E*), a region around the

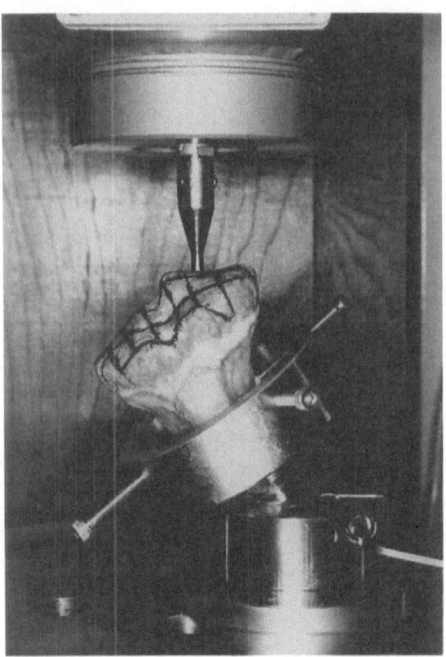

Fig. 3. Each 7.5 × 7.5 mm square of the grid is loaded with an indenter in a loading machine. A specimen holder which can tilt a maximum of 55 degrees allowed the indenter to contact the surface of the subchondral plate perpendicularly

free-edge of the meniscus (F), a meniscus-covered region (M), a meniscal-horn region (H), and a periarticular region (P).

The specimen was mounted on a holder in the loading machine. The holder could tilt up to 55 degrees, which allowed the indenter to contact the surface of the subchondral plate perpendicularly (Fig. 3). The indentation stiffness was measured by the indentation test at each square on the subchondral plate. A variance analysis on the measured values was performed among the squares and among the six anatomical regions.

Results

The indentation tests, during the initial 0.3 mm compression, showed approximately a straight load-deformation curve and never reached a yielding point. In the preliminary experiment, 66 sets of measurements were performed. An indentation test on a bony plate and a compression test on a bone plug constituted one set of measurements. The maximum load, the maximum slope of the load-deformation curve during the initial 0.3 mm compression in indentation tests, the maximum strength, and the elastic modulus measured in the bone plugs were statistically correlated with each other ($P < 0.01$).

In studying the topographic variation of the indentation stiffness of the tibial subchondral plate, a highly significant correlation was observed between the maximum load and the maximum slope of the load-deformation curve during the initial 0.3 mm compression ($P < 0.01$, $r = 0.94$). Therefore, the maximum

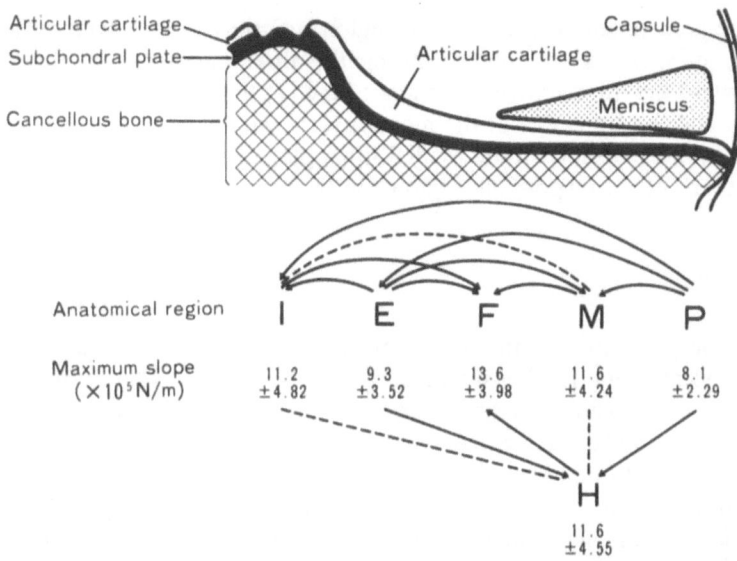

Fig. 4. Results of *t*-tests on average indentation stiffness between two anatomical regions. An *arrow* is drawn from the larger value to the smaller, and a *dashed line* connects two values with no statistically significant difference between them

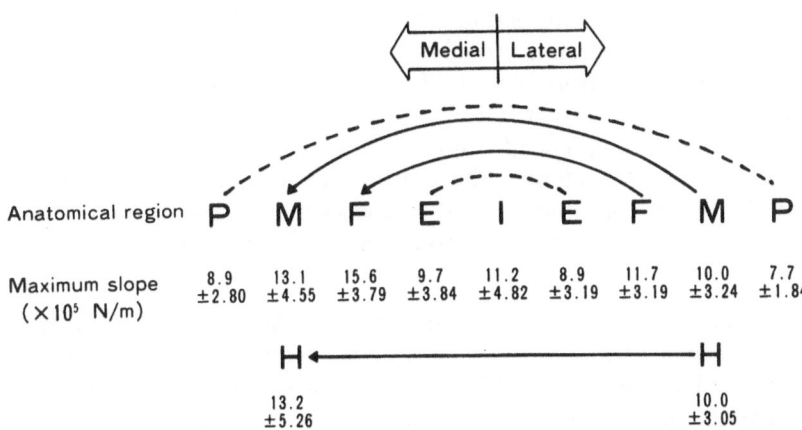

Fig. 5. Results of *t*-tests between corresponding medial and lateral anatomical regions. *Arrows* and *dashed lines* are drawn as in Fig. 4

slope of the load-deformation curve was used to represent indentation stiffness in the following analysis. Variance analysis showed that indentation stiffness varied significantly not only with the location of the 45 squares but also with the six anatomical regions ($P < 0.01$). The results of the Student's *t*-test done between two of the six anatomical regions are shown in Fig. 4. Indentation stiffness of the tibial subchondral plate listed by region in descending magnitude was (a) the region around the free-edge of the meniscus, (b) the meniscus-covered region, (c) the meniscal-horn region, (d) the intercondylar region (no difference in stiffness was noted among the three regions), (e) the sloping region toward the intercondylar eminence, and (f) the periarticular region. The region around the free-edge of the meniscus, the meniscus-covered region, and the meniscal-horn region on the medial tibial plateau had a greater indentation stiffness than the corresponding regions on the lateral tibial plateau (Fig. 5).

Discussion

Topographic variation of the mechanical properties of cancellous bone in the knee joint were previously investigated by compression tests on cubic specimens [5, 6], cylindrical specimens [4, 7, 8], by penetrating tests [9], and indentation tests. There have been various types of indentation tests described. Lereim et al. [10] employed a steel ball 5 mm in diameter against cancellous bone. Bargren et al. [11] used a indenter with a conical end of 120 degrees. Johnson et al. [12] used a flat-ended cylindrical indenter. The indentation test is considered to have several advantages compared with other tests on cut-out bone specimens:

1. It allows a detailed and precise mapping of bone stiffness at various levels below the articular surface
2. Indentation can be performed at each site with the bone specimen *in situ* retaining the natural constraints provided by the surrounding bone
3. It can be a non-destructive test, if the loading is performed within the yielding point, as in the present study

This study used a flat-ended 4 mm-diameter indenter as described by Johnson et al. [12]. We adopted the criterion of an initial 0.3 mm indentation to minimize the contribution of the subchondral cancellous bone and to avoid reaching a yielding point. Johnson et al. [12] analyzed the first 0.5 mm indentation.

The subchondral plate is a well defined cortical shell between the calcified layer of the articular cartilage and the medullary cavity [13]. It is constructed of both cortical and lamellar parts [14]. Most of the mechanical studies on bone in the proximal tibia have been performed after removal of the subchondral plate and the overlying articular cartilage. We have not been able to find a previous report on the mechanical testing of the subchondral plate in the proximal tibia.

In the present study, the indentation stiffness of the subchondral plate varied with the anatomical region. This topographic variation was almost consistent with the previously reported stiffness distribution of the tibial subchondral cancellous bone [7, 8, 9]. The regions with greater amounts of indentation stiffness were around the free-edge of the meniscus and beneath the meniscus. These regions correspond to the high pressure area in the loaded knee joint [15, 16, 17]. Noble and Alexander [18] reported a pyramidal profile of the subchondral bone density in the proximal tibia. The subchondral bone density reached a maximum just mesial to the meniscal inner border and then fell sharply when moving both toward the tibial spine and toward the outermost part of the tibial plateau, where it was lowest. These variations of the subchondral bone density correspond well with the topographic variation of indentation stiffness in the subchondral plate observed in the present study.

References

1. Morrison JB (1970) The mechanics of the knee joint in relation to normal walking. J Biomech 3:51–61
2. Takei T, Akizuki S, Wada T, Kobayashi S, Okuhara T (1985) Microstructural response of articular cartilage to compressive load. Orthop Trans 9:301–302
3. Kobayashi S, Takei T (1987) The effect of subchondral bone on maintenance of fluid film between articular surfaces and deformation of articular cartilage under static load. Orthop Trans 11:305
4. Kobayashi S, Takei T (1990) Mechanical property distribution of cancellous bone in the knee joint. In: Nather A and Goh JCH (eds) Proceedings of the 6th International Congress on Biomedical Engineering, pp 483–485
5. Behrens JC, Walker PS, Shoji H (1974) Variations in strength and structure of cancellous bone at the knee. J Biomech 7:201–207
6. Kobayashi S, Takei T, Okuhara T, Wada T (1984) The role of subchondral bone in maintaining fluid film in the loaded joint spaces. On the influence of mechanical properties of subchondral bone (in Japanese). Proceedies of 1984 Annual Meeting of Japanese Society for Orthopaedic Biomechanics 6:123–130
7. Goldstein SA, Wilson DL, Sonstegard DA, Matthews LS (1983) The mechanical properties of human tibial trabecular bone as a function of metaphyseal location. J Biomech 16:965–969
8. Hvid I, Jensen J (1984) Cancellous bone strength at the proximal human tibia. Eng Med 13:21–25
9. Hvid I, Hansen SL (1985) Trabecular bone strength patterns at the proximal tibial epiphysis. J Orthop Res 3:464–472
10. Lereim P, Goldie I, Dahlberg E (1974) Hardness of the subchondral bone of the tibial condyles in the normal state and in osteoarthritis and rheumatoid arthritis. Acta Orthop Scand 45:614–627
11. Bargren JH, Day WH, Freeman MAR, Swanson SAV (1978) Mechanical tests on the tibial components of non-hinged knee prostheses. J Bone Joint Surg [Br] 60:256–261
12. Johnson JA, Krug WH, Nahon D, Miller JE, Ahmed AM (1983) An evaluation of the load bearing capability of the cancellous proximal tibia with special interest in the design of knee implants. Trans ORS 8:403
13. Mankin HJ (1963) Localization of tritiated thymidine in articular cartilage of rabbits. III. Mature articular cartilage. J Bone Joint Surg [Am] 45:529–540
14. Lemperg R (1971) The subchondral bone plate of the femoral head in adult rabbits. I. Spontaneous remodelling studied by microradiography and tetracycline labelling. Virchows Arch [A] 352:1–13
15. Walker PS, Erkman MJ (1975) The role of the menisci in force transmission across the knee. Clin Orthop 109:184–192
16. Fukubayashi T, Kurosawa H (1980) The contact area and pressure distribution pattern of the knee. Acta Orthop Scand 51:871–879

17. Ahmed AM, Burke DL (1983) In vitro measurement of static pressure distribution in synovial joints — Part I: Tibial surface of the knee. J Biomech Eng 105:216–225
18. Noble J, Alexander K (1985) Studies of tibial subchondral bone density and its significance. J Bone Joint Surg [Am] 67:295–302

The Effect of Flexibility and Central Stem of the Cementless Total Knee Tibial Tray on Initial Fixation

Ichiro Yoshii[1] and Leo A. Whiteside[2]

Summary. The effect of material properties, central stem and its length on tibial tray micromovements were investigated using 100 preserved cadaver tibias. Both axial and shear tests were performed on independent specimens. Axial compressive loads of 100 kg were applied to the anterolateral portion of tibial trays in axial loading tests. Subsidence on the loaded side, lift-off on the contralateral side, micromotions on both sides, and bending of the tray were measured. There was significantly less subsidence, micromotion, and lift-off with the softer Ti6Al4V trays than with the harder CoCr trays. Less subsidence and lift-off were also demonstrated in trays with a long central stem when compared to the other stem groups. For shear loading tests, shear loads of 25 kg were applied to the central portion of the posterior rim of the tray anteriorly. Subsidence and micromotion were measured on the medial and lateral side of the trays, and the mean values were statistically determined for each group. The presence of a central stem significantly reduced both subsidence and micromotion regardless of stem length. Therefore, we concluded that the flexible Ti6Al4V tibial tray remains in better contact with the cut tibial bone surface than the more rigid CoCr tray during eccentric loading. We also concluded that the tibial tray with a long stem can achieve better initial fixation of the implant to bone when compared to the short stem and no stem.

Key words. TKA — Initial fixation — CoCr — Ti6Al4V — Central stem — Stem length — Tibial component — Micromovement

[1] Yoshii Orthopaedic Hospital, 6-7-5 Ohashidori, Nakamura, Kochi, 787 Japan
[2] DePaul Biomechanical Research Laboratory, 3165 McKelvey Road, Suite 240, Bridgeton, Missouri 63044, USA

Introduction

In recent years, the use of porous coatings as a means of prosthetic joint fixation has gained popularity in cementless total knee arthroplasty (TKA), and the clinical importance of achieving rigid fixation of the tibial component in TKA is well accepted. Recent laboratory studies comparing the rigidity of initial fixation in different designs of tibial components concluded that tibial trays secured with cancellous screws achieve superior fixation compared to those secured with stem and pegs only [1, 2]. However, screws alone may not be enough to facilitate the desired degree of bone ingrowth [3].

Material properties such as the modulus of elasticity have not been evaluated to determine their effect on fixation of total knee components. Flexibility of the tibial tray may have a major effect on micromovement and load transfer between the metallic structure and bone. Recently, Ti6Al4V has been extensively evaluated as an acceptable material for orthopaedic implants. This material has an elastic modulus of 100 GPa, which is approximately one half that of CoCr.

A central stem may also influence rigidity of initial fixation. Since the stem has proven benefical in cemented TKA, the same effect would be expected in cementless cases. However, reports available in the literature are contradictory, and the effects of the central stem in cementless TKR are unclear [4–6, 7, 8, 9]. The length of the central stem is also considered to affect the fixation of the tibial tray. A finite element model developed by Crowninshield, et al. predicted that a long central stem would reduce levels of stress applied to underlying bone [5]. However, an experimental model developed by Volz et al. showed little effect of the tibial stem [2]. Although the use of long stems on the tibial component has been advocated for some clinical situations [10, 11], the effect of the tibial stem length in cementless applications has not yet been established experimentally or clinically. The goal of this study is to evaluate the effect of flexibility of the material on tibial tray micromovement by comparing CoCr and Ti6Al4V tibial trays, and to evaluate the effect of a central stem and its length on micromovement for both axial and shear loads.

Materials and Methods

One hundred phenol-preserved cadaver tibial specimens were used in this study. Twenty eight specimens were used for the comparison between CoCr and Ti6Al4V implants. The component design was identical for each material group, and it was secured only with four cancellous screws. The rest of the specimens were used for the evaluation of the effect of the central stem, in which medium sized Whiteside Ortholoc Modular Knee System tibial component (Down Corning Wright, Arlington, TN) was used. In addition to the cancellous screws, this tray was designed to be secured with or without a 7.6 cm-long central stem (the short stem) or a 15.2 cm-long stem (the long stem) (Fig. 1).

Fig. 1. Dimensions of tibial trays used in this study. This design allowed the assessment of the effectiveness of the stem and its length

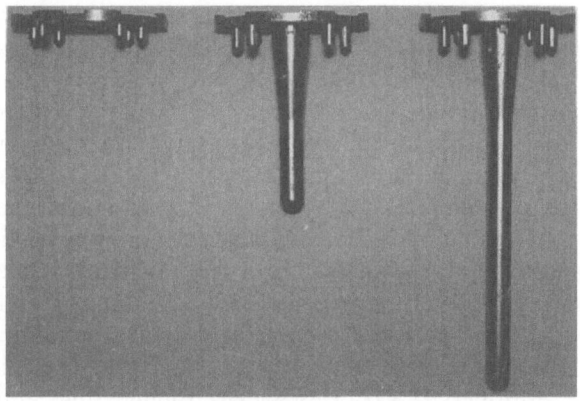

After freeing the bones of surrounding muscles, periosteum, and soft tissues, tibial trays were implanted using standard surgical techniques. The tibia was cut perpendicularly to the long axis of the tibial shaft 10 mm distally to the lateral surface of the tibial plateau so that the bone was under-resected on the medial side and evenly resected on lateral side. The tibial surface was then planed with a tibial surface planer to create a flat and highly regular surface. Tibial specimens were chosen so that the peripheral rim of the tibial tray could be aligned with the circumference of the tibial cortex. All specimens in which circumferential peripheral contact could not be achieved were discarded. After planing, the tibia was cut 20.3 cm below the resected surface and mounted in a 5.1 cm-long PVC pipe with dental cement so that the resected surface was horizontally oriented and the tibia was exposed 15.2 cm below the cut surface. Rigid initial fixation was achieved with the use of four 6.5 mm cancellous screws angled divergently at 10° through the peripheral tibial cortex.

Before implant, a simple penetration test was performed to assess underlying cancellous bone quality. This method was used in Miura's study, which was adapted from Hvid's work [1, 12]. The specimens were mounted in a servohydraulic testing machine (Instron Corporation, Canton, MA), A 2.5 mm-diameter rod was inserted 5 mm deep into the cancellous bone perpendicular to the resected surface at a speed of 0.5 mm/sec, while the applied load was monitored on a chart recorder. The insertion points were identical to the location of the four screw holes on the tibial tray. Peak load on each of the four points was measured and the mean value was used for the evaluation of cancellous bone quality.

After the penetration test, the tibial trays were implanted and tested using both axial and shear loading protocols.

Axial Loading Test

A compressive axial load was applied to the tibial trays by means of the Instron machine. Specimens were mounted in the machine so that the tibial tray was oriented perpendicularly to the loading axis. A compressive axial load was applied to 70 specimens, 14 for each group. In that, 28 specimens were used to compare Ti6Al4V and CoCr. The other 42 specimens were used for the evaluation of the effect of the central stem. These specimens consist of (a) a stemless group (group without stem), (b) a short stem group (group with the short stem), and (c) a long stem group (group with the long stem). Specimens were mounted in the Instron machine horizontally so that the tibial cut surface was perpendicular to the load. An eccentric load with haversine waveform from a 5 kg preload to a 100 kg peak load was applied to the same point on the anterolateral peripheral rim of the tray in order to apply the load to the area of softest bone on the cut surface of the tibia. This loading point simulated the worst case scenario of eccentric loading conditions. This was repeated for 1000 cycles at a frequency of 1 Hz and three linearly variable differential transformers (LVDT) (Schaevitz Engineering, Pensauken, NJ) were used for measuring micromovements. One LVDT was mounted on the anterolateral region of the tibial bone as close to the cut surface as possible. This LVDT was referenced to the bottom surface of a flag which was attached to the anterolateral surface of the tray and measured the vertical motion of the tray on the loaded side. Another LVDT was mounted to the center portion of the tibial tray and was referenced to the top surface of the same anterolateral flag. This LVDT was used to measure the amount of displacement due to metal bending in the tray. The top and bottom reference positions on the anterolateral flag were identically located 13 mm from the loading point. A third LVDT was mounted on the posteromedial region of the tibial bone, also

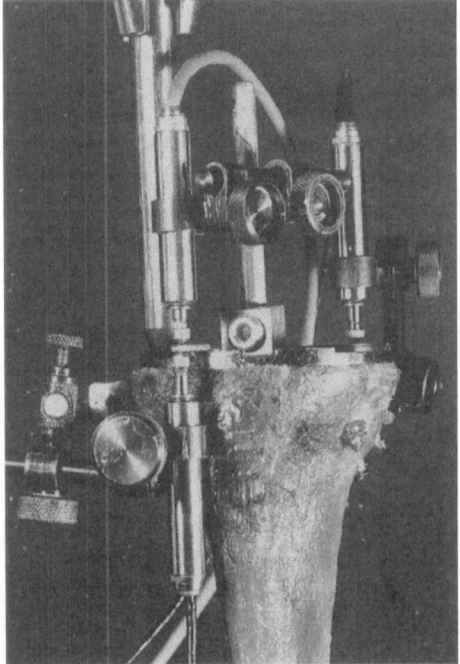

Fig. 2. Location of the LVDTs and the loading position in axial loading test. These points were identical for all test specimens. *LVDT*, linearly variable differential transformers

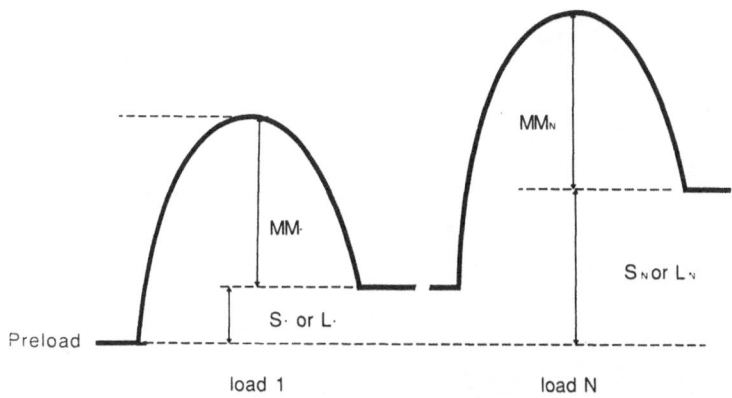

Fig. 3. Displacement measurements. MM, micromotion; S, total subsidence; L, total lift-off; MM_1, micromotion at 1st load; MM_N, micromotion at Nth load; S_N, total subsidence at Nth load; L_N, total lift-off at Nth load

as close to the cut surface as possible. This LVDT was referenced to the posteromedial rim of the tray and measured the vertical movement of the tray on the unloaded side. All three LVDTs were placed parallel to the loading axis and the location of each LVDT was identical for each test (Fig. 2).

Micromovements were recorded using a chart recorder (Astro-Med, West Warwick, RI). The relationship between the displacement measurements and loading is shown in Fig. 3. Micromotion is defined as the recoverable displacement of the tibial tray from one peak load to the next preload for each measured cycle. Subsidence is defined as the permanent sinking displacement of the tray relative to the bone surface at preloaded phase. Lift-off is the permanent lifting displacement of the tray relative to the bone surface at the preloaded phase. Metal bending is defined as the recoverable magnitude of micromotion measured by the LVDT attached to the center of the tray from one peak load to the next preload. All the measurements were calculated for each specimen at initial load and every 100 cycles thereafter.

Mean values and standard deviations of each measurement were determined for each material and statistical significance of apparent differences between material groups was determined using the analysis of variance technique (ANOVA). The relationships between displacement measurements, bone quality, and loading cycles were displayed with the use of three-dimensional graphs for the comparison of the two materials. The graphs were formed using the SYSTAT statistics program and their regression lines for each cycle were determined. The statistical significance of the differences among these regression lines for each material were determined using the analysis of covariance technique (ANCOVA). In order to determine how much metal bending contributed to micromotion, correlation coefficients between metal bending and micromotion were calculated at every 100 cycles for each material,

Fig. 4. Location of the LVDTs and the loading position in shear loading test. These points were identical for all test specimens. *LVDT*, linearly variable differential transformers

and the statistical significance of the correlation coefficients between the two materials was determined by the two-tail student *t*-test.

As well as the material groups, mean values and standard deviations of each set of measurements were also determined for each stem group and the statistical significance was determined by ANOVA. The relationship between measured displacement and bone quality were evaluated with the use of regression line graphs for each measured cycle, and the statistical significance of these regression lines for each stem length group was determined by ANCOVA.

Posteroanterior Shear Loading Test

Shear testing was done with 30 specimens, 10 for each stem group. Specimens were mounted in the Instron machine so that the tibial tray was oriented parallel to the direction of applied load. A shear load in the posterior-to-anterior direction was applied with a loading device that was made to facilitate application of an even force to the central tibial notch between the medial and lateral condyles of the tray. Haversine waveform loads were applied from 2–25 kg for 1000 cycles at a frequency of 1 Hz. Two LVDTs were symmetrically placed on the medial and lateral posterior rims of the tibial tray. They were oriented parallel to the loading direction in order to measure micromovement of the tibial tray relative to the proximal tibia in the posteroanterior direction (Fig. 4).

Because of the difficulty in applying even shear loads medially and laterally, mean values of medial and lateral displacements were calculated for both subsidence and micromotion in every specimen at the initial cycle and every 100 cycles thereafter.

Mean values and standard deviations of subsidence and micromotion were determined for each group at each measured cycle and the statistical significance of differences among the groups was determined by ANOVA. The relationships between displacement measurements and bone hardness were

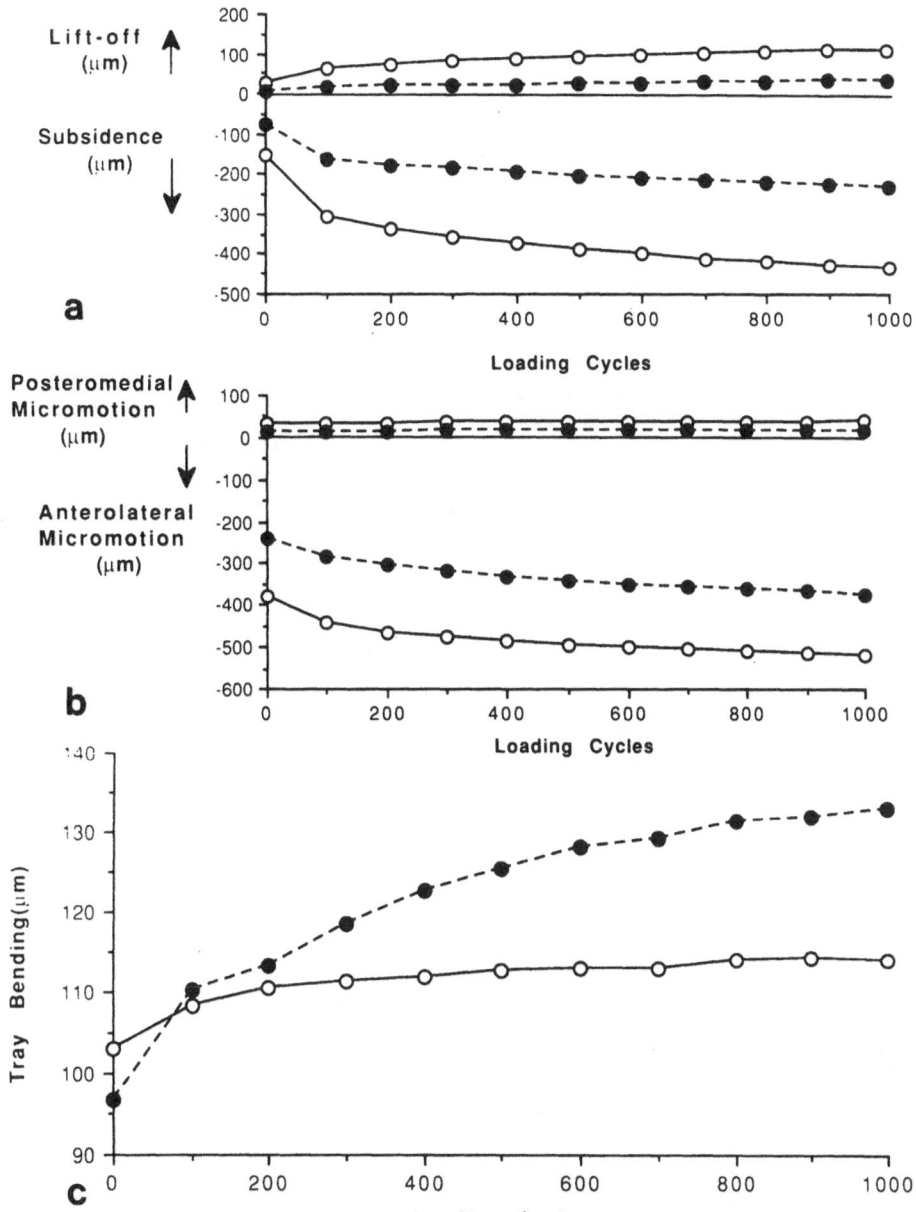

Fig. 5a–c. Graphs of displacement measurements for each material group. **a** Total subsidence and total lift-off for each material group. Positive direction represents lifting movement. Negative direction represents sinking movement. All the measurements for the Ti6Al4V tray demonstrate significantly less movement than for the CoCr tray ($P <$ 0.05). **b** Anterior and posterior micromotion for each material. Positive direction represents lifting movement. Negative direction represents sinking movement. All the measurements for the Ti6Al4V tray demonstrate significantly less movement than for the CoCr tray ($P <$ 0.05). **c** Bending of the trays for each material. There are no significant differences demonstrated between the two material groups. *Open circles,* CoCr; *solid circles,* Ti6Al4V

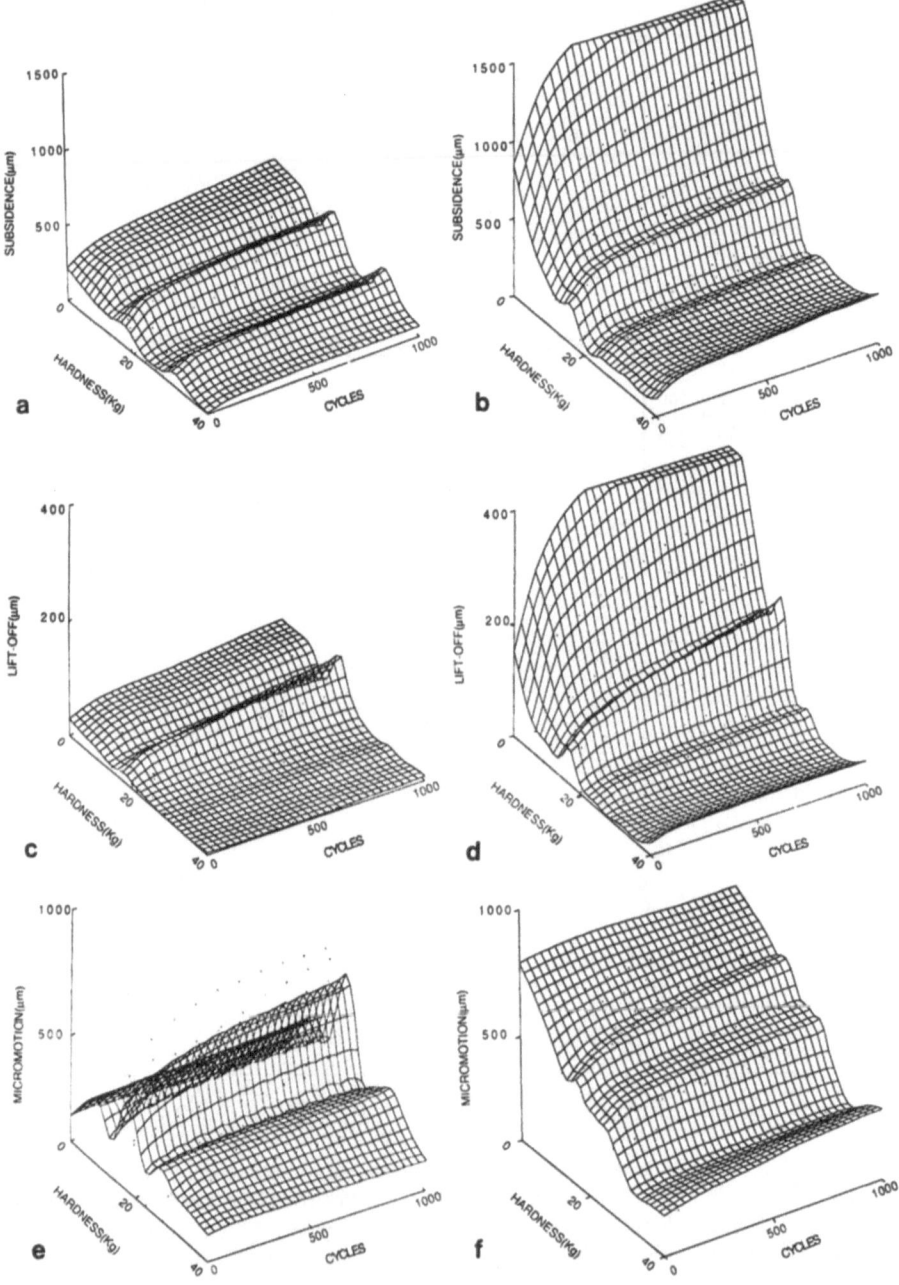

evaluated with the use of regression line graphs for each measured cycle determined by ANCOVA.

Results

Subsidence ($P < 0.05$), anterolateral micromotion ($P < 0.05$), lift-off ($P < 0.05$), and posteromedial micromotion ($P < 0.05$) were significantly less with Ti6Al4V than with CoCr at every measured cycle except the initial cycle (Fig. 5a,b). Metal bending was greater with Ti6Al4V, but no statistically significant differences were demonstrated (Fig. 5c). Three-dimensional graphs of displacement measurement such as total subsidence, total lift-off, and anterolateral micromotion versus the mean value of the penetration test and cycles are shown in Fig. 6a–f. Subsidence and lift-off were positively correlated with the number of cycles and inversely correlated with the mean value of the penetration test (inversely correlated with bone hardness). This was seen in both material types, but the rate of increase of these displacements was greater with CoCr than with Ti6Al4V. The regression lines of these displacements vs. bone hardness were significantly different between the two materials (for subsidence $P = 0.006$ and for lift-off $P = 0.004$). When the CoCr trays were tested, micromotion increased as the bone hardness decreased; but in the Ti6Al4V trays, the largest micromotion occurred at the middle level of bone hardness. Despite this difference in appearance of the graphs, no significant differences of the regression lines were demonstrated between the two materials ($P = 0.154$).

Correlation coefficients between metal bending and anterolateral micromotion for each material group are significantly higher for Ti6Al4V than for CoCr ($P < 0.0001$). These values were R-square ≥ 0.75 for Ti6Al4V, and R-square ≥ 0.075 for CoCr.

When comparing the magnitude of displacement measurements in the stem groups, subsidence in the long stem group was significantly less than those

---

Fig. 6a–f. Three-dimensional graphs of displacement measurements, bone hardness and loading cycles. **a** Three-dimensional graph of total subsidence for the Ti6Al4V tray. **b** Three-dimensional graph of total subsidence for the CoCr tray. As bone hardness decreases and as loading cycles increases, total subsidence increases, but this trend was more prominent with the rigid CoCr tray than with the flexible Ti6Al4V tray. **c** Three-dimensional graph of total lift-off for the Ti6Al4V tray. **d** Three-dimensional graph of total lift-off for the CoCr tray. Total lift-off also increases as bone hardness decreases and as loading cycle increases, but the increasing rate is greater in the CoCr tray. **e** Three-dimensional graph of anterolateral micromotion for the Ti6Al4V tray. The greatest amount of micromotion occurred with bone of intermediate hardness. **f** Three-dimensional graph of anterolateral micromotion for the CoCr tray. Micromotion increased progressively as the bone hardness decreased. The pattern of the CoCr graph was different from that of Ti6Al4V

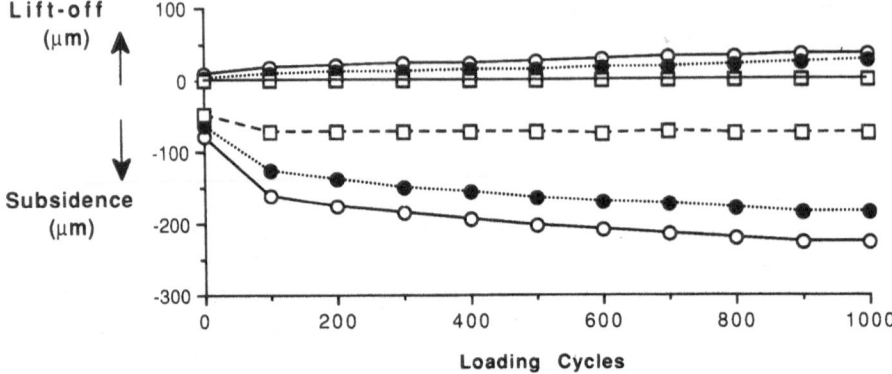

Fig. 7. Graphs of subsidence, lift-off for each group in axial loading test at every measured cycle. Positive direction represents lifting movement. Negative direction represents sinking movement. The long-stem group (*open squares*) shows the least subsidence and lift-off at every measured cycle ($P < 0.05$). The short-stem group (*solid circles*) shows no significant differences from the stemless group (*open circles*) both for subsidence and lift-off

of both the stemless and the 7.6 cm-stem groups at every measured cycle except the initial cycle ($P < 0.05$). Mean lift-off of the long-stem group was significantly less than that of the stemless group and the short-stem group. The short-stem group appeared to be less than the stemless group but did not show any significant difference in either subsidence or lift-off (Fig. 7). There were no statistically significant differences in anterolateral micromotion, posteromedial micromotion, or metal bending among any of the groups.

Regression lines for micromotion, subsidence, and lift-off versus bone hardness are shown in Fig. 8a,b. All three groups show an inverse relationship between bone hardness and the amount of subsidence and micromotion. This trend was also seen in lift-off for the stemless group and the short-stem group, however, the long-stem group showed lift-off in only one case. Thus, despite the increase in subsidence in the long-stem group with softer bone, this tendency was less than that seen in the other two stem groups. The regression line of subsidence vs. bone hardness of the long-stem was significantly different from that of the other two groups at every measured cycle ($P < 0.0001$). The regression line of lift-off vs. bone hardness of the long-stem group appeared to be different from the other groups, but no significant difference was demonstrated ($P = 0.084$). However, it is remarkable that none of the long stem trays lifted off more than 10 μm, even in the softest bone. The other parameters such as anterolateral and posteromedial micromotion, and metal bending showed no significant differences among all stem groups.

Posteroanterior Shear Loading Test

Magnitudes of subsidence and micromotion for groups with stems were significantly less than for the stemless group at every measured cycle ($P < 0.05$),

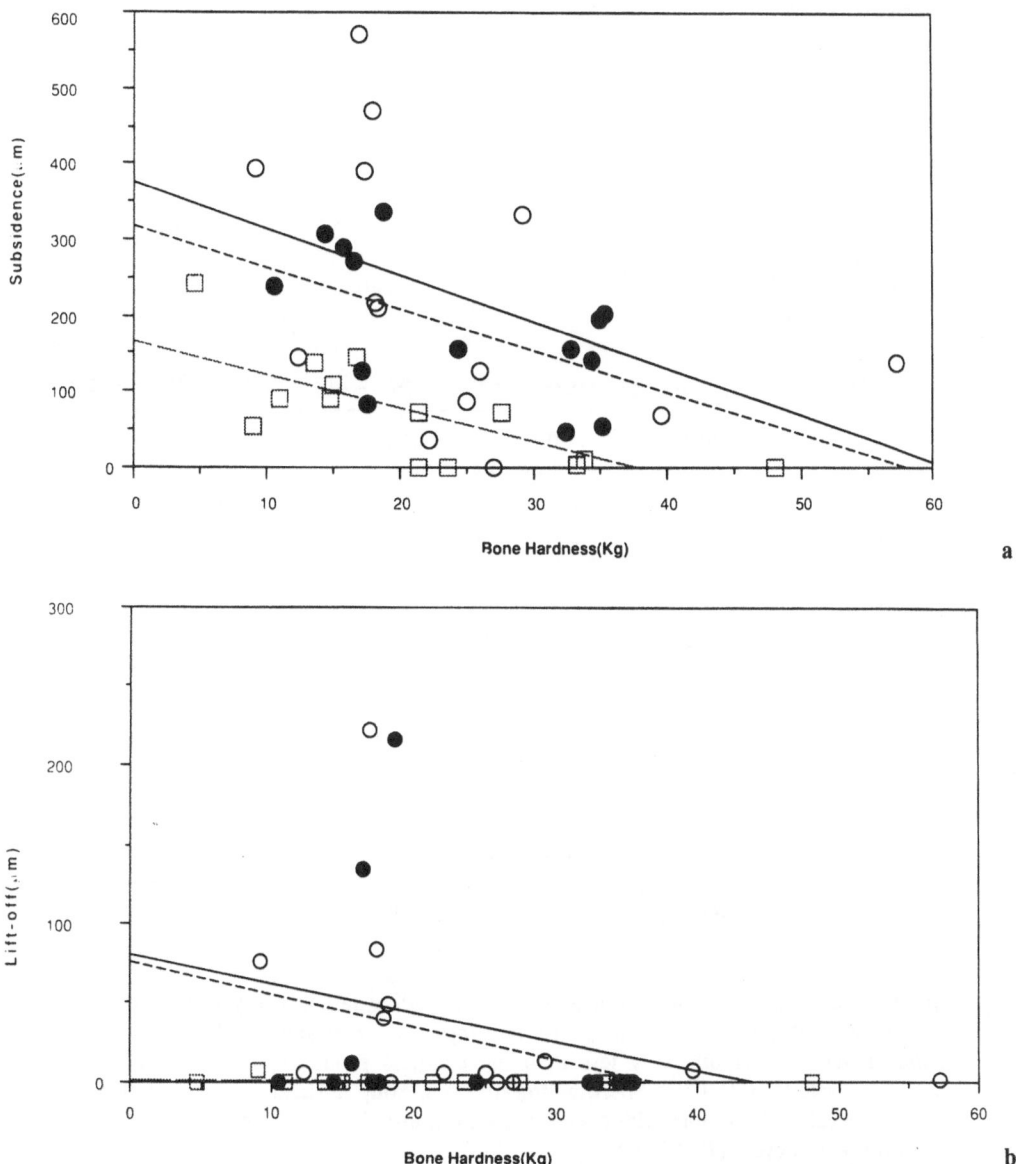

Fig. 8a,b. a Graph of subsidence and bone hardness in axial loading test at 1000th cycle with regression lines. The long-stem group showed a significantly different regression line from the other two groups ($P < 0.0001$). **b** Graph of lift-off and bone hardness in axial loading test at 1000th cycle with regression lines. A regression line for the long-stem group appeared to be different from the other two groups, but no significant difference is demonstrated ($P = 0.084$). *Open squares*, long stem; *open circles*, stemless; *solid circles*, short stem

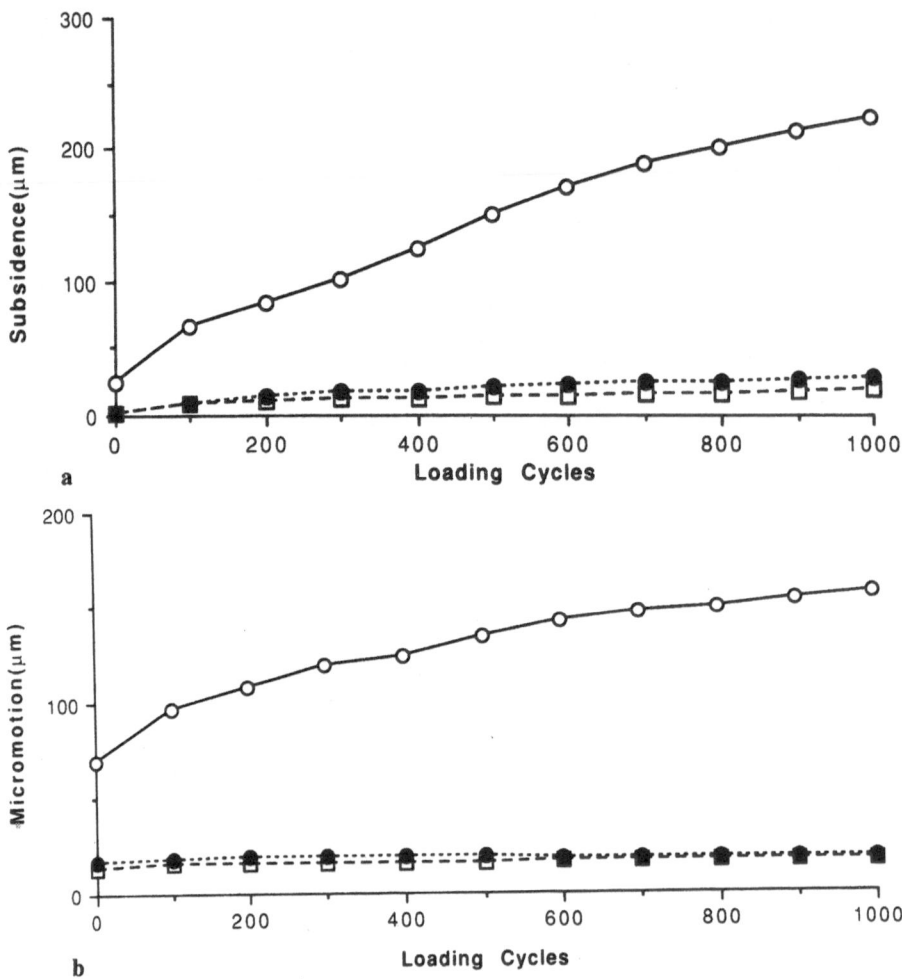

Fig. 9a,b. a Subsidence for each group in shear loading test at every measured cycle. The groups with a stem, the long-stem group (*open squares*) and the short-stem group (*solid circles*), are significantly less than the stemless group (*open circles*) at every measured cycle ($P < 0.05$). **b** Micromotion for each group in shear loading test at every measured cycle. The groups with a stem are significantly less than the stemless group at every measured cycle ($P < 0.05$)

while there showed no significant difference between the groups with stems (Fig. 9a,b).

Regression lines for micromotion and subsidence versus bone hardness are shown in Fig. 10a,b. As seen in the axial loading test, as bone hardness increased, subsidence and micromotion values decreased. The slope of the regression lines for the short- and long-stem groups were significantly smaller than that determined for the stemless group (for subsidence, $P = 0.008$; for micromotion, $P = 0.02$). In comparing the two stem groups, there was no significant difference.

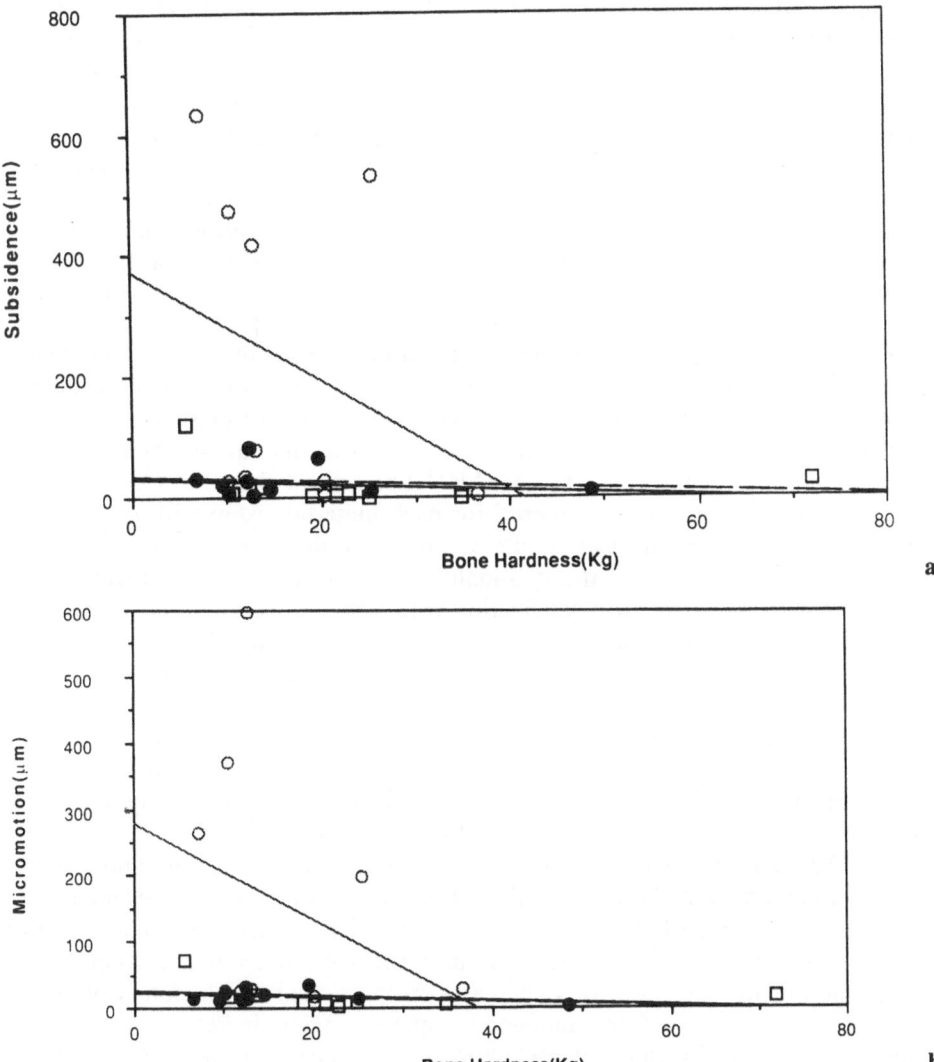

Fig. 10a,b. a Graph of subsidence and bone hardness in shear loading test at 1000th cycle with regression lines. The regression line of the stemless group is significantly different from that of the other two groups ($P = 0.008$). **b** Graph of micromotion and bone hardness in shear loading test at 1000th cycle with regression lines. The regression line of the stemless group is significantly different from the other two groups ($P = 0.02$). There is no significant difference between the short- and the long-stem groups. *Open square*, long stem; *solid circle*, short stem; *open circles*, stemless

Discussion

Bone ingrowth in cementless TKA is generally accepted as necessary to achieve acceptable long term results [13]. Although many factors influence bone ingrowth, the single most important factor is commonly believed to be micromovement between porous coated surfaces and the underlying cancellous bone. This is directly related to the degree of initial fixation. This axial loading test was designed so that loads twice the normal body weight were applied to the weakest region of the tibial surface in order to simulate the worst-case scenario [14]. Despite the inability to exactly duplicate clinical simulations, the fact that measurement points were consistently located allows comparisons to be made between implant materials. Although some biomechanical studies have shown that the medial side is more heavily loaded during single-legged stance, gait analysis studies have shown that the anterolateral surface also sees heavy loads during heel strike [15]. Anterolateral sinking has also been reported in clinical series of cementless tibial components [11].

Bone hardness was evaluated by a penetration test in this study. The peak load represents a rough estimate of hardness of the underlying cancellous bone, although this does not express exact bone strength or elasticity.

Displacement measurements were statistically evaluated and all the measurements were significantly less for Ti6Al4V tibial trays than for CoCr trays. These measurements were compared for each material group with regression lines of displacement measurements vs. bone hardness and they were significantly different between the two materials. These results demonstrate that the Ti6Al4V tibial tray provides fixation with less relative micromovement of the implant than a more rigid CoCr tray of the same dimensions. This seems to be due to the greater flexibility of the Ti6Al4V trays. This trend was seen regardless of bone hardness.

Correlation coefficients were determined in order to assess the relationship between metal bending and micromotion. The degree of correlation was much greater for Ti6Al4V than for CoCr. This may be explained by a higher percentage of micromotion contributed by metal bending for Ti6Al4V than for CoCr. The harder CoCr tray tends to tilt more during eccentric loading which may weaken screw fixation. Once tilting has occurred, the load is transferred to the loaded portion of the tibia through a smaller surface area, and this in turn increases the tendency to sink into the cancellous bone. Therefore, micromotion of the CoCr tray increases proportionally with bone softness. Furthermore, lift-off of the unloaded opposite region increases, worsening ipsilateral sinking. Because of its low modulus of elasticity, the Ti6Al4V tray bends more with eccentric loading and distributes the load more evenly across the tibial cut surface, so the tilting magnitude is smaller. This effect occurs even when the underlying bone is soft, so that micromotion does not increase proportionally as bone hardness decreases when a Ti6Al4V tray is used. Therefore, the three-dimensional graph such as those seen in Fig. 6e,f showed different patterns between the two materials. This effect of flexibility seems to minimize migration of the tray. As load is repeated, this effect becomes more apparent. When repetitive load is applied, screw fixation weakened even with the flexible Ti6Al4V tray, but the Ti6Al4V tray still compensates tilting with bending. Thus, the Ti6Al4V tray showed less increase of subsidence and lift-off than the CoCr tray corresponding to the loading cycles.

The results of the axial loading test comparing the two stem groups and the stemless group demonstrated that a central stem helped to reduce subsidence of the loaded portion and tilting of the tray. Miura et al. suggested that the effect of screws was to prevent lift-off on the unloaded portion of the tibial tray but they were unable to demonstrate a major effect of thickening of the 7.6 cm stem [1]. However, the present study suggests that a longer central stem had an additive effect over that of screw fixation. This effect may be due to actual cortical contact distally, or possibly better metaphyseal bone contact proximally. Since this additive effect was not seen consistently when the short stem was used, it seems that metaphyseal cancellous contact alone is insufficient to significantly improve fixation of the tray. These findings agree with those of Volz et al., who reported that a short stem was ineffective in improving fixation of a cementless tibial component [2]. The regression line graph of subsidence vs. bone hardness demonstrates in the case of soft bone that some of the load is applied through the stem to the endosteal surface, thus protecting the cut tibial surface from overload in compression. Part of this effect seen in soft bone may be due to the stem's resistance to tilting, and the resultant effect on load distribution under the tibial tray. When tilting and lift-off are minimized, the tilting axis remains distant from the point of application of the load, and the area through which the load is applied is much greater than that under a tibial component that is allowed to tilt [1].

The question of shear loading of the tibial implant is seldom addressed, but there is one study on the effect of screw fixation which suggests that the additive effect of a stem in the presence of screws is very small when the tibial component is fixed to hard bone. Although the present study agrees with this finding in the case of hard bone, our data suggests a further conclusion that screw fixation alone is insufficient when the tibial bone is soft. The addition of a stem greatly improves tibial fixation against shear forces with both 7.6-cm and 15.2-cm stems, which suggests that the resistance is generated in the proximal metaphyseal cancellous bone.

The results of this study suggest that tibial trays made with Ti6Al4V provide more stable and rigid fixation than CoCr trays and that the central stem can play a major role in fixation of the tibial component in cementless TKR even when the implant has been fixed to the tibia with four cancellous screws. These effects may have some advantages in clinical applications.

References

1. Miura H, Whiteside LA, Easley JC, Amador DD (1990) Effects of screws and sleeve on initial fixation in uncemented total knee tibial component. Clin Orthop Rel Res 259:160–168
2. Volz RG, Nisbet JK, Lee RW, McMurtry MC (1988) The mechanical stability of various noncemented tibial components. Clin Orthop Rel Res 226:38–42
3. Sumner DR, Jacobs JJ, Turner TM, Urban RM, Galante JO (1989) The amount and distribution of bone ingrowth in tibial components retrieved from human patients. Trans ORS 14:375

 4. Bartel DL, Burstein AH, Santavicca EA, Insall JN (1982) Performance of the tibial component in total knee replacement. J Bone Joint Surg [Am] 64:1026
 5. Crowninshield RD, Murase K, Pedersen DR (1983) An analysis of tibial component design in total knee arthroplasty. J Biomech 16:13
 6. Dempsey AJ, Finlay JB, Bourne RB, Rorabeck CH, Scott MA, Millman JC (1989) Stability and anchorage considerations for cementless tibial components. J Arthroplasty 4:223
 7. Manley MT, Stulberg BN, Stern LS, Watson JT (1987) Direct observation of micromotion at the implant-bone interface with cemented and non-cemented tibial components. Trans Orthop Res Soc 12:436
 8. Samuelson K (1988) Bone grafting and noncemented revision arthroplasty of the knee. Clin Orthop Rel Res 226:93
 9. Shimagaki H, Bechtold JE, Sherman R, Gustilo RB (1988) Initial stability of tibial components in cementless total knee arthroplasty. Trans ORS 13:477
10. Strickland AB, Chan KH, Andriacchi TP, Miller J (1988) The initial fixation of porous coated tibial components evaluated by the study of rigid body motion under static load. Trans ORS 13:476
11. Whiteside LA, Pafford J (1989) Load characteristics of a non-cemented total knee arthroplasty. Clin Orthop Rel Res 239:168–177
12. Hvid I, Hansen SL (1985) Trabecular bone strength patterns at the proximal tibial epiphysis. J Orthop Res 3:464–472
13. Ryd L, Lindstrand A, Rosenquist R, Selvik G (1986) Tibial component fixation in total knee arthroplasty. Clin Orthop Rel Res 213:141–149
14. Murase K, Crowninshield RD, Pedersen DR, Chang T (1983) An analysis of tibial component design in total knee arthroplasty. J Biomech 16:13–22
15. Andriacchi TP, Galante JO, Fermier RW (1982) The influence of total knee replacement design on walking and stair climbing. J Bone Joint Surg [Am] 64: 1328–1335

Stress Analysis of the Proximal Tibia After Total Knee Arthroplasty with the Finite Element Method

N. Yoshino, N. Inoue, Y. Watanabe, F. Yamashita, Y. Hirasawa,[1]
T. Hirai, and T. Katayama[2]

Summary. Two-dimensional finite element analysis, considering heterogeneous and anisotropic properties of cancellous bone, was applied to the analysis of stress distribution in the natural proximal tibia before and after total knee arthroplasty. The detailed material properties of the proximal tibia were estimated by image analysis of soft X-rays. In the natural proximal tibia under bi-condylar uniform loads, which is regarded as ideal, the stress concentration was located in the middle portion between the condyles when the model was based on homogeniety and isotropy of cancellous bone. On the other hand, stress concentration was not found in analysis assuming heterogeniety and anisotropy. The stress distribution in the proximal tibia with the model of heterogeneity and anisotropy of cancellous bone was shown to be close to the actual clinical outcome. The finite element analysis employed in the present study may be beneficial in the evaluation of tibial component prosthetic designs.

Key words. Stress analysis — Finite element method — Heterogeneity — Anisotropy — Image analysis — Cancellous bone — Total knee arthroplasty

Introduction

Since the Finite Element Method (FEM) was first utilized for analysis of the structure of aircraft by Turner et al. in 1954, FEM has been employed for many mechanical analyses of materials and structures. In the orthopaedic field, FEM has recently been employed to evaluate the design of prostheses and

[1] Department of Orthopaedic Surgery, Kyoto Prefectural University of Medicine, Kyoto, Japan
[2] Department of Mechanical Engineering, Doshisha University, Kyoto, Japan

to examine the causes of prosthetic mechanical problems such as loosening and sinking of the tibial component [1–6]. However, the investigators in those studies did not demonstrate which design produces the minimum stress concentration at the prosthesis/bone interface. A variety of results have been reported stemming from the differences of several conditions, such as geometries of the models, material properties, loading conditions, boundary conditions, and dimensions. For the application of FEM to the stress analysis around the prosthesis/bone interface, especially surface replacement type prostheses, the material properties of the cancellous bone should be detailed because cancellous bone has heterogeneous and anisotropic properties [7, 8]. However, cancellous bone has been treated as homogeneous and isotropic in the majority of previous finite element analyses because the effective methods to evaluate the detail distribution of the material property of the cancellous bone have not been established so far.

The objective of the present study was to asses the effectiveness of consideration of the heterogeneous and anisotropic material properties of cancellous bone in the proximal tibia using FEM. To this end, the heterogeneous and anisotropic material properties of the cancellous bone were first estimated by image analysis. Second, the hetrogeneous anisotropic model was compared with the homogeneous isotropic model for the stress analyses in the natural proximal tibia. Finally, the results of the finite element analyses of the tibia after total knee arthroplasty (TKA) were verified by comparison with X-ray images taken after TKA.

Methods

Estimation of the Elastic Modulus of Cancellous Bone in the Natural Proximal Tibia

The proximal tibia of an adult female was sliced to a 3 mm thickness in the mid-frontal plane, and then a soft X-ray image was taken (Fig. 1a). Image analysis quantified the volume fraction of the trabeculae and trabecular orientation for each of the 117 5 mm × 5 mm regions (Fig. 1b) on the soft X-ray image.

The images were digitized (256 × 256 pixels, 64 gray levels) and input into a microcomputer. The volume fraction of the trabeculae, V_f, was obtained from:

$$V_f = \sum_x \sum_y G_{can}(x, y) \Big/ \sum_x \sum_y G_{cor}(x, y)$$

where Gcan(x, y) was the gray level of the digital image of cancellous bone and Gcor(x, y) was that of the cortical bone.

To quantify the trabecular orientation, a two-dimensional Fourier transformation was applied to a digitized image of the soft X-ray. To detect the directionality, information contained in the power spectrum, p(x, y), was transformed from the Cartesian system (x, y) into a polar system (r, θ) (Fig. 2).

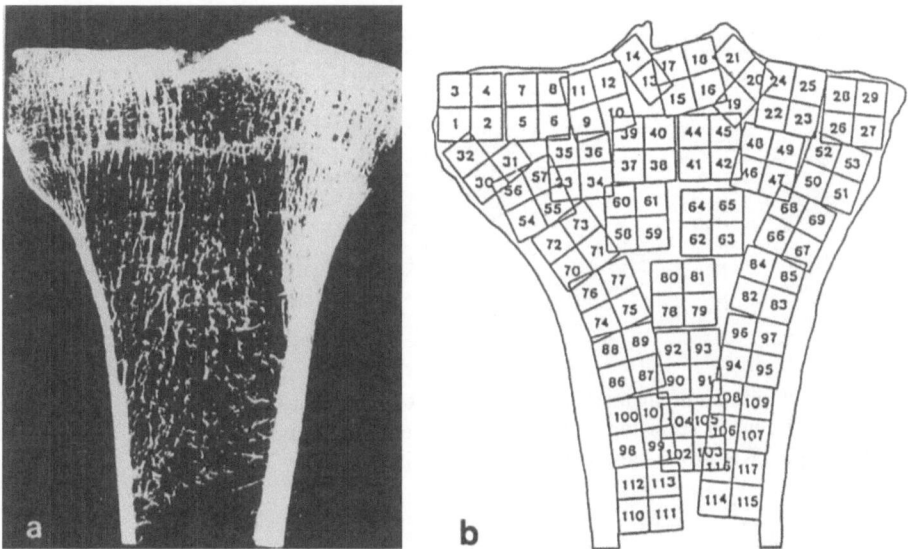

Fig. 1a,b. a Soft X-ray image of the sliced proximal tibia. **b** Area for quantification of trabecular volume fraction and orientation

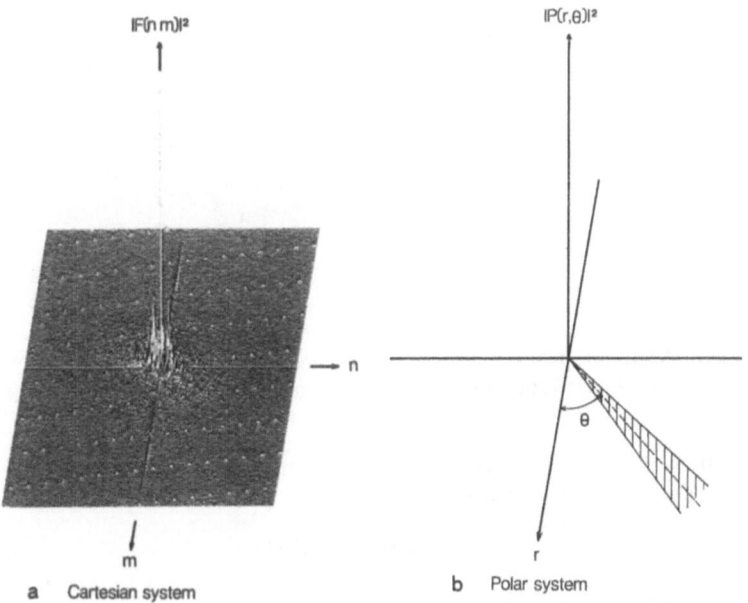

a   Cartesian system   b   Polar system

Fig. 2a,b. Measurement of trabecular orientation. **a** $|F(n, m)|^2$, Power spectrum of digitized soft X-ray image of cancellous bone in the Cartesian system; **b** $|P(r, \theta)|^2$, Power spectrum of digitized soft X-ray image of cancellous bone in the polar system. The intensity of orientation of the prime image was revealed as the energy in the fan-shaped area

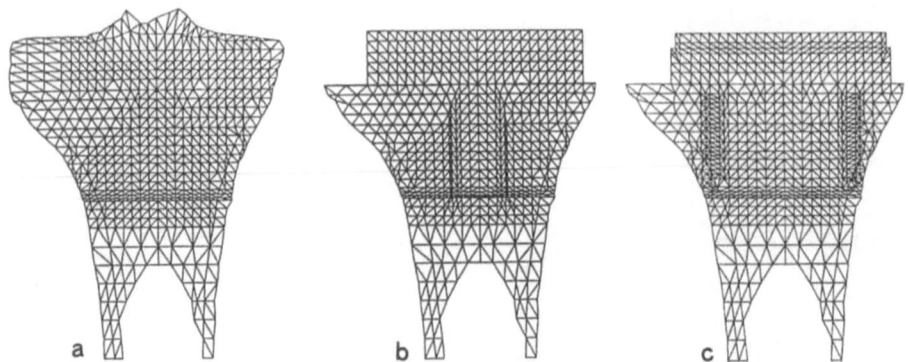

Fig. 3a–c. Finite-element mesh **a** of natural proximal tibia (1337 elements, 722 nodal points) **b** of central stem type TKR (1241 elements, 673 nodal points) and **c** of screw fixation type TKR (1306 elements, 707 nodal points)

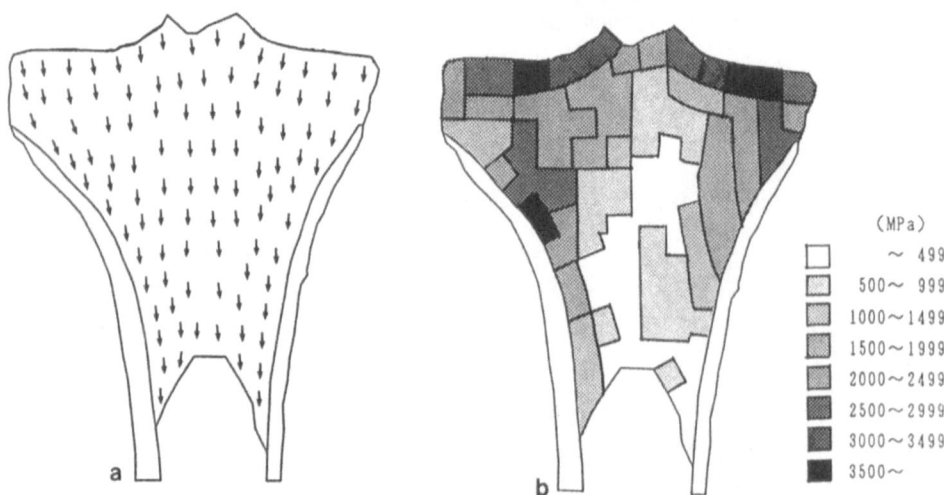

(MPa)

☐ ~ 499
▨ 500~ 999
▨ 1000~1499
▨ 1500~1999
▨ 2000~2499
▨ 2500~2999
▨ 3000~3499
■ 3500~

Fig. 4a,b. Material properties of cancellous bone according to **a** mean trabecular orientation and **b** distribution of the Young's modulus of the cancellous bone

The function, $P(r, \theta)$, was decomposed into $P_\theta(r)$ for each θ, and $P_r(\theta)$ for each r. The orientation for each θ was:

$$F(\theta) = \sum_r P_\theta(r).$$

The distribution function $f(\theta)$, which indicated the orientation factor of the trabeculae for each angle θ, was obtained from:

$$f(\theta) = F(\theta) \Big/ \sum_\theta F(\theta).$$

The mean orientation angle, θ_m, was determined from $f(\theta)$.

Inoue et al. [9, 10] demonstrated that elastic moduli can be estimated from the volume fraction of the trabeculae and the distribution function of the trabecular orientation when cancellous bone is assumed to have homogeneous and anisotropic properties using the following simplified assumptions:

$$E_L = \alpha_1 E_f V_f, \qquad E_T = \alpha_2 E_f V_f \qquad G_{LT} = 4077\ V_f\ 0.5358(\alpha_1/\alpha_2)$$

$$\alpha_1 = \sum_{\theta=0}^{180} \cos^4(\theta - \theta m)f(\theta), \qquad \alpha_2 = \sum_{\theta=0}^{180} \sin^4(\theta - \theta m)f(\theta)$$

E_f = modulus of the cortical bone
V_f = volume fraction of the trabeculae
$f(\theta)$ = distribution function of the trabecular orientation
θ_m = mean angle of the trabecular orientation

The procedure mentioned above provided the direction of principal axis and the elastic moduli on the principal axis for each 5 mm × 5 mm region. Heterogeneity of the proximal tibia could be represented by combining the values for each of the 117 regions.

Finite Element Analysis

The two-dimensional FEM was performed to study the stress distribution in the proximal tibia under loaded conditions. Figures 3a–c show FEM models for a natural proximal tibia, which had 1337 elements among 722 nodal points, and two proximal tibias after TKA (central stem type had 1241 elements among 673 nodal points, and screw fixation type had 1306 elements among 707 nodal points). Figure 4a shows the mean trabecular orientation and Fig. 4b shows the distribution of Young's modulus in mean trabecular orientation angle of cancellous bone. The material properties of cancellous bone, cortical bone, ultra high molecular weight polyethylene (UHMWPE), and metal (Titanium)

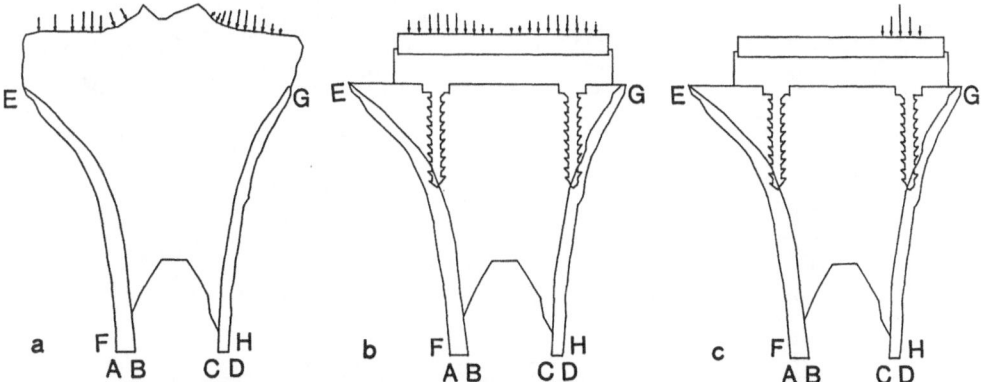

Fig. 5a–c. Loading and boundary conditions **a** of bi-condylar uniform loading for the natural proximal tibia **b** of bi-condylar loading for proximal tibia after TKA and **c** of uni-condylar loading for proximal tibia after TKA

Table 1. Material properties

Young's moduli	(MPa)
Cancellous bone	131~4,059
Cortical bone	12,000
UHMWPE	500
Titanium	132,500
Poisson's ratio	0.3

are shown in Table 1. Figure 5 shows the loading and boundary conditions for the models of the natural proximal tibia and the proximal tibia after TKA. The tibia is assumed to be supported in the vertical and horizontal directions at A–B and C–D, and in the horizontal direction at E–F and G–H.

Results

Stress Distribution in the Natural Proximal Tibia

Figures 6a,b show the distribution of von Mises' equivalent stress divided by bone density, and the volume fraction of the trabeculae. The stress concentration was found to be in the middle portion of the bilateral condyle by an analysis when cancellous bone was assumed to be homogeneous and isotropic (Fig. 6a). However, the stress concentration was not found when cancellous bone was assumed to be heterogeneous and anisotropic (Fig. 6b).

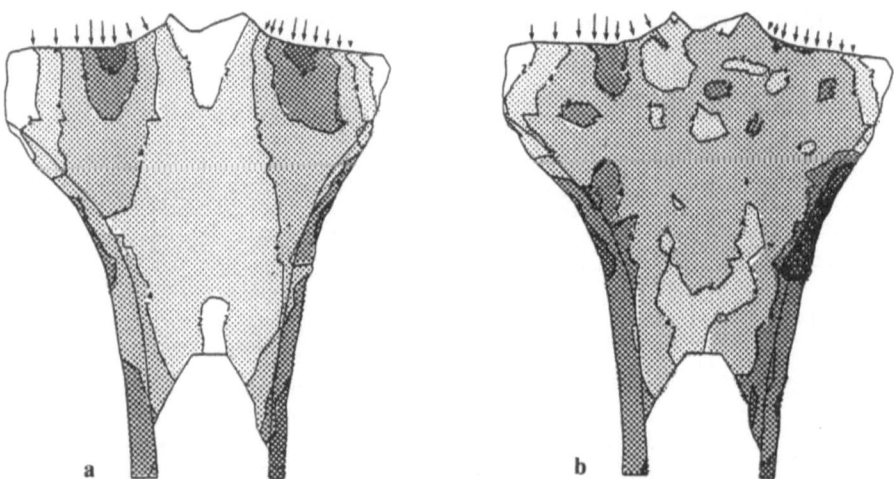

Fig. 6a,b. Von Mises' equivalent stress distribution per given load and per volume fraction of trabeculae for the natural proximal tibia **a** assuming heterogeneity and anisotropy and **b** assuming homogeneity and isotropy

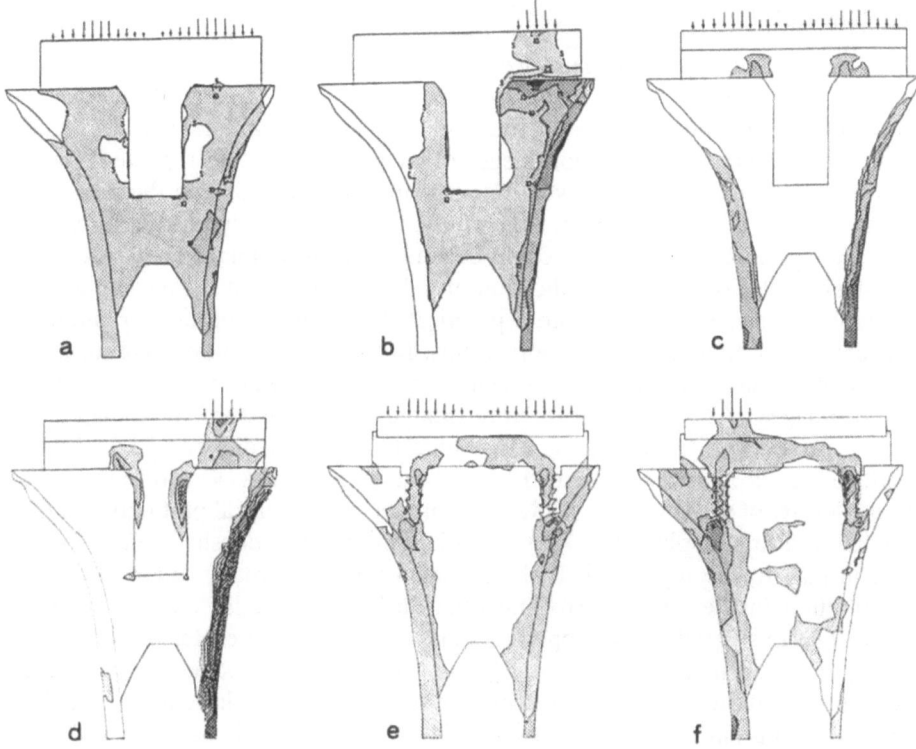

Fig. 7a–f. Von Mises' equivalent stress distribution per given load and per volume fraction of trabeculae for the proximal tibia after TKA with **a** bi-condylar loading for UHMWPE central stem type TKA **b** uni-condylar loading for UHMWPE central stem type TKA **c** bi-condylar loading for metal-backed central stem type TKA **d** uni-condylar loading for metal-backed central stem type TKA **e** bi-condylar loading for screw fixation type TKA **f** uni-condylar loading for screw fixation type TKA

Stress Distribution in the Proximal Tibia after TKA

Figures 7a–f show von Mises' equivalent stress distribution per given load and per volume fraction of trabeculae for the proximal tibia after TKA. Though stress concentration was not found in the UHMWPE central stem type TKA model when the bi-condylar load was uniform (Fig. 6a), it was found to be concentrated at the prosthesis/bone interface under a uni-condylar load (Fig. 7b). Though stress concentration was found in the lateral portion of the central stem under uni-condylar load, the load was found to be transmitted to the cortical bone in the metal-backed UHMWPE stem type TKA model (Fig. 7c,d). Conversely, stress concentration was found at the screws and in the surrounding cancellous bone, and the load was found to be transmitted to the cortical bone through the screws in the screw fixation type TKA model (Fig. 7e,f).

Discussions

The material properties of the cancellous bone for the FEM in many previous studies were based on the data from compression tests. Therefore, elastic moduli of cancellous bone could be obtained in only limited portions, and trabecular orientation was not assessed. Askew et al. [5] reported stress analysis including heterogeneity and anisotropy, but they divided the proximal tibia into four regions. In the present study, the elastic moduli of the proximal tibia divided into small regions could be estimated using image analysis of the mid-frontal section. The quantification of volume fraction of bone using image analysis has recently become more practical. However, trabecular orientation has not been obtained from those methods. In the present study, quantification of the trabecular orientation was achieved utilizing Fourier analysis of soft X-ray images. Analysis of soft X-ray images of the natural proximal tibia demonstrated that trabeculae were oriented from the medial and lateral tibial plateaus to the cortical bone of the tibial metaphysis (Fig. 4). Moreover, the bone density of these areas was higher than that of the central part between the medial and lateral plateaus. These localized high bone density areas may be due in part to the high load. If the human knee is optimally designed, stress distribution in the natural proximal tibia during bi-condylar uniform loading could be considered to be approximately even. The finite element analysis, taking heterogeneity and anisotropy of the cancellous bone into account, results in more uniform stress distribution under a bi-condylar uniform loading than the analysis which assumes the homogeneous and isotropic material properties. These results were found to be closer to the actual clinical outcome than any previous model. The importance of this increased uniformity has implications in defining which mechanical properties cancellous bone must have in order to exhibit a natural stress distribution. Therefore, our numerical analysis system using a two-dimensional model would be more beneficial for analyzing the stress distribution of the tibia after TKA, and would have applications in tibial component prosthetic design.

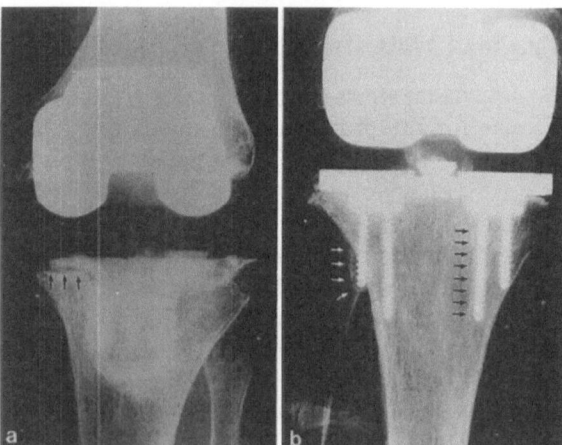

Fig. 8a,b. X-ray image of tibia after TKA with **a** central stem type TKA with radiolucent zone under medial tray and **b** screw fixation type TKA with linear sclerotic and osteolytic change around the screws

In the UHMWPE central stem type TKA, the stress concentration was focused at the middle of the prosthesis/bone interface under a uni-condylar load, and a radiolucent zone was clinically observed at the same area in X-ray images (Fig. 8a). In the screw fixation type TKA, the stress concentration was localized around the screws under both bi-condylar and uni-condylar loads, and changes such as linear sclerosis or osteolysis in the surrounding bone are clinically observed in X-ray images (Fig. 8b). When cancellous bone is not considered to be homogeneous and isotropic, but rather heterogeneous and anisotropic, FEM is more useful as a model.

However, there are many factors to be defined with finite element analysis. In the present analysis, material connections such as prosthesis/bone interfaces, UHMWPE/metal tray junctions, and metal tray/screw junctions were assumed to be rigidly bonded. Loading conditions were also assumed to be static. These conditions must be simplified for the purpose of creating a model, and they are only an approximation of actual conditions. Boundary and loading conditions will have to be established in future studies.

Conclusions

1. Two-dimensional finite element analysis considering heterogenous and anisotropic properties of cancellous bone was applied to analysis of stress distribution in the proximal tibia.
2. The stress distribution in the proximal tibia with the model of heterogeneity and anisotropy of cancellous bone showed to be close to the clinical outcome.
3. The finite element analysis employed in this study may be beneficial in the evaluation of tibial component prosthetic design.

References

1. Garg A, Walker PS (1986) The effect of the interface on the bone stress beneath tibial components. J Biomech 19:957–967
2. Lewis JL, Askew MJ, Jaycox DP (1982) A comparative evaluation of tibial component designs of total knee prostheses. J Bone Joint Surg [Am] 64:129–135
3. Murase K, Crownishield RD, Pedersen D, Pedersen DR (1982) An analysis of tibial component design in total knee arthroplasty. J Biomech 16:13–22
4. Vasu R, Carter DR, Shurman DJ, Beanpre GS (1986) Epiphysial-based designs for tibial plateau components-1. Stress analysis in the frontal plane. J Biomech 19: 647–662
5. Askew MJ, Lewis JL (1981) Analysis of model variables and fixation post length effects on stress around a prosthesis in the proximal tibia. J Biomech Eng 103: 239–245
6. Bartel DL, Burstein AH, Santavicca EA (1982) Performance of tibial component in total knee replacement. J Bone Joint Surg [Am] 64:1026–1033

7. Li K (1977) Study on trabecular architecture of the upper end of the tibia (in Japanese). J Jpn Orthop Ass 51:291–303
8. Kapandi IA (1974) The physiology of the joints. In: Lower limb, vol 2. Churchill Livingstone, Edinburgh and London, pp 73–135
9. Inoue N (1987) Mechanical properties of cancellous bone: the dependence of strength and elastic modulus on trabecular orientation (in Japanese). J Kyoto Pref Univ Med 97:847–865
10. Inoue N, Sakakida K, Yamashita F, Hirai T, Katayama (1986) The elastic modulus of cancellous bone-dependence on trabecular orientation. Biomech Basic Appl Res 207–212

Intraoperative Evaluation of Component Stability in Total Knee Arthroplasty

Kenjiro Yoshida, Kanji Asada, Akira Shimazu, and Hiroshi Sakane[1]

Summary. Information concerning component stability immediately following total knee arthroplasty (TKA) must be reliable. To this end, we developed a measurement system for tibial component, which is thought to be the most frequent loosening component, using three non-contact displacement transducers fixed on the tibial cortex. The measurement system consists of three parts: A device which is connected to the tibial component and has three steel plates, a sensor unit, and an external fixator which connects the sensor unit to the tibial shaft. A load is applied with a push-pull gauge from which a voltage output is supplied to the A–D converter. Sampled data are calculated as displacements by a personal computer that displays the tibial component micromotion data on a cathode ray tube (CRT).

Clinical application of this measurement began in March 1990. Between that date and September 1990, 11 joints in 9 cases were measured. System accuracy is within 10 µm. Even though the measurement takes 10–15 minutes, the quantitative information obtained from this system seems to be useful for the evaluation of the quality of component fixation.

Key words. Total knee arthroplasty — Quantitative measurement — Micromotion — Displacement transducer — Subsidence and lift-up

Introduction

The stability of a total knee replacement mainly depends upon the under-lying individual bone strength and surgical skill, assuming the same design of components. Comparisons of the stability between different component designs applied on cadaveric bone were performed to evaluate the best artificial joint

---

[1] Osaka City University Medical School, Department of Orthopaedics, Osaka, Japan

design [1], but comparison of all kinds of components now available is nearly impossible. If this test could determine the most stable component, other factors such as minimizing the amount of resected bone might be a controversial subject. There must be an optimum selection of the design of surface replacement of the knee for the individual bone strength and activities. If bone strength could be evaluated with computer tomography to determine the severity of the osteoporosis of the rheumatoid arthritis cases, pre-operative assessment for the selection of the design of the artificial bone might be possible. It is hard to evaluate the relation between the surgical skill and the final result such as loosening of the component or deformation of the bone. Therefore, information concerning the quality of component stability immediately after operation must be reliable. Direct measurement of micromotion of the component is a most reliable method for the evaluation of the strength of the prepared bone surface for replacement. The strength of the bone could be defined as being the amount of component displacement by the load applied on it. We developed a measurement system for tibial component, which is thought to be the most frequent loosening component, using three non-contact displacement transducers fixed on the tibial cortex.

Method

Micromotion of tibial component is caused by external forces applied in daily activities to the component. Application of forces in the same amount and direction as in daily activities is extremely difficult to perform during an operation. Even though a skilled surgeon can manually detect the stability of component fixation, direct measurement of micromotion of the component by forces applied manually may be very helpful information. The amount of force used by a surgeon to push the component is around 20–30 kg. Maximum displacement of the component may be observed where the force is applied, but direct measurement of component micromotion precisely at the point of application of force is difficult to perform. Therefore, the area where the force is applied and the place where the displacement is measured should be physically separate.

The measurement system consists of three parts: A device which is connected to the tibial component and has three steel plates (JIS SS41), a sensor unit, and an external fixator which connects the sensor unit to the tibial shaft. The fixation apparatus is designed for porous coated anatomical (PCA) modular total knee system (Howmedica Inc, Rutherford, New Jersey). The sensor (Keyence Inc. Osaka, Japan) detects the distance between the sensor and the steel plate without contact to minimize the influence of its fixation and has a theoretical resolution of 1 micron with a 12-bit analog digital converter. The sensors must be set a minimum of 50 mm apart to avoid interference between the sensors. For this purpose two opposite steel plates must be set outside the tibial component. A load is applied with a load cell (Kyowa, Tokyo, Japan) from which a voltage output is supplied to the A–D

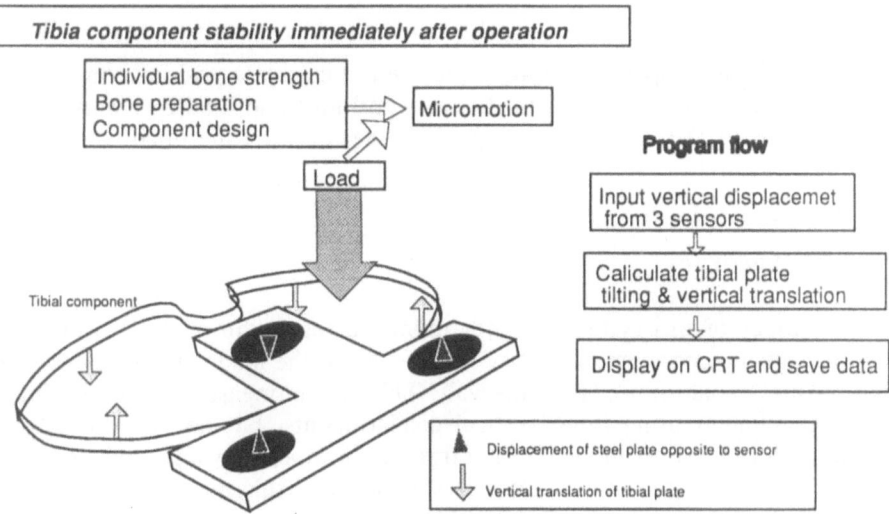

Fig. 1. Outline of the micromotion measurement system

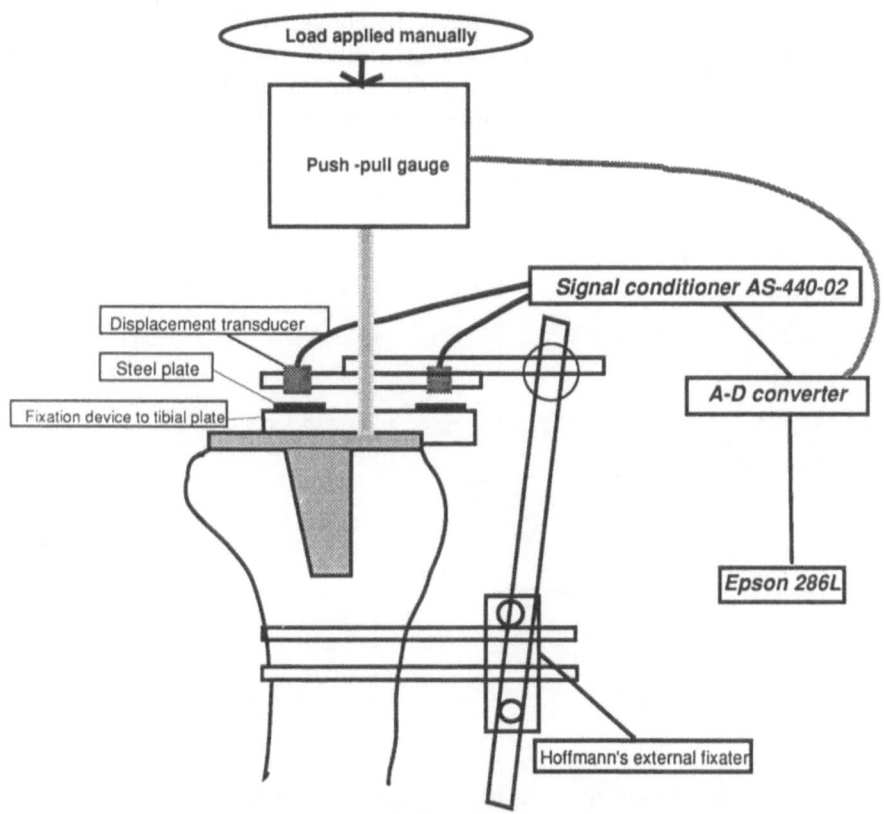

Fig. 2. Hardware requirements for the measurement of tibial component micromotion. Micromotion can be detected by the sensors without contact

converter. Four sample voltage data are used to calculate position coordinates (the number of millimeters from the zero position) and load(Kg) with a personal computer (NEC 9801 or compatible) that displays the tibial component micromotion and the load on the CRT (Figs. 1, 2).

System Accuracy

The accuracy of this system depends upon the resolution of the sensor itself and the tibial component deformity in response to the load. The sensor we use has a 1 μm resolution at 2 mm intervals. This system measures only the tilting tibial component using three vertical displacements, but not other displacements such as rotation and horizontal translation which may affect the accuracy of the vertical displacement. Comparison between this measurement system output and the directly measured value that was detected by a strain-gauge distant sensor on the tibial component mounted on a plastic bone model was performed. The linearity of the measurement system seemed to be good enough, ranging between ±10 μm that is mainly generated by the noises from the cables between the sensors and the A–D converter (Fig. 3).

This measurement system calculated the vertical translation from different parts of the component, consequently deformation of the tibial component itself is one of the major factors which affected the measurement accuracy.

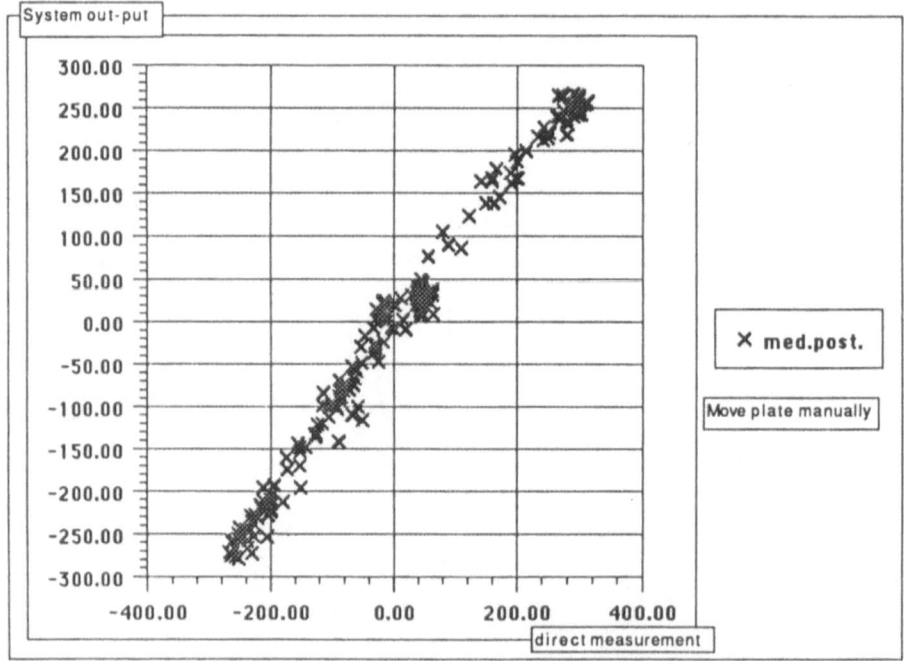

Fig. 3. System Linearity, system output, and direct measurement. Axes units in μm

Bending tests on the tibial component revealed that the deformation of the component is under $9\,\mu m/10\,kg$. The tibial component is considered to be floating on the tibial cancellous bone and the load applied on the component may not always act as the bending force of the component.

Procedures During Operation

The sensor unit and the fixation devices are sterilized with ethylene oxide gas, and the calibration procedure is carried out preoperatively. When the final setting of the tibial component has been completed, the sensor unit is fixed to the tibial bone, and the component fixation device with steel plates is set on the tibial component which is connected transiently to the sensor unit with a 1 mm plastic separator. This sensor unit is securely fixed with a Hoffman's external fixation device to the tibial cortex 40 mm away from the tibial baseplate to avoid jamming of the component stem. When all cables from the sensors are connected to the microprocessor, the transient separator is removed, and the tibial component is loaded with the push-pull gauge manually, up to 20–30 kg of force, as vertical to the component as possible. Anteromedial, posteromedial, anterolateral, and posterolateral quadrants of the tibial component were pushed separately, and the measured displacements of the component and the force are recorded automatically by the computer.

Materials

Clinical application of this measurement began in March 1990. Between that date and September 1990, 11 joints in 9 cases were measured to determine micromotions of the tibial component immediately after fixation. In two osteoarthritis cases and one rheumatoid arthritis case, the component was fixed without bone cement and the measurement was also carried out before

Table 1. Cases

OA		
	OA-C case1	72 female
	OA-C case2	68 female
	OA-C case3-R	72 female
	OA-C case3-L	72 female
	OA-UC case4	68 male
	OA-UC case5	68 male
RA		
	RA-C case1-R	68 female
	RA-C case1-L	68 female
	RA-C case2	47 female
	RA-C case3	65 female
	RA-UC case4	56 male

OA, osteoarthritis; RA, rheumatoid arthritis; C, cemented; UC, uncemented; R, right knee; L, left knee

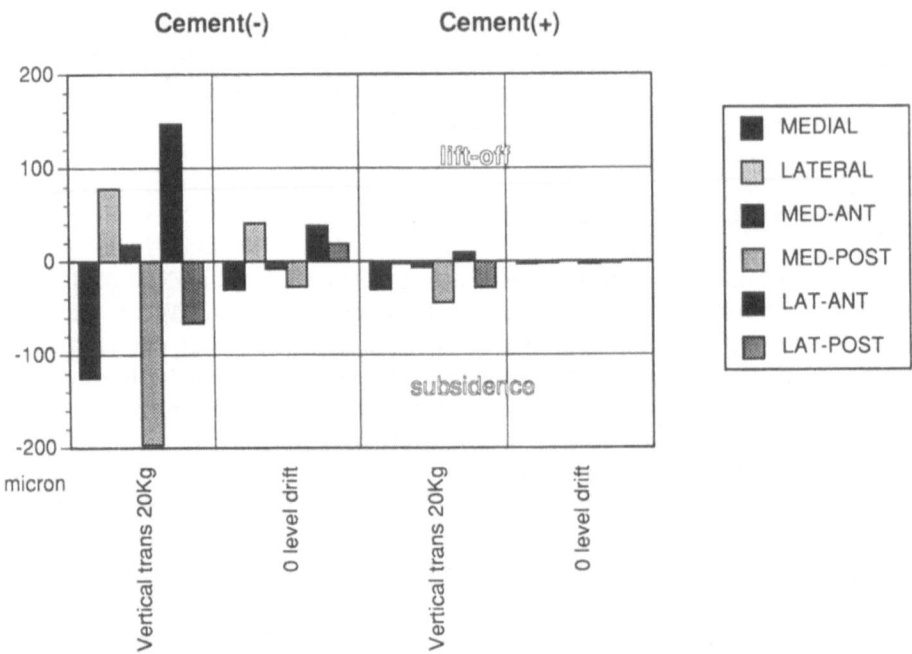

Fig. 4. Average tibial component motion on Osteoarthritis cases with medial posterior quadrant 20 kg load applied

cementing (Table 1). We analyzed tibial micromotions when the load was 20 kg perpendicular to the tibial component in the posteromedial quadrant.

Result

The measurement usually takes 5–10 minutes which includes setting the component attachment, and the sensor unit with the external fixation device and cables. Although one measurement takes only one minute, 5 minutes or more are required because we need more time for retightening the screws or trying other fixation methods to obtain a better fixation of the component. Visualization was prevented by the plastic shield around the head used as a protection against bacterial contamination in the operation field, but it was easy to see the magnified image on the CRT.

The average subsidence of osteoarthritis cases was about 200 μm on the posteromedial quadrant where a 20 kg load was applied. A 140-μm lift-up was observed on the anterolateral edge of the component. When the 20 kg load was removed, a 40-μm lift-up was still noticed, but when the component was cemented there was a small amount of subsidence and a negligible lift-up when loaded. However, there was almost no residual lift-up of the component when the load was removed (Fig. 4).

Tibia plate motion

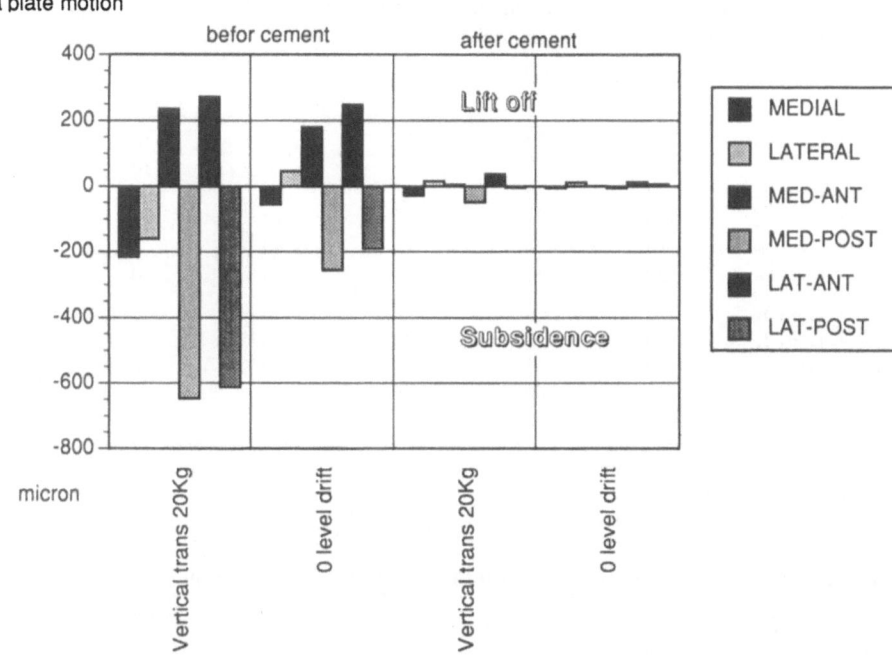

Fig. 5. Average tibial plate movement on RA tibia. *RA*, rheumatoid arthritis

The average displacement in rheumatoid arthritis (RA) cases before cementing was apparently larger than in osteoarthritis (OA) cases, exceeding 200 µm of subsidence which was usually a visible amount, and the residual displacement when load was removed exceeded 200 µm in subsidence and lift-off. When cemented, the displacement of the tibial component was securely fixed as in OA cases with under 20 µm of subsidence when loaded and minimal when the load was removed (Fig. 5).

To discuss the stability of the components individually, the final measurement of micromotion is important. The maximum subsidence of the last measurement did not exceed 200 µm, and the lift-up was under 150 µm (Table 2). All cemented cases except one RA case were well stabilized, and especially when the load was removed the residual deformity of the component was hard to detect with our system. However, in uncemented cases and in one RA cemented case, we observed a significant residual displacement of the component, 10–80 µm of subsidence and 0–60 µm of lift-off (Table 3).

Discussion

Various kind of efforts to reduce migration of components such as the addition of a metal backing and a long stem to the standard polyethylene component significantly reduced both migration and inducible displacement [2]. Different

Table 2. Residual displacement of tibial plate after 20 kg load off. OA cemented cases were very stable with minimum residual displacement, but RA cemented case3 was still unstable after cement. (Units = μm)

Residual Def.	Medial	Lateral	Med-ant	Med-post	Lat-ant	Lat-post
OA-C case1	0	−10	0	0	−10	0
OA-C case2	0	0	0	0	10	0
OA-C case3-R	0	−10	0	0	0	−10
OA-C case3-L	−10	0	0	−10	0	0
OA-UC case4	10	−50	40	−40	0	−80
OA-UC case5	−20	30	0	−20	30	10
RA-C case1-R	0	0	0	0	0	0
RA-C case1-L	0	10	10	0	10	10
RA-C case2	0	0	0	0	0	0
RA-C case3	−40	60	−10	−30	50	30
RA-UC case4	0	0	40	−40	40	−40

OA, osteoarthritis; RA, rheumatoid arthritis; C, cemented; UC, uncemented; R, right knee; L, left knee; −, subsidence; +, lift-off

Table 3. Stability of tibia. Medial-posterior 20 kg load. (Units μm)

Final stability	Medial	Lateral	Med-ant	Med-post	Lat-ant	Lat-post
OA-C case1	−30	−10	−20	−30	−10	0
OA-C case2	−60	30	0	−100	60	−30
OA-C case3-R	20	−50	30	−20	−10	−60
OA-C case3-L	−60	30	−20	−60	30	0
OA-UC case4	20	−50	120	−110	80	−150
OA-UC case5	−50	50	20	−90	80	−30
RA-C case1-R	0	−20	20	−20	10	−30
RA-C case1-L	−30	0	0	−40	10	−20
RA-C case2	−20	−30	10	−60	10	−60
RA-C case3	−120	130	−20	−130	140	30
RA-UC case4	−20	−40	100	−160	90	−170

OA, osteoarthritis; RA, rheumatoid arthritis; C, cemented; UC, uncemented; R, right knee; L, left knee; −, subsidence; +, lift off

fixation systems have different levels of fixation [3]. Measurable migration and inducible displacement may be the rule rather than the exception in total knee arthroplasty. Accordingly, absolute rigid fixation would not be necessary for successful function of a TKA [4].

Subsidence of the tibial component was noted in cases of obese patients taking systemic corticosteroids and in whom there was inadequate coverage of the cut tibial plateau surface [5]. However, it is difficult to be sure whether or not the fixation was absolutely rigid and how to know how to control the stability. The roentgen-photogrametric method to detect micromotion could not measure whether fixation immediately after operative procedure and its accuracy was under 100 μm. Hvid [6] measured the trabecular bone strength during operation. The axial strength of trabecular bone at the knee is critical for the maintenance of support and fixation of the prosthetic components after total surface knee arthroplasty. The resistance of trabecular bone to penetration was measured in 150 consecutive total knee arthroplasties. Tibial bone strength was lower in rheumatoid than in osteoarthritic knees. From this point of view, micromotion immediately after operative procedures must be a reliable factor for individual cases that could receive better treatment with our measurement system.

Friedman has said in 1981 [7] that the choice of prosthesis must be individualized according to the patient's goals and activity level. The perfect prosthesis is not yet available, but research in biomechanics, materials, and fixation may resolve some of the current limitations. This situation still holds true in total knee arthroplasty, because the mechanical characteristics and daily activities of the patients are extremely different from one patient to another. We could not find a perfectly designed artificial joint but came up with the best selection for the patients.

Our measurement system works very nicely during operation, but it takes 10–15 minutes in restricted operation time. The most time consuming procedure of this measurement system is the fixation of the sensor unit to the tibial cortex with an external fixation device, maintaining a 2-mm gap between sensors and placing opposite steel plates which are fixed on the tibial component. Hoffman's external fixation device which we are using now might help reduce the time required due to its easy setting. The measurement itself takes only 5 minutes with a real-time data acquisition and display system. Requiring 15 minutes may not disturb operative procedure completion within one pneumatic tourniquet time.

The program language we used was a BASIC compiler that makes the data sampling rate and handling of data too slow to draw the load deformation curve accurately. If a detailed load deformation curve can be obtained, the measurement could be performed within the limit of elastic deformation as a non-destructive measurement.

The displacement of the tibial component relative to the tibial bone is three-dimensional, but the displacement caused by the vertical load is directed either downward or upward. This measurement system has only three vertical displacement sensors that may be acceptable for the detection of major displacements of the component when loaded vertically. Theoretically, our measurement system can be modified to detect other directions, but the time needed for this might extend and disturb the operative procedure.

The load applied for the measurement is restricted by the method that is used in the operative procedure, and we chose a manual application of the load. Simulation of the load of a walking cycle is extremely difficult to obtain with manual loading. We applied the load in the medioanterior, medial-posterior, lateral-anterior, and lateral-posterior quadrants of the tibial component. Under this unstable stress, the tibial component is forced to lean down much more than with a balanced normal loading that may reveal more responsivenes of the quality of the underlying cancellous bone to the component displacement.

Even though the subsidence of the component was exceeded and reached 170 µm, and the lift-up was 100 µm, we selected cementless fixation for a young rheumatoid arthritis case. The quality of fixation of the tibial component with cement immediately after the procedure is apparently better than without cement, but a micromotion smaller than 150 µm may be acceptable for bone ingrowth. An almost absolutely rigid fixation was obtained with cement except on osteoporotic cancellous bone in a rheumatoid arthritis patient who used a cane to control the weight-bearing, and there was no roentgenographic change in the cementbone interface. The residual deformity is a result of the cancellous bone micro-fracture, accumulation of which might cause sinking or loosening of the component. One cemented RA case and cementless OA cases had apparently more residual deformity than the cemented cases. These cases should be treated with gentler rehabilitation programs and should be directed to use a cane or crutches to control the weight-bearing.

The load applied in the measurement is an eccentric force applied on specified quadrants of the component. Pitching and yawing of the tibial component may be well controlled by a long stem, similar to the keel of a yacht, but a round stem may not stabilize the component against rotational stress. Pegs, stems, and screws can control the quality of the fixation that we selected during the measuring procedure with our system.

Even though the measurement takes a little bit of time, the quantitative information obtained from this system seem to be useful for the evaluation of the quality of component fixation. Comparison of this data with the follow up X-ray files may determine optimum fixation of the component.

Reference

1. Shimagaki H, Bechtold JE, Sherman RE, Gustilo RB (1990) Stability of initial fixation of the tibial component in cementless total knee arthroplasty. J Orthop Res 8:64–71
2. Albrektsson BE, Ryd L, Carlson LV, Freeman MA, Herberts P, Regner L, Selvik G (1990) The effect of a stem on the tibial component of knee arthroplasty: A roentgen stereophotogrammetric study of uncemented tibial components in the Freeman-Samuelson knee arthroplasty. J Bone Joint Surg [Br] 72:252–8
3. Dempsey AJ, Finlay JB, Bourne RB, Rorabeck CH, Scott MA, Millman JC (1989) Stability and anchorage considerations for cementless tibial components. J Arthroplasty 4:223–30
4. Ryd L, Lindstrand A, Rosenquist R, Selvik G (1986) Tibial component fixation in knee arthroplasty. Clin Orthop 213:141–9
5. Laskin RS (1988) Tricon-M uncemented total knee arthroplasty: A review of 96 knees followed for longer than 2 years. J Arthroplasty 3:27–38
6. Hvid I (1988) Trabecular bone strength at the knee. Clin Orthop 227:220–21
7. Friedman RJ (1981) Current status of total knee arthroplasty. Postgrad Med 70: 207–14

Gait Analysis

Three-Dimensional Measurement of Pelvic Rotation

KENJI KAWATE, YUTAKA OHNEDA, and SUSUMU TAMAI[1]

Summary. The movement of the pelvis during ambulation shows a 3-dimensional rotational pattern. In order to clarify the gait characteristics of patients with coxarthrosis, the rotation of the pelvis during ambulation was measured on the frontal, sagittal, and horizontal planes using three gas-rate-sensors. Then we drew waveforms using a microcomputer with an A–D converter. Thirty-six normal subjects and 46 patients with coxarthrosis were studied.

In the normal subjects, the pelvis dropped on the swing phase side, and Trendelenburg phenomenon was observed in all cases on the frontal plane. The waveforms were regular, symmetrical and highly reproducible on repeated gait cycles. In contrast, the waveforms of the coxarthrosis patients were irregular, asymmetrical and poorly reproducible. The pelvis was in anteversion at heel strike in each normal subject on the sagittal plane, while in most of the patients, the pelvis inclined backward at heel strike of the affected side. The waveform was flat around heel strike of the affected side on the horizontal plane in the majority of patients. Using the Fourier Transform technique, objective proof was provided that the waveforms of the patients with coxarthrosis are irregular, asymmetrical and poorly reproducible as compared with those of normal subjects.

The mean rotational angle of the patients was smaller than that of the normal subjects on the frontal plane, and larger on the sagittal plane. We were able to distinguish the patient from the normal subject by observation of the ratio of the rotational angle on the sagittal plane to that of the frontal plane.

Key words. Pelvic rotation — Three-dimensional measurement — Gas-rate-sensor — Patient with coxarthrosis — Waveform — Fourier transform — Ratio of the rotational angle

---

[1] Department of Orthopedic Surgery, Nara Medical University, 840 Shijo-cho, Kashihara, Nara, 634 Japan

Introduction

The movement of the Pelvis during ambulation shows a three-dimensional pattern which is altered by a variety of disease processes affecting the hip joint. The movement has been studied by various techniques. Levens et al. [1] took photographs of pins drilled into the cortices of bones using three cameras. Inman et al. [2] measured pelvic movements by photographing interrupted lights. Stokes et al. [3] used a two-camera SELSPOT system interfaced with an HP1000 minicomputer. Thurston et al. [4] observed the movements using a television/computer system. Leeuwen et al. [5] used precision potentiometers attached to a firmly strapped external pelvis girdle. In the present study, the rotation of the pelvis during ambulation was measured on the frontal, sagittal, and horizontal planes using three gas-rate-sensors, and the results of normal subjects were compared with the results obtained from patients with coxarthrosis.

Materials and Methods

Thirty-six normal subjects (7 male and 29 female) and 46 patients with coxarthrosis (4 male and 42 female) were studied. Eleven and 35 of the 46 patients showed bilateral and unilateral involvement, respectively. The patients were further divided into 2 groups consisting of 17 early stage and 29 advanced/end stage cases.

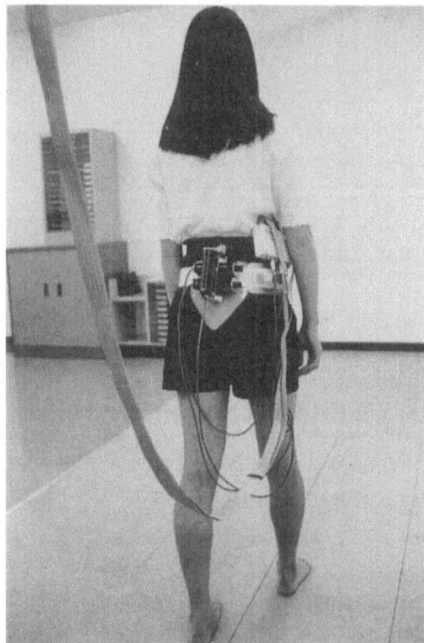

Fig. 1. A subject is fitted with the pelvic girdle on which three gas-rate-sensors are installed

Three gas-rate-sensors (Rotational Angle Measurement system G-2210, Anima Inc., 3-65-1 Shimoishihara, Chofu-shi, Tokyo JAPAN) were firmly attached to a pelvis girdle, one in each dimension (Fig. 1). A gas-rate-sensor is an angular accelerometer using hydrogen gas.

The normal subjects and patients then performed a total of 6 trials walking along an 8-meter path into which was incorporated a 250 cm-long force platform. The 3-planar rotation of the pelvis and perpendicular component of the floor reaction force were simultaneously measured. The analog data obtained was converted to digital form, and each individual waveform was drawn on the display using a microcomputer. A measuring program was made using C language with a sampling time of 3 msec.

Results

Frontal Plane Pelvic Rotation

In the normal subjects, the pelvis dropped on the swing phase side during normal ambulation, with the so-called Trendelenburg phenomenon observed in all cases. The waveforms were regular, symmetrical, and highly reproducible on repeated gait cycles, and showed 2-point peaks almost synchronously with the perpendicular component of the floor reaction force (Fig. 2). In contrast, the waveforms of the advanced/end stage patients were irregular, asymmetrical, and poorly reproducible on repeated gait cycles as compared with the normal subjects. The waveforms in these patients did not show 2-point peaks but rather numerous small peaks (Fig. 3).

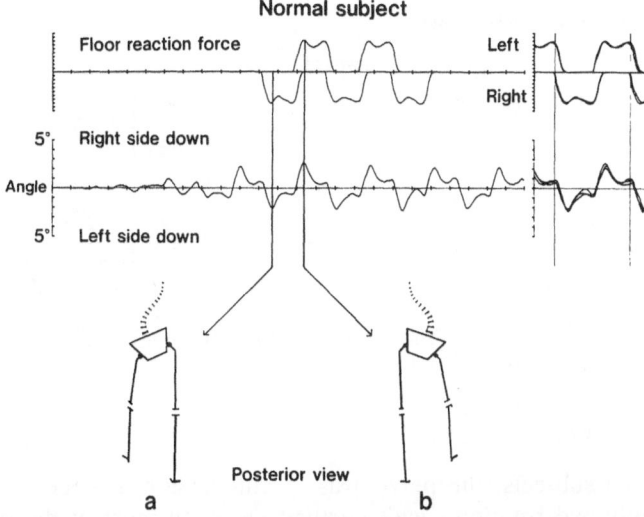

Fig. 2a,b. Typical waveform of a normal subject on the frontal plane. The pelvis drops on the swing phase side. **a** Right stance phase. **b** Left stance phase

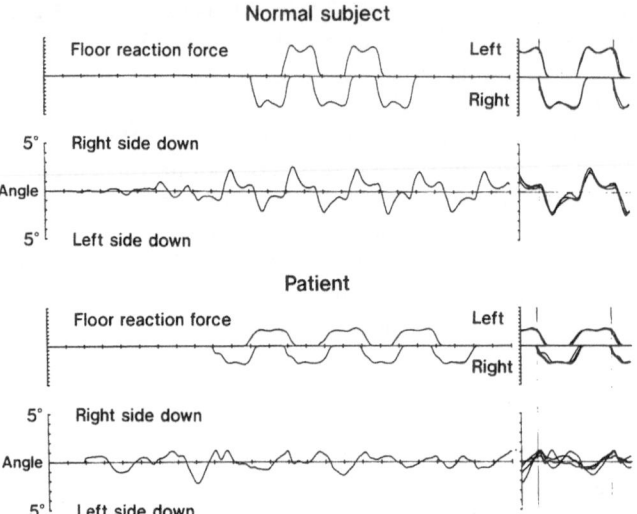

Fig. 3. Typical waveforms of the normal subject (*above*) and patient with right coxarthrosis (*below*) on the frontal plane

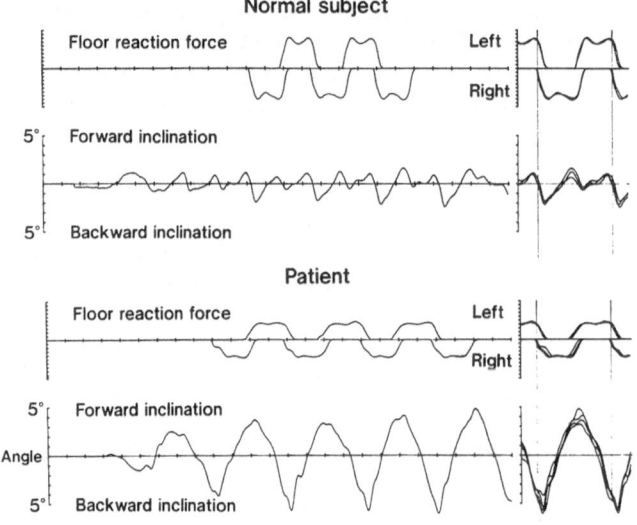

Fig. 4. Typical waveforms of the normal subject (*above*) and patient with right coxarthrosis (*below*) on the sagittal plane

Sagittal Plane Pelvic Rotation

In the normal subjects, the pelvis was in anteversion at heel strike and had approximately two rotation cycles per gait cycle. In most of the patients, the pelvis inclined backward at heel strike of the affected side, with some patients showing one rotation cycle per gait cycle (Fig. 4).

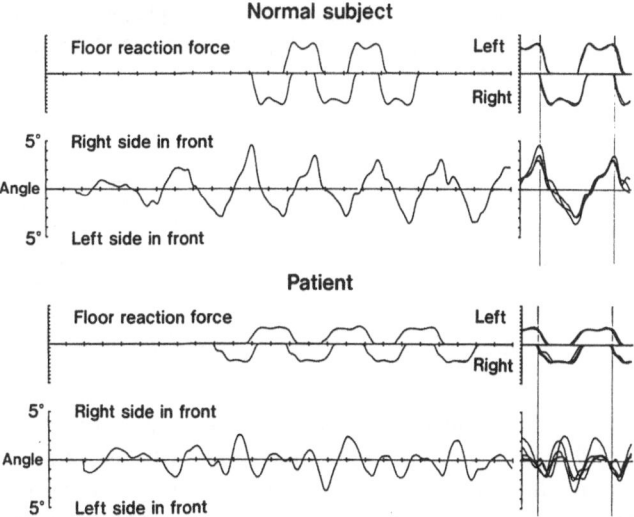

Fig. 5. Typical waveforms of the normal subject (*above*) and patient with right coxarthrosis (*below*) on the horizontal plane

Horizontal Plane Pelvic Rotation

In the normal subjects, the waveforms were symmetrical and peaked at heel strike, whereas in the majority of patients, the waveforms were asymmetrical and peaked late at mid-stance on the affected side (Fig. 5).

Rotational Angle

On the frontal plane, the mean angle in the normal subjects was $6.9 \pm 1.7°$, in contrast to $4.9 \pm 1.5°$ and $3.8 \pm 1.2°$ in the early stage and advanced/end stage patients, respectively, with significant differences found between the normal subjects and both patient groups. On the sagittal plane, the mean angle of the normal subjects was $3.6 \pm 1.2°$, which was significantly smaller than that of the advanced/end stage patients ($7.0 \pm 2.5°$). On the horizontal plane, the mean angle of the normal subjects was $8.3 \pm 3.5°$, which did not differ significantly from that of the early stage ($7.5 \pm 3.1°$) and advanced/end stage patients ($7.2 \pm 2.9°$) (Table 1).

Discussion

Using high-accuracy gas-rate-sensors, it was found that the pelvis of all of the normal subjects dropped on the swing phase side on the frontal plane, which is known as Trendelenburg phenomenon. It is thought that this phenomenon is invariably found in all normal subjects.

Table 1. Mean angles of the pelvic rotation on the three planes. The mean rotational angle of the patients was smaller than that of the normal subjects on the frontal plane, and larger on the sagittal plane

	Normal subjects (n = 36)	Patients	
		Early stage (n = 17)	Advanced and end stage (n = 29)
Frontal plane	6.9 ± 1.7°	4.9 ± 1.5°*	3.8 ± 1.2°*
Sagittal plane	3.6 ± 1.2°	4.5 ± 2.0°	7.0 ± 2.5°*
Horizontal plane	8.3 ± 3.5°	7.5 ± 3.1°	7.2 ± 2.9°

(mean ± SD, * $P < 0.001$)

On the frontal plane, the waveforms of the normal subjects were regular, symmetrical and highly reproducible, and demonstrated 2-point peaks. In the patient group, however, the waveforms were irregular, asymmetrical and poorly reproducible on repeated gait cycles. These differences are thought to be attributable to an attempt by the patient to minimize pain by adjusting the position of his/her legs at impact. On the sagittal plane in the patient group, it was found that the pelvis inclined backward at heel strike of the affected side. This is thought to be due to pelvic compensation for the altered movement of the disabled hip joint. On the horizontal plane in the patient group, the wave peaked late at mid-stance on the affected side. This was also thought to represent an attempt to mitigate sudden impact.

Then we calculated symmetry, reproduction, and abnormality indices using the Fourier Transform technique about the waveforms of the frontal plane. Abnormality is defined as the difference between an individual's wave pattern

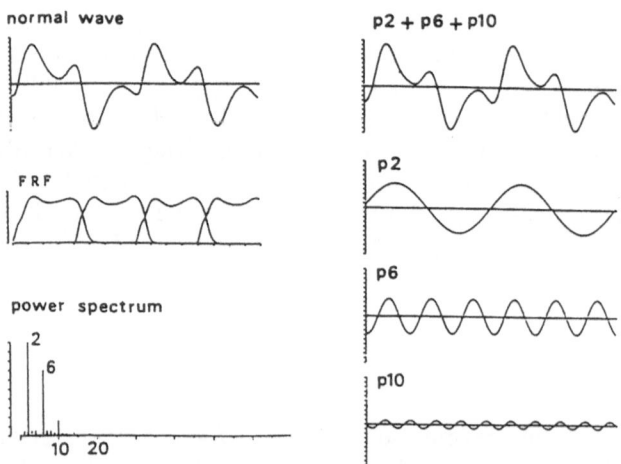

Fig. 6. Basic waveform which consists of only 2, 6, and 10 frequency cycles. *FRF*, floor reaction force

and the basic waveform which consists of only 2, 6, and 10 frequency cycles (Fig. 6). The three indices are calculated as follows:

1. Reproduction index $= \sum_{n=1}^{10} P_{2n-1}/P_2$

2. Symmetry index $= \sum_{n=1}^{5} P_{4n}/P_2$

3. Abnormality index $= \left(\sum_{n=1}^{20} P_n - S \right) \Big/ S$

$$P_n = \mathrm{sqr}(A^2_n + B^2_n) \quad S = P_2 + P_6 + P_{10}$$

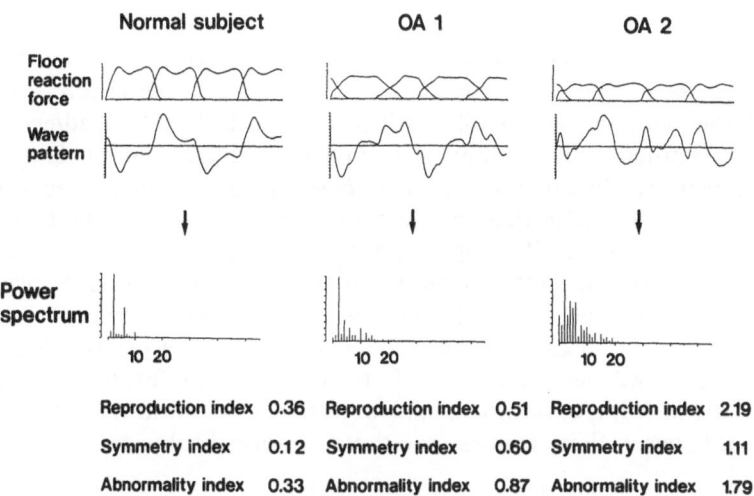

Normal subject	OA 1	OA 2
Reproduction index 0.36	Reproduction index 0.51	Reproduction index 2.19
Symmetry index 0.12	Symmetry index 0.60	Symmetry index 1.11
Abnormality index 0.33	Abnormality index 0.87	Abnormality index 1.79

Fig. 7. Fourier Transform of a normal subject and two patients (OA 1 and OA 2). *OA,* osteoarthritis

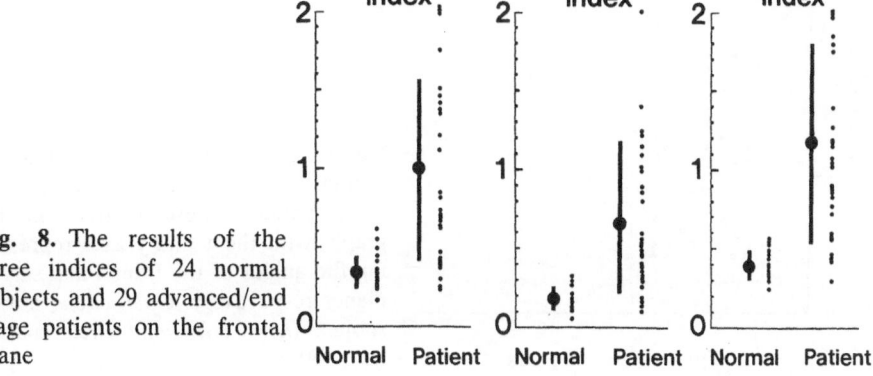

Fig. 8. The results of the three indices of 24 normal subjects and 29 advanced/end stage patients on the frontal plane

Lower number values denote the ability of the subject to perform tasks effectively. When these indices were calculated for each case, significant differences ($P < 0.001$) in the waveforms of the normal subjects and advanced/end stage patients were seen. In the normal subjects, the waveform consisted mainly of 2, 6, and 10 frequency cycles, and the waveform was easily reproducible. In the patient group, however, the waveform consisted of various frequency cycles and showed very poor indices (Fig. 7). By plotting the results of these 3 indices for the 24 normal subjects and 29 advanced/end stage patients on a graph, objective proof was provided that the waveforms of patients with coxarthrosis are irregular, asymmetrical and poorly reproducible as compared with those of normal subjects (Fig. 8).

Next, the mean rotational angle of the patients was smaller than that of the normal subjects on the frontal plane, and larger on the sagittal plane. To better appreciate these findings, the results of the 36 normal subjects and 29 advanced/end stage patients were plotted on a graph, with the X and Y axes representing the angles of the frontal and sagittal planes, respectively. As seen in Fig. 9, the two groups are easily distinguished, with the advanced/end stage patients and normal subjects found in the upper-left side and lower-right of the graph respectively. In other words, it is possible to differentiate to a large extent the two groups according to whether the rotational angle is larger on the frontal or sagittal plane. We can therefore distinguish the patient from the normal subject using the ratio of the rotational angle on the sagittal plane to that on the frontal plane. A ratio larger than one is considered to be abnormal.

In conclusion, by measuring the rotational angle of the pelvis using three gas-rate-sensors, we were able to clarify the characteristic features of ambulation in patients with coxarthrosis and objectively demonstrate considerable differences between these features and those of normal subjects.

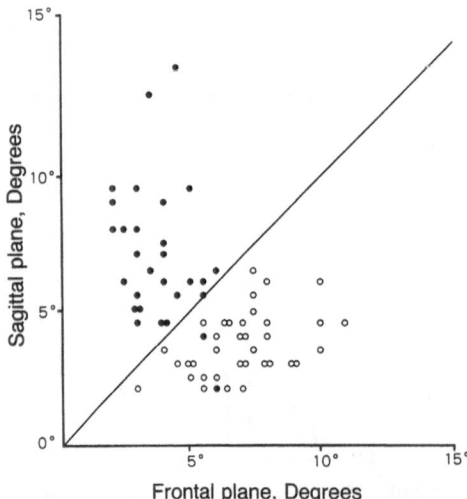

Fig. 9. The rotational angles of the 36 normal subjects and 29 advanced/end stage patients were plotted on this graph, with the x and y axes representing the angles of the frontal and sagittal planes respectively. *Open circle*, normal subject ($n = 36$); *solid circle*, patient ($n = 29$)

References

1. Levens AS, Berkeley CE, Inman VT, Blosser JA (1948) Transverse rotation of the segments of the lower extremity in locomotion. J Bone Joint Surg 30A:859–872
2. Inman VT, Ralston HJ, Todd F (1981) Human walking. Williams and Wilkins, Baltimore, pp 22–61
3. Stokes VP, Andersson C, Forssberg H (1989) Rotational and translational movement features of the pelvis and thorax during adult human locomotion. J Biomech 22:43–50
4. Thurston AJ, Harris JD (1983) Normal kinematics of the lumbar spine and pelvis. Spine 8:199–205
5. Leeuwen JLV, Vink P, Spoor CW, Deegenaars WC, Fraterman H, Verbout AJ (1988) A technique for measuring pelvic rotations during walking on a treadmill. IEEE Trans Biomed Eng 35:485–488

Miscellaneous

Biomechanics of Natural and Artificial Joints

MASANORI OKA, KEN IKEUCHI, SADAMI TSUTSUMI,[1] TAKAO YAMAMURO, TAKASHI NOGUCHI[2], and TAKASHI NAKAMURA[1]

Summary. We investigated and compared biomechanics of natural and artificial joints. The most important functions of the joints rely on excellent lubrication and uniform distribution of impact loads on to the underlying bones. As to the lubrication, we measured dynamic changes of the joint space optically and verified the existence of fluid film lubrication in natural joints. From the aspect of fluid film lubrication, ultra high molecular weight polyethylene (UHMWPE) is not as good a material as polyvinyl alcohol hydrogel (PVA-H) and articular cartilage. The momentary stress transmitted through the specimens revealed that subchondral cancellous bones played the most important role and that UHMWPE had a higher peak stress and a shorter duration of sustained stress than articular cartilage and PVA-H, suggesting a worse damping effect. From the results of finite element method (FEM) analysis, intramedullary stem fixation might not avoid stress shielding. In artificial joints in the future, it is desirable to preserve as much subchondral cancellous bone as possible and to replace the involved joint surface with materials whose mechanical properties are similar to those of articular cartilage. We also reported biocompatibility, wear resistant properties of PVA-H, and attachment of this material to underlying bones. Although some problems still remain to be solved, PVA-H seems to be a very interesting and promising material which meets the requirements of artificial articular cartilage.

Key words. Joints — Lubrication — Load-bearing function — Artificial articular cartilage — Biocompatibility

---

[1] Research Center for Biomedical Engineering, Dept. of Artificial Organs, Division of Artificial Locomotive Systems, Kyoto University, Kyoto, Japan
[2] Department of Orthopaedic Surgery, Kyoto University

Introduction

The clinical success of artificial joints has given remarkable benefit to patients with joint disorders. It cannot be denied, however, that the complications of total replacement surgery, particularly "loosening", have frequently occurred and that revision cases have been increasing in number every year [1, 2]. The etiologies of aseptic loosening have been discussed, but no conclusion has yet been reached.

In this article, we attempt to investigate and compare the biomechanics of natural and artificial joints with special attention to the causes of aseptic loosening. The most important functions of joints rely on excellent lubrication and uniform distribution of impact loads onto the underlying bones.

If we jump down from a heighth of 3 m, for example, our hip joints are subjected to a great impact load approximately seven times our body weight. Usually, in healthy joints, these impact loads are well attenuated, and articular cartilage and subchondral bones play a very important role in reducing the load. Furthermore, subchondral cancellous bones remodel themselves constantly, according to Wolff's law, and transmit the load most effectively to the distal bone structures. In current artificial joints, however, all these important ,structures, including healthy portions are resected, causing loss of the shock-absorbing effect as well as of bone stock. Remarkable foreign-body reactions due to polyethylene (PE) wear debris suggest that it may not be sensible to rely only on this material in future.

For the purpose of preserving as much cancellous bones as possible, we have attempted to develop an artificial articular cartilage, PVA-H, a rubber-like gel. We have improved the mechanical properties of this gel through a new synthetic process.

The first objective of this article is to investigate lubrication of UHMWPE, in comparison with that of articular cartilage and of artificial articular cartilage. Secondly, we describe load-bearing functions of various joint materials including articular cartilage with particular reference to shock-absorption. Thirdly, we briefly describe the prospects for the artificial articular cartilage which we are developing.

Materials and Methods

Materials used were UHMWPE for medical use and osteochondral specimens taken from the femoral condyle of pigs. Cylindrical test pieces 5 mm in diameter with a 5 mm thickness were used in every test. In the case of osteochondral specimens, the thickness of articular cartilage was approximately 1 mm.

Another material used was PVA-H which we chose as an artificial articular cartilage. PVA is atactic and has a viscosity-average degree of polymerization of 7000 (its molecular weight is 308,000). The polymer was dissolved in a solvent of water and dimethyl sulfoxide (DMSO). The resulting PVA solution

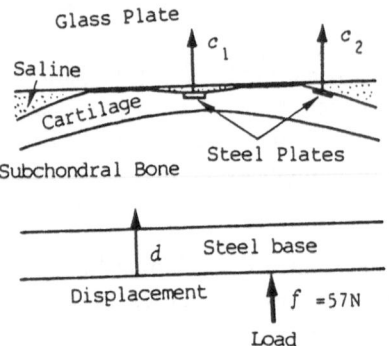

Fig. 1. Cross-sectional view of the array of a glass plate and an osteochondral specimen. The specimen was pressed to the glass plate. Two thin steel plates (C1, C2) were attached to the cartilage surface as target for the laser beam (From [13] with permission)

was cooled to below room temperature to obtain a transparent gel which was formed as a consequence of crystallization of the PVA molecules. The PVA-H gel was dried using a vacuum dehydration machine for 24 hours at room temperature, followed by heat treatment in a silicone oil bath for one hour at 140°C. The resultant PVA-H gel was placed in water until the maximum level of water absorption was reached. The saturated water content was around 20 wt%.

Lubrication

To investigate the lubrication properties of the materials, we examined the time-dependent changes of thickness of the fluid films formed between the

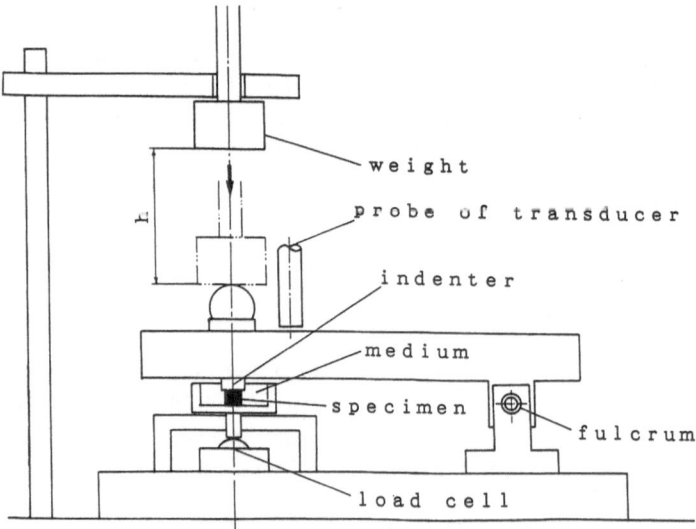

Fig. 2. Diagram of the apparatus. A weight (27 N) was dropped from 10 mm height on the specimen in fluid. The stress transmitted is measured by a load cell set up under the specimen (From [13] with permission)

materials and a glass plate. As shown in Fig. 1, test specimen in saline was pressed to a glass plate under a load of 57 N. Steel plates (10 μm thick) were attached to the material surface and the vertical displacement of a laser beam spot focused on the steel plate was detected by an obliquely fixed charge-coupled-device (CCD) through the glass plate. In the case of osteochondral specimen, two steel plates, one located centrally (C1), and another located peripherally (C2) were attached to the cartilage surface. Central fluid pressure through a very thin hole in the glass was simultaneously measured by a diaphragm-type transducer.

Damping Effects

We measured the momentary stresses transmitted through the materials by a load cell set up under the specimen (Fig. 2). The impact load was applied by dropping a weight (27 N) from a height of 10 mm onto the specimen soaked in saline. Time-dependent deformation of the specimen was also measured by a transducer probe.

Load Bearing Function of the Hip Joint

To investigate the differences of stress distribution between natural and artificial hip joints, we carried out FEM analysis. The computer program used was COSMOS/M which made possible the analysis of the transmission of stresses through the gap between two soft materials.

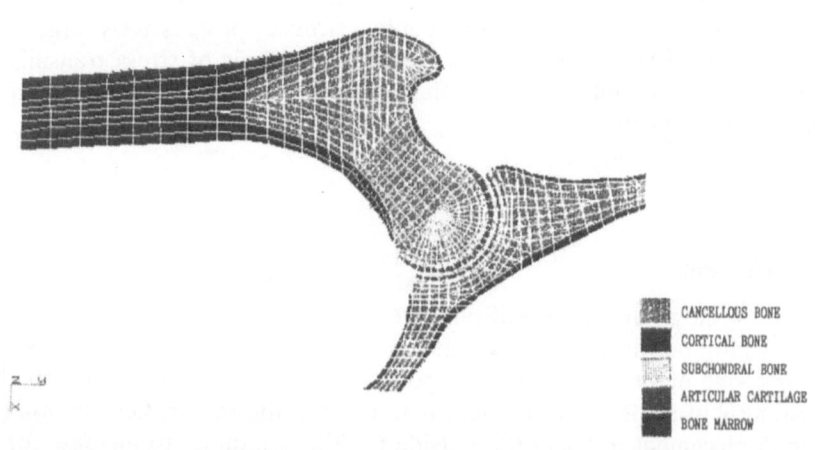

CANCELLOUS BONE
CORTICAL BONE
SUBCHONDRAL BONE
ARTICULAR CARTILAGE
BONE MARROW

Fig. 3. Hip joint model for FEM analysis: Five zones are divided based on anatomical and mechanical characteristics

Table 1. Material characteristics

Material	Young's modulus (MPa)	Poisson's ratio
Cortical bone	7,000	0.25
Articular cartilage	10	0.40
Subchondral bone plate	3,000	0.25
Cancellous bone	1,000	0.30
PVA-Hydrogel	30	0.40
Titanium fiber	30,000	0.30
UHMWPE	500	0.40
Bone cement	2,000	0.30
Stainless steel	100,000	0.30
Alumina	350,000	0.19
Spring element	0.001	0.30

The human hip joint consists of a total of 910 nodes and 880 elements. The material structure of the hip joint was manually zoned into 5 region (Fig. 3). An artificial hip joint consists of 740 nodes and 709 elements. Its femoral stem is composed of stainless steel, the femoral head of Alumina, and the socket of UHMWPE. Each component is fixed to the bones with acrylic cement. The material characteristics are shown in Table 1. A load of 1764 N is applied to the pelvic bone with an angle of 16 degree from the vertical axis.

In both models, a 1 mm gap between the surfaces is assumed to exist before loading. In order to avoid the instability of pelvic bones due to the establishment of the joint gap, the distal femur end is constrained completely. For the same reason, the superior and medial pelvic bone ends are stabilized by weak spring elements (0.001 MPa) in Young's modulus. As the load increases the initial gap decreases and a greater portion of the apposing articular surfaces progressively not separate, come into contact. Concurrent with the increase of intra-articular contact stresses, progressively larger stresses are developed within both elements. In the analysis of stress transmission in the femur, we adopted the side-plate theory in order to perform quasi three-dimensional analysis.

Results

Lubrication

Figure 4 shows the surface displacement corresponding to the reduction of the gap width between the glass and the cartilage. The upward displacement (narrowing of gap) of the central portion (C1) increases until 20 seconds after loading, but stops when the fluid is trapped in the recess. On the other hand, the displacement of C2 at the outside position continues to increase for a long time. Figure 5 gives the central fluid pressure (Ps). After a rapid increase in the initial stage, Ps continues to increase during the creep deformation period.

Fig. 4. Upward surface displacement of C_1 and C_2 which corresponds to the reduction of gap width between the glass plate and the cartilage surface under loading of 57 N. Absissa indicates the time after loading in seconds (From [13] with permission)

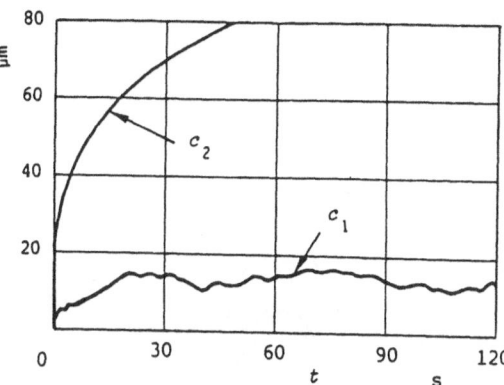

Fig. 5. History of central fluid pressure under loading of 57 N. Ordinate indicates fluid pressure in KPa (From [13] with permission)

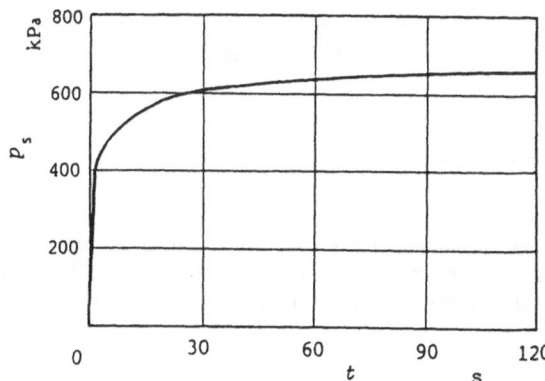

Fig. 6. Time-dependent changes of the gap width between the glass plate and materials under loading of 57 N. Ordinate indicates decrease of the gap. In the case of PVA-H20, polyvinyl alcohol hydrogel, the gap width increases until 50 seconds after loading and decreases thereafter (From [13] with permission)

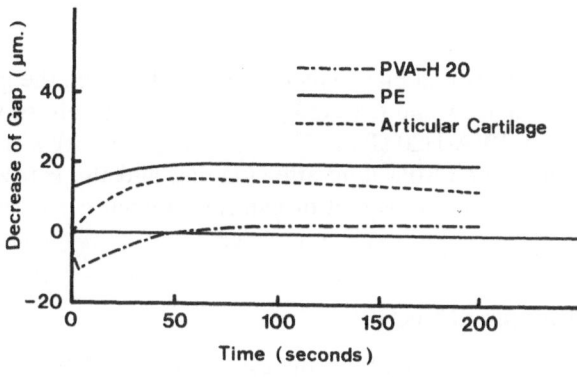

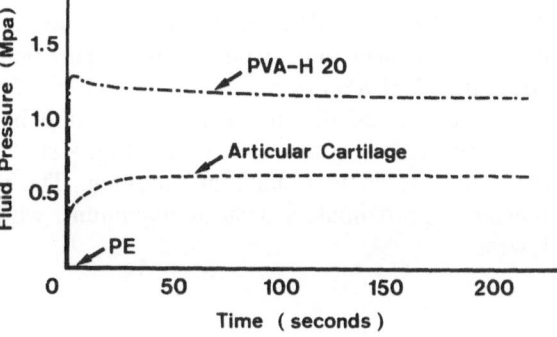

Fig. 7. Changes of the fluid pressures in gap. In the case of UHMWPE, the fluid pressure becomes zero soon after loading (From [13] with permission)

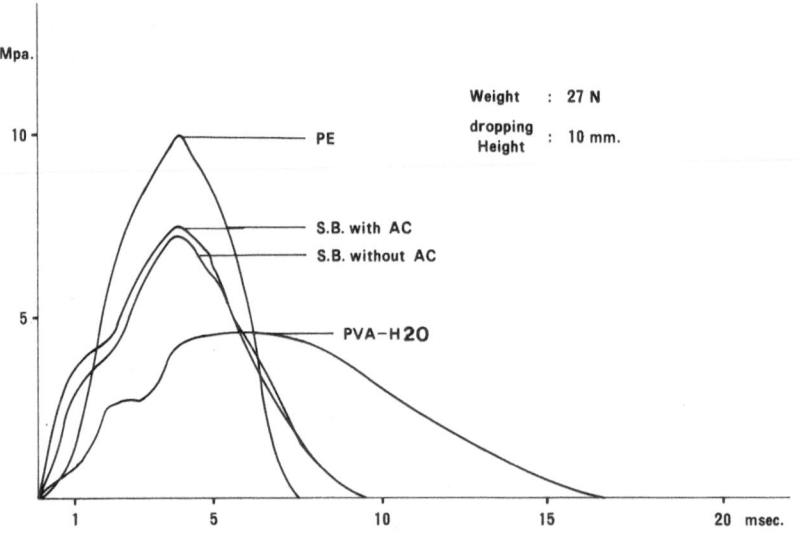

Fig. 8. Impact stresses transmited through materials. Abscissa indicates the time after loading while ordinate indicates the stress. *PVA-H20*, polyvinyl alcohol hydrogel; *PE*, polyethylene; *S.B.*, subchondral bone; *AC*, articular cartilage (From [13] with permission)

As shown in Figs. 6 nad 7, the gap between the glass and UHMWPE become very narrow and the fluid pressure became zero soon after loading. Conversely, when PVA-H20 (PVA-H 20% water content) was used, the fluid pressure was kept over 1 MPa long after loading and the gap width was maintained to the same thickness as that of natural articular cartilage.

Damping Effects

As shown in Fig. 8, in the case of UHMWPE, the peak stress reached 10 MPa and the duration of sustained stress was as short as 7 msec. On the other hand, the peak stress for PVA-H-20 was much lower (4.5 MPa) and the stress was maintained longer than 16 msec. This means that PVA-H has a better damping effect than UHMWPE.

Such a marked difference was not found between the subchondral bones with and without articular cartilage. Fig. 9 shows time-dependent deformation of each material in momentary loading. The strain of PVA-H was highest, reaching approximately 35% at maximum, while that of UHMWPE was the lowest.

Fig. 9. Deformation of materials under momentary stresses. Ordinate indicates strain. *PVA-H20*, polyvinyl alcohol hydrogel; *PE*, polyethylene; *S.B.*, subchondral bone; *AC*, articular cartilage

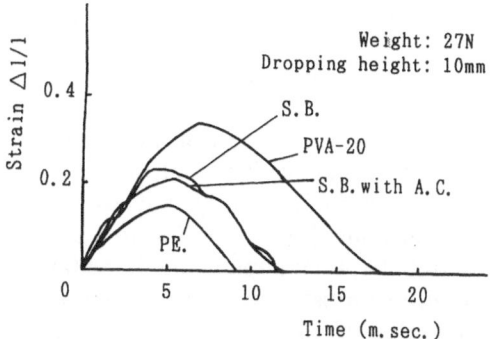

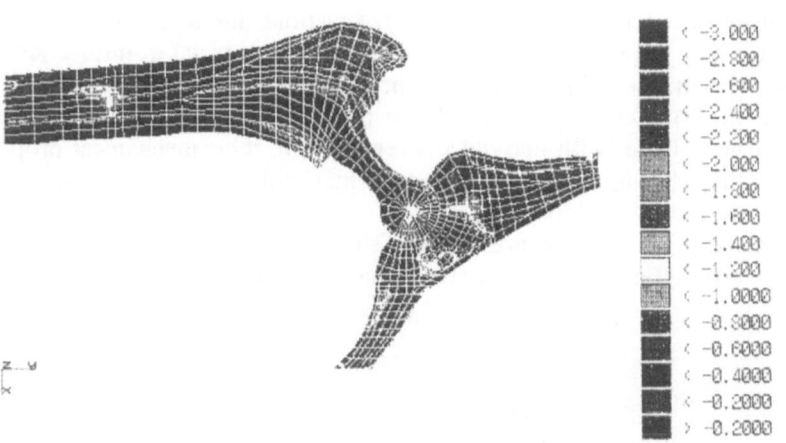

Fig. 10. Stress patterns for artificial hip joint. Intensities of stresses from .3 Mpa to 0.2 MPa are shown by 16 colors, as shown on right column (in this figure black and white)

Load-Bearing Functions of Hip Joints

FEM analyses of the human hip joint reveals that stresses of approximately 1.8 MPa are evenly distributed in subchondral cancellous bones of the femoral head and that on the medial femoral cortex, stresses of about 5 MPa are transmitted. On the other hand, the stresses in the medial femoral cortex of artificial hip joints show a very low value particularly in the medial neck region where only 0.6 MPa stress was found, thus indicating stress-shielding. On the medial stem-cement interface, compressive stress of 25 MPa were found and on the lateral interface, tensile stresses of 22 MPa were found (Fig. 10).

Discussion

The most serious and the greatest complication in current artificial joints is aseptic loosening. The etiologies of aseptic loosening have been discussed in various ways. Since most of the loosening occur at the interface between bone and acrylic cement, it was attributed to the usage of acrylic bone cement [1]. The incidence of loosening, however, has not been eliminated using cementless artificial joints [3]. According to Lennox [4], granulomatous reactions at the interface between implant and bone take place not only in cemented prostheses but also with tissue in-growth press-fit prostheses.

Histological examinations of the scar tissues surrounding the reoperated joints revealed marked macrophages and foreign-body giant cell reactions around the numerous wear particles of PE.

Recently, Howie et al. [5] have confirmed experimentally that the resorption of bone and the formation of a granulomatous membrane at the interface between acrylic cement and bone are mainly induced by particles of high-density PE. Remarkable foreign-body reactions due to PE wear debris suggest that it may not be sensible to rely only on this material in future. As shown in Fig. 11, current artificial hip joints involve five interfaces; that is, bone to bone cement, bone cement to PE, PE to metal, metal to bone cement, and bone cement to bone, with marked discrepancies in their mechanical properties. In fact, these five interfaces, each bonded mechanically by different materials, will become a cause of loosening.

As the history of hip arthroplasty shows, it was not until the appearance of endoprosthetic replacement that massive intact cancellous bones of the

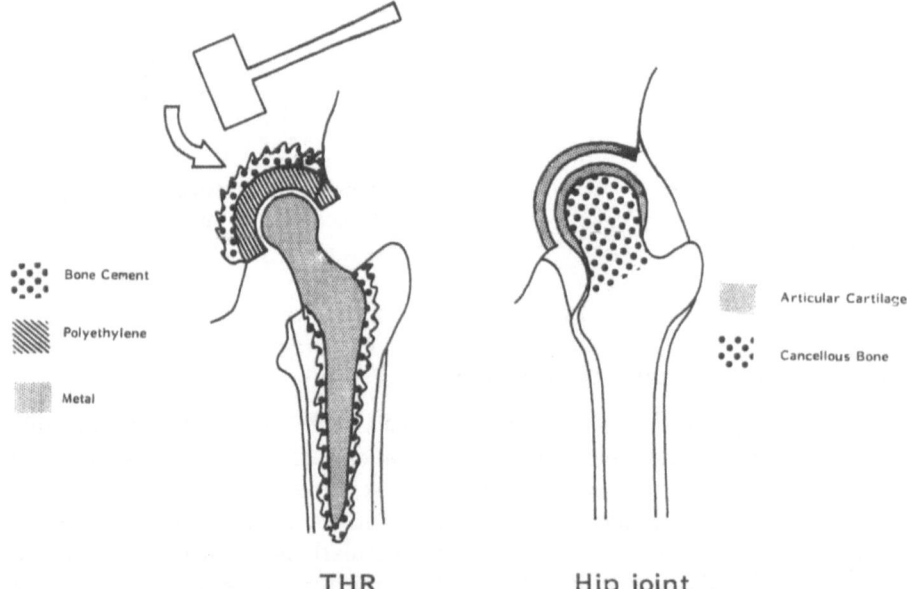

Bone Cement

Polyethylene

Metal

Articular Cartilage

Cancellous Bone

THR   Hip joint

Fig. 11. Shock absorbing effects of natural hip joints and total hip replacements (THR)

proximal femur were resected. As illustrated in Fig. 11, the cancellous bone of the proximal femur plays a very important role in the shock-absorbancy of joints [6]. Nevertheless, in current artificial joints, too much healthy bone is resected.

Considering the problems with current artificial joints, we have attempted to develop an artificial articular cartilage with which to replace the involved joint surface, and to spare healthy subchondral cancellous bones. In this paper, we investigate and compare biomechanics of natural and artificial joints.

Lubrication

A possible mechanism of the surprisingly low friction in animal joints may be explained in terms of fluid film lubrication in which minimal contact occurs between both joint surfaces and almost all the load is borne by the fluid film. To clarify this lubrication mechanism of joints, it is necessary to know both the width of the joint space and the undulation of the joint surface. However, in the available literature very few papers exist which describe the space within animal joints. Takei [7] examined serial sections of human knee joints which were instantaneously frozen after the application of a load (120 kg), and reported that, under normal conditions (without osteoarthritic changes), the space between the articular surfaces was never less than 100 µm. This finding is interesting, although the examination was not carried out under a physiologically dynamic condition.

We optically measured the dynamic changes in gap width between the glass and the articular cartilage under a loading condition. From these experimental results, one can say that the pressure of fluid film trapped in the dented portion of articular cartilage remains continuously high, and that the joint space maintains a minimum heighth after loading. These results strongly support the existence of fluid film lubrication in animal joints [8, 9].

On the other hand, in the case of UHMWPE, the gap width was very narrow and the fluid pressure in the gap became zero soon after loading. These results suggest that UHMWPE is not as good a material as articular cartilage at least in terms of fluid film lubrication.

Regarding the lubrication mode of artificial joints, Charnley [10] himself was of the opinion that boundary lubrication is mainly operative. This is the reason why Charnley chose a small femoral head of 22 mm diameter for his study.

Medley et al. [11] calculated the film thickness of normal human ankle joint to be about 0.7 µm, and suggested that transient elastohydrodynamic lubrication played a role in synovial joint lubrication.

As McCutchen proposed [12], "weeping lubrication", whereby the fluid film exuded from cartilage seems to play a very important role, not only in lubrication but also in weight bearing of joints. Inspection of our experimental results and a numerical analysis of squeeze film effect revealed that soft materials like articular cartilage can bear a large part of load via the fluid film generated between the surfaces by creep deformation. The numerical analysis of squeeze film effect disclosed that while the articular cartilage supports only 10% of the load by direct contact, such load bearing does not occur in PE which supports approximately 90% of the load directly [8, 9]. Based on these biomechanical considerations, we are trying to develop a new biomaterial for a better artificial articular cartilage [13].

In the case of PVA-H20, the fluid pressure was maintained over 1 MPa long after loading and the gap width was maintained, suggesting a better lubrication than UHMWPE.

Damping Effects

As regards the damping effect of the artificial joints, surprisingly little work has been published despite the fact that almost all surgeons use UHMWPE as the soft component against the hard counterpart. We have studied the damping effects of various prosthetic materials, using the test apparatus shown in Fig. 8. We measured the load transmitted through the materials by a load cell set up under the specimen. These results indicate that UHMWPE is inferior in damping to both the natural joint and PVA-H20. Furthermore, one can state that it is desirable to preserve as much subchondral cancellous bone as possible and to use a biomaterial with properties similar to articular cartilage for artificial joints in the future [13].

Usually, the biomaterials are used in combination with some other materials such as acrylic bone cement, because the materials must be attached firmly to

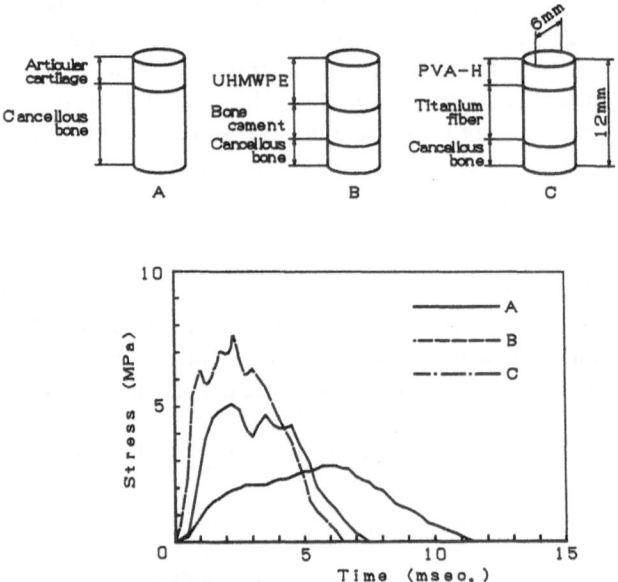

Fig. 12. Shock absorbing effects of various composite materials in specimens *A*, *B*, and *C* which are composed of various materials including cancellous bones were subjected to tests for shock absorbancy. Specimen *B* showed the highest peak stress. *PVA-H20*, polyvinyl alcohol hydrogel; *UHMWPE*, ultra high molecular weight polyethylene

the underlying bone. Therefore, we investigated damping of various combinations of materials using the same apparatus.

As shown in Fig. 12, specimen A with UHMWPE indicated the highest peak stress, while specimen C with PVA-H20, the lowest. Specimen C was composed of Titanium fiber mesh with PVA infiltrated into the pores.

Load-Bearing Mechanism of Joints

FEM analyses of natural hip joints revealed neither stress concentration nor stress shielding in any region of the bones. The stresses are evenly distributed to the underlying bones. In artificial hip joints, on the other hand, marked stress-concentration was found in the interface on the medial side of the stem and bone cement, while marked stress shielding was found on medial femoral cortex. As Huiskes [14] reported, intramedullary fixation might not eliminate all stress shielding on cortical bones.

Using the same FEM program, we analyzed the stress transmission through a hip joint with both surfaces replaced by a 6 mm thick artificial osteochondral composite material made of Titanium fiber mesh and PVA-H. As shown in Fig. 13, stress distribution in the proximal femur is very similar to that of natural hip joints, indicating neither stress concentration nor stress shielding. These results might support the idea to revive a surface arthroplasty by using new biomaterial whose mechanical properties are similar to that of natural articular cartilage.

As biocompatibility is very important, we will briefly describe the biocompatibility of PVA-H.

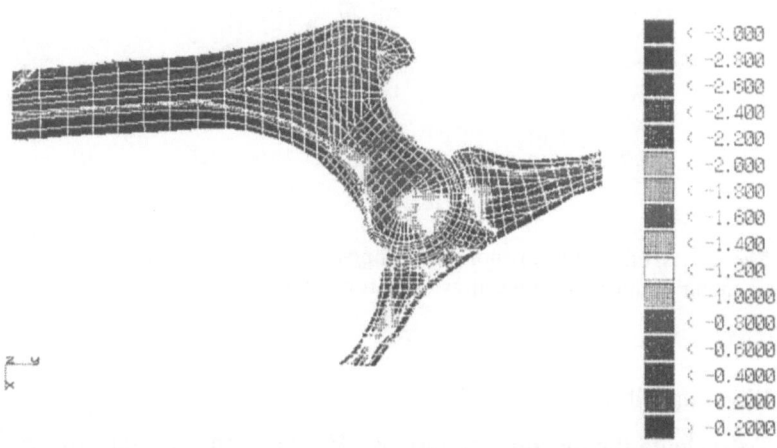

Fig. 13. Stress patterns for surface arthroplasty. Stresses are evenly distributed to the underlying bones, similar to those of natural hip joint

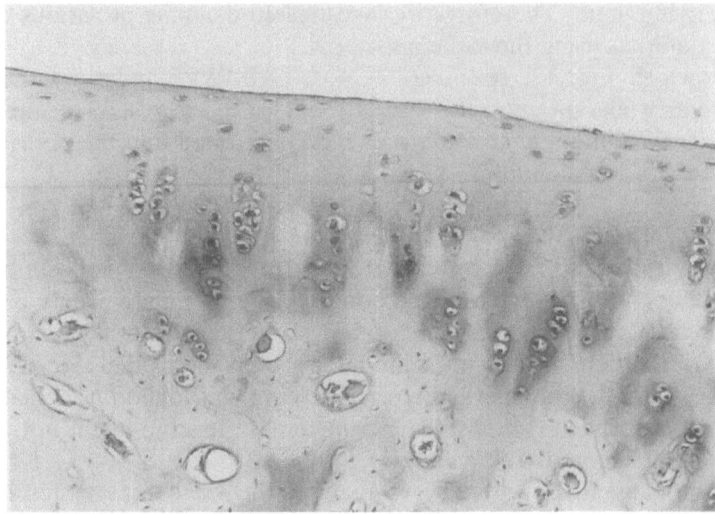

Fig. 14. Photomicrograph showing articular cartilage of the rabbit femur implanted with PVA-H20 at 52 weeks after operation (Safranin-0, original magnification ×100)

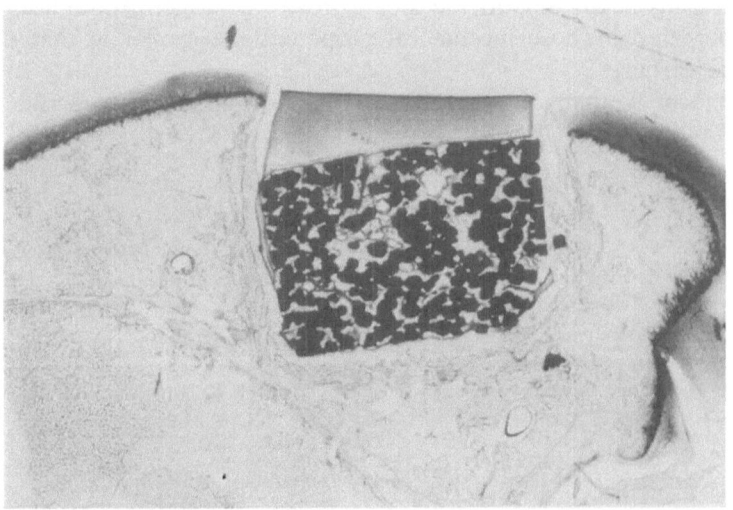

Fig. 15. Artificial Osteo-chondral-composite material implanted in the patellar groove of rabbit femur at 4 weeks after operation. (Giemsa stain ×10)

Biocompatibility of PVA-H

To evaluate the biocompatibility of PVA-H as an artificial cartilage, a series of in vivo tests within the intra-articular and the intramuscular environment were conducted. The details of this study will be published elsewhere [15]. Briefly, the PVA-H was shaped into cylinders 4 mm in diameter and 4 mm in height for implantation. The disinfected implant plug was inserted into an osteochondral defect in the patellar groove of the femur of 24 white rabbits.

Tissue reactions of cartilage, bone, synovium, and muscle with the implanted PVA-H were studied histologically at 2, 4, 8, 26, and 56 weeks postoperatively. In the experimental group in which PVA-H was implanted, inflammatory reactions were very slight for all these tissues. Histological pictures are shown in Fig. 14 for the articular cartilage of the femur 52 weeks after the operation. It was observed that the structure of cartilage is maintained without any pathological changes and that stainability with Safranin-0 is almost normal for the PVA-H implant group.

It is widely accepted that although a block of UHMWPE causes minimal inflammation its wear particles do cause irritation. We introduced small particles (50–300 µm) of PVA-H and UHMWPE into the knee joints of the same rats. Histological findings of the synovial membranes and the articular cartilage two months after operation showed that the UHMWPE particles caused intensive inflammation in comparison with the PVA-H particles.

Attachment of the Material to the Bone

In the above experiments, the PVA-H specimens were implanted directly into the osteochondral defects by press-fitting, with no means of attaching them on to the underlying bone. To use prosthetic materials for artificial joints, we must fix the materials firmly to the bone. This is another important problem to be addressed. To fix the implant firmly, it is necessary to combine the PVA-H with another material which, chemically or mechanically, binds with the bone

Fig. 16. New bone ingrowth from the *right* side into the pores of alumina at 8 weeks after operation. (Giemsa stain ×100)

tissue. In an attempt to strengthen the bonding, we have devised an implant composed of PVA-H and a porous alumina ceramic.

To this end, we first infiltrate a PVA solution into the pores of porous alumina and bind these two materials by gelling the PVA. The shear force between the PVA-H with a water-content of 20% and alumina with a porosity of 40% and with a pore size of 250 μm was about 2.2 MPa. We implanted this artificial osteochondral composite material into the patellar groove of a rabbit femur (Fig. 15). The histological findings of the undecalcified tissue sections revealed considerable new bone ingrowth into the pores of alumina, just two weeks after the operation. These new bone trabeculas in the alumina pores increased in volume and matured time-dependently. Figure 16 shows the bone having grown into the pores 26 weeks after operation.

We are currently trying to use an artificial osteochondral composite material made of a titanium fiber mesh (porosity 50%) whose pores were filled with PVA, because the mechanical properties of this material are better than those of the brittle, porous alumina, particularly in respect to toughness.

Wear Resistance

No matter how excellent the lubrication of materials is, the material cannot be used as a prosthetic material unless the wear resistance is sufficient.

We investigated wear resistance of PVA-H in comparison with that of UHMWPE using "pin on disc" device. A pin from the polymer specimen with a diameter of 4 mm was pressed against an alumina disc running at a speed of 50 mm per second under a stress of 3 MPa in distilled water. The wear volume was estimated from the decrease in height of the polymer pin measured by a displacement transducer. The wear properties are expressed by the following wear factor:

$$\text{Wear factor} = \frac{\text{Wear volume (mm}^3)}{\text{Load (N) sliding distance (m)}}$$

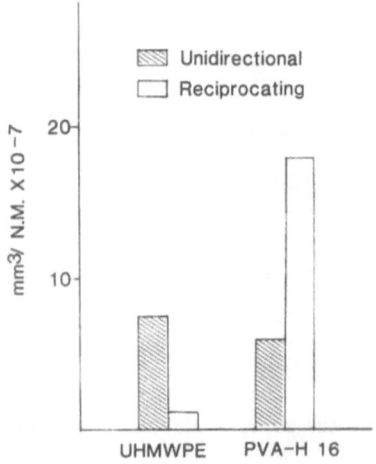

Fig. 17. Wear factor of polyvinyl alcohol hydrogel (*PVA-H16*) and UHMWPE. Ordinate indicates wear factor mm^3/(Nm × 10^{-7})

As shown in Fig. 17, wear factor of PVA-H (16% water content) obtained by unidirectional pin on disc test, measure up with that of UHMWPE.

Wear factor by reciprocating test, however, ranged approximately 18 times that of UHMWPE. In the reciprocating test, the sliding speed becomes zero at the end of each stroke. As the coefficient of friction of PVA-H becomes higher, such as at lower sliding speeds, these bad results reflect the fact that PVA-H is still weak at low sliding speeds. To improve wear resistance, we are now improving the mechanical properties by using higher molecular weight PVA and adopting a new annealing process.

The surface replacement arthroplasty of hips which was developed from the abovementioned concepts and which had often been performed [16, 17], seems to have declined at present because this operation has presented a great many complications such as early loosening. Unfortunately, the advantage of preserving the femoral head was not utilized effectively. The reason for this is rather simple: The materials used hitherto were not of a new type, but PE, metals, and bone cements which are completely different mechanically from the surrounding biological tissues. Consequently, stress shielding occurs for the femoral head, causing bone resorption and loosening.

For reviving a new surface arthroplasty, it is essential to develop a new biomaterial whose mechanical properties are similar to that of natural articular cartilage. For this purpose, we have attempted to develop an artificial articular cartilage and have chosen the PVA-H as such. Various mechanical properties of PVA-H such as lubrication and shock-absorbancy were indicated to be superior to those of UHMWPE. The excellent biocompatibility of PVA-H found in the animal joint experiments encourages us to use this as a new joint material [13, 15].

Although we still have various problems to solve, the PVA-hydrogel seems to us to be a very interesting and promising material for surface arthroplasty.

Conclusion

By optically measuring dynamic changes of the joint space, the existence of fluid film lubrication in natural joint was verified. From the aspect of fluid film lubrication, UHMWPE is not as good a material as PVA-H and articular cartilage.

In the shock absorbing effects of joints, subchondral cancellous bones play the most important role. The damping of UHMWPE is worse than that of osteochondral specimen and PVA-H.

Based on the results of FEM analysis, intramedullary stem fixation might not eliminate stress shielding, while the stresses are evenly distributed to the bones in natural hip joint.

In the future, it is desirable to preserve as much subchondral cancellous bone as possible, and to use materials whose mechanical properties are similar to that of natural articular cartilage.

We reported biocompatibility, wear resistant properties of PVA-H, and attachment of this material to underlying bone. From these results, PVA-H seems to be a very interesting and promising material for surface arthroplasty.

References

1. Ranawat CS, Atkinson RE, Salvati EA, Wilson PD (1984) Conventional total hip arthroplasty for degenerative joint disease in patients between the ages of forty and sixty years. J Bone Joint Surg [Am] 66:745–52
2. Stauffer RN (1982) Ten year follow-up study of one hundred consecutive Muller curved stem total hip replacement arthroplasties. J Bone Joint Surg [Am] 64: 983–90
3. Engh CA, Massin P (1989) Cementless total hip arthroplasty using the anatomic medullary lock system: Results showing a survivorship analysis. Clin Orthop 249:141–58
4. Lennox DW, Schofield BH, McDonald DF, Riley LH (1987) A histological comparison of aseptic loosening of cemented, press-fit, and biologic ingrowth prostheses. Clin Orthop 225:171–91
5. Howie DW, Vernon-Roberts B, Oakeshott R, Manthey B (1988) A rat model of resorption of bone at the cement bone interface in the presence of polythylene wear particles. J Bone Joint Surg [Am] 70:257–64
6. Radin E, Paul O (1971) Response of joints to impact loading: In vitro wear. Arthritis Rheum 14:356–62
7. Takei T (1979) Deformation of the articular cartilage and joint space of the human knee joint under static load. J Jpn Orthop Ass 53:577–93
8. Ikeuchi K, Oka M (1988) A study of cartilage deformation with squeeze film effect. In: Tsuruta T (ed) Transactions of 3rd World Biomaterials Congress. Business Center for Academic Societies, Tokyo, p 270
9. Ikeuchi K, Oka M (1989) An experimental study of deformation and squeeze film effect in a synovial joint. Trans Jpn Soc Mech Eng 55,516C:2123–30
10. Charnley J (1979) Low Friction Arthroplasty of the Hip. Springer Verlag, Berlin, Heidelberg, New York, pp 3–15
11. Medley JB, Dowson D, Wright V (1984) Transient elastohydrodynamic lubrication models for the human ankle joint. Engineer in Medicine, pp 137–151
12. McCutchen W (1969) More on weeping lubrication. In: Wright V (ed) Lubrications and wear in joints. Sector, London, pp 117–121
13. Oka M, Noguchi T, Kumar P, Ikeuchi K, Yamamuro T, Hyon SH, Ikada Y (1990) Development of an artificial articular cartilage. Clin Materials 6:361–381
14. Huiskes R (1990) The various stress patterns of press-fit, in grown, and cemented femoral stems. Clin Orthop 261:27–38
15. Noguchi T, Yamamuro T, Oka M, Kumar P, Kotoura Y, Hyon SH, Ikada Y (1991) Poly(vinyl Alcohol) Hydrogel as an artificial articular cartilage: Evaluation of biocompatibility. J Applied Biomaterials 2:101–107
16. Freeman MAR, Brown GC, (1978) ICHL cemented double cup hip replacement. Arch Orthop Trauma Surg 92:191–8
17. Amstutz HC, Kabo M, Hermens K, Herzmens K, O'Carroll PF, Kilgns D (1987) Porous surface replacement of the hip with chamfer cylinder design. Clin Orthop 222:140–60

Validation Study of the Rigid Body Spring Model Using the Metacarpophalangeal Joint

TAKATOSHI IDE, YASUHIRO YAMAMOTO, and SHIGERU TATSUGI[1]

Summary. Estimation of the force distribution on the metacarpophalangeal joint surface was studied by using sixteen cadaveric metacarpophalangeal joints, and the results were compared with the stress analysis calculated by computer simulation using Kawai's Rigid Body Spring Model (RBSM). According to the results of analysis based on the two-dimensional RBSM (2D-RBSM) computer simulation, displacement was 6.7 times greater than that of the experiment while the peak pressure was only 0.6 times greater. The results of the two-dimensional RBSM modified to take into consideration the depth of the joint, showed that it was comparable with the experimental studies in terms of peak pressure and displacements. The elastic modulus of the cartilage was estimated to be 25.8 MPa and the spring constant of the cartilage was estimated to be 22.6 N/mm/unit area after displacement matching by the modified two-dimensonal RBSM.

Key words. Rigid body spring model (RBSM) — Computer simulation — Spring constant — Stress analysis

Introduction

It was clearly shown that Kawai's Rigid body Spring Model (RBSM) [1] is extremely useful in performing numerical qualitative analysis of problems related to direct contact of body joints [2, 3, 4, 5]. For this analysis, load distribution tests were carried out using cadaveric metacarpophalangeal joints, and comparative studies were done with the stress distribution calculated using computer simulation.

---

[1] Department of Orthopedic Surgery, Yamanashi Medical College, 1110 Shimokato, Tamaho-cho, Yamanashi, 409-38 Japan

Materials and Methods

Experimental Study Using Cadaveric Specimens

Sixteen fresh frozen human cadaveric metacarpophalangeal joint specimens (MP joint) of the index, middle and ring fingers were selected. After removing all soft tissues including the capsule, the proximal and distal ends were embedded in a special fixation device using bone cement on the X-Y table which permitted two-dimensional free movement. Flexion and extension movement of the MP joint was also allowed to keep an axis-symmetrical situation of the joint contact area with an axial loading force perpendicular to the joint surface. The joint position was carefully adjusted to align for an axis-symmetrical orientation. This was verified by monitoring X-axis, Y-axis, and angular displacements when a perpendicular load was applied to the joint surface temporally. After the most natural position was obtained, the flexion and extension angular movement was fixed throughout the experiment to maintain axis-symmetry. During this procedure, slippage of the joint was kept to a minimum and, to avoid concentrating the weight in one place after setting the MP joint, the loads were added and the position was readjusted. For this experiment, the Instron all-purpose test machine (Instron 1321) was used and the extensometer was used to measure the displacement (Fig. 1). Moreover, the contact area and pressures of the MP joint surface were measured with a pressure sensor (Fuji Photo Company) and in addition, a carbon-particle-impregnated rubber pressure transducer was also used. Weight was applied to a maximum load of 20 lbs at a rate of 2 lbs/sec and assumed elastic modulus were calculated from a load-displacement curve and the joint contact area.

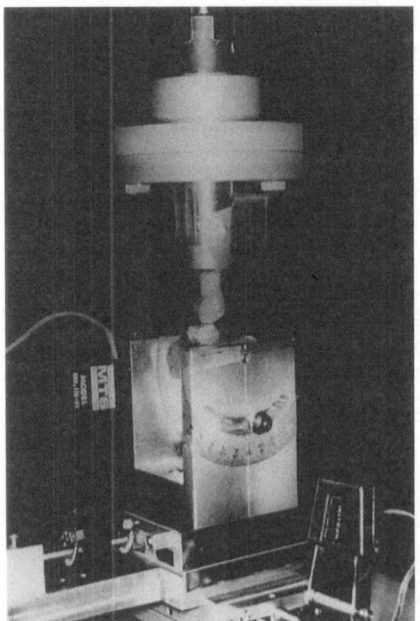

Fig. 1. A mechanical testing was performed using a universal test machine (Instron 1321) and the cartilage deformation was measured by an extensometer. The peak pressure and contact area of the articulating surfaces were measured by a Fuji prescale and a thin carbon-particle-impregnated rubber pressure transducer

Fig. 2. A computer simulation of gravel structures was performed utilizing a new analytic method which can be describe as "Rigid-Body Spring Model" (RBSM) by Kawai. The RBSM is formulated by considering the rigid bodies to be in a state of equilibrium with external loading

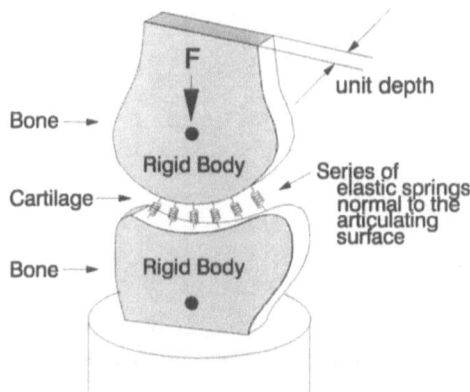

Computer Simulation Using Kawai's Rigid Body Spring Model

The Rigid Body Spring Model is formulated by considering the rigid body to be in equilibrium with external loading. Reaction forces between adjacent bodies are produced by the spring system distributed over the possible contact surface between the adjacent bodies. The weight which is added to these rigid bodies is transmitted through these springs. Displacements of the simulated compressive springs can be described as functions of the displacement of the centroid of the associated rigid bodies. Strain energy due to relative displacement of the spring system can then be formulated in the quadratic function of the displacement vectors of the centroid of adjacent rigid bodies. Using the RBSM method, generally bone segments are assumed to be rigid, while articulating surfaces and musculo-ligamentous structures are simulated by springs (Fig. 2). The distribution and density of spring elements at the articulating surfaces depend upon the contact area and the combined stiffness property of the cartilage and subchondral bone. RBSM must rely upon an iterative process in order to achieve the equilibrium condition due to the difficulties encountered in contact stress problems, and such a requirement will not be easily met if Finite Element Method (FEM) is to be used.

Two-Dimensional RBSM and the Spring Constant

A section of the specimen at the metacarpophalangeal joint was used as the model and the springs were positioned 0.1 mm apart along the joint between the proximal phalanx and the metacarpal bone. The stiffness equation used for RBSM calculations is the same as the standard formula function of the Finite Element Methods:

$$KU = F \tag{1}$$

where K is the stiffness matrix, U the displacement vector, and F the reaction force vector.

Whereas the FEM calls for a change in position of each contact point, the RBSM calls for the change in position of the rigid bodies center of gravity. The

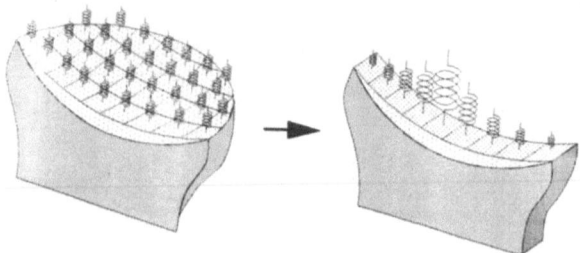

Fig. 3. The joint contact area was divided into small unit sub-squares and multiplied in the transverse direction. The number of units of the springs created non-uniform two-dimensional springs across the length of contact line

spring constant, with specific assumptions made regarding the cartilage surface flatness and the stiffness is formulated below:

$$Kd = E/h = (F/dh \times h/A) \times 1/h \qquad (2)$$

where E is the elastic modulus, F the loading force, h the initial cartilage thickness, A the contact area, and dh the deformation of cartilage.

The elastic modulus can be estimated by the following formula:

$$E = (F/A)/(dh/h) = F/dh \times h/A \qquad (3)$$

where F/dh is the slope of the load-deformation curve in material testing. Thus the spring constant can be estimated with a slope of load-deformation curve and with the contact area.

Modified Two-Dimensional RBSM and Spring Constant

As for the numerical value of the model taken from the metacarpus and the first phalanx, it assumes a two-dimensional model which is deep at the center of the joint and shallow at the periphery. If many of the simulated springs are lined up along the depth of the joint, the spring constant can be thought of as those springs bundled up and calculated together (Fig. 3). In this model, it is necessary to measure the joint's depth direction in advance.

The spring constant in the modified RBSM was assumed as given in the formula below:

$$Kdm = Kde \times (Y \times \text{unit area}) \qquad (5)$$

where Kdm is the spring constant of each spring in the modified 2D-RBSM, Kde the spring constant from the experimental study, Y the joint depth, and the unit area is $1\,\text{mm}^2$.

Results

Load Test

Sixteen specimens were attached to the Instron and a total of 49 measurements were taken. The aim was a load of 88.96 N, but in the actual test, scattered dispersion was noticed and the average was 95.7 N. Then according to that load, the cartilage's change in position was shown to be 0.117 mm. Further, the slope calculated from the load-deformation curve was an average of 846.5 N/mm. According to the pressure-sensitive film, the average peak pressure was measured at 2.1 MPa and 1.9 MPa at the carbon-particle-impregnated rubber pressure transducer.

From the experiment, the average joint cartilage spring constant was estimated to be 16.5 N/mm and the elastic modulus was 19.2 MPa.

Measurement of the MP Joint's Geometry and the Material Constant

The area of the MP joint's contact surface was an average of 49.6 mm^2, the thickness of the joint cartilage was 1.18 mm, and the maximum depth was 7.99 mm. It was determined that the outline of the joint contact surface was similar to the following formula with geometrical measurements:

$$Y = [1 - ABS(X^3)] \times Dm$$

where Y refers to every X-coordinate on the Y-coordinate contact surface line segment (joint depth), ABS the absolute value, Dm the maximum depth of the

MP joint, and X is defined as

$$X = \frac{\text{X-coordinate of the joint surface line segment}}{\text{length of the contact surface line segment}}.$$

Results of the Computer Simulation

Using the weight load and spring constant derived from the load experiment, each of the 16 specimens were analyzed according to the two-dimensional RBSM model. The results were an average displacement of 0.789 mm, and a peak pressure of 1.26 MPa, 6.7 and 0.6 times more than the experimental results, respectively. Moreover, according to the modified two-dimensional RBSM, analysis showed an average peak pressure of 1.95 MPa, and an average displacement of 0.137 mm. A stress distribution comparison of the two-dimensional and the modified two-dimensional RBSM is shown in Fig. 4, and a comparison of peak pressures is shown in Fig. 5.

Corrections for the Cartilage Spring Constant

In order for the displacement of the cartilage obtained in the load experiment to reach that of the displacement calculated from the rigid bodies analysis of the modified two-dimensional RBSM experiment, the revised spring constant was estimated with displacement matching by the following formula:

$$Kd = \frac{\text{Computational displacement}}{\text{experimental displacement}} \times Kdm \qquad (6)$$

Where Kdm is the spring constant of each spring in modified 2D-RBSM estimated by experimental study.

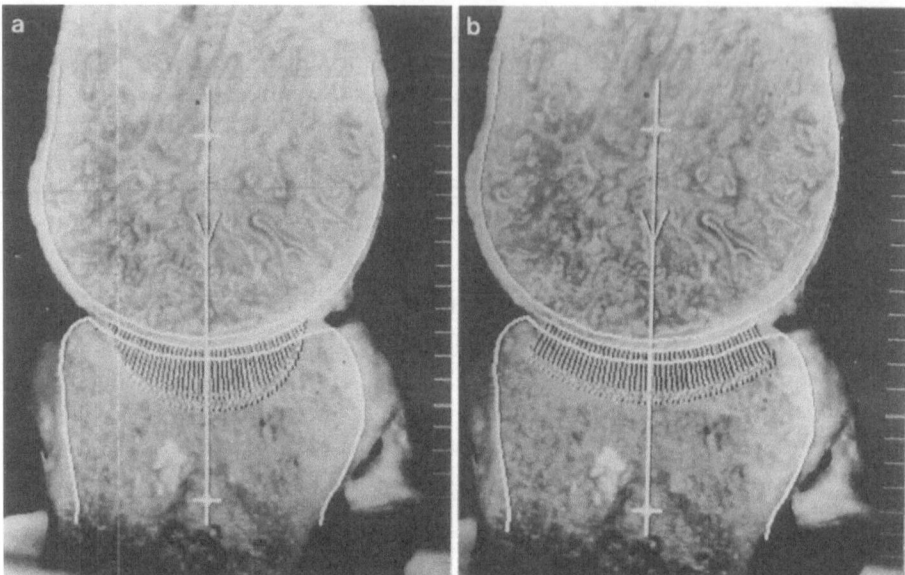

Fig. 4a,b. Stress distribution measured utilizing **a** the Modified 2D-RBSM (A) shows a more realistic pattern as compared with **b** the original 2D-RBSM (B) with the unit depth across the length of the contact line

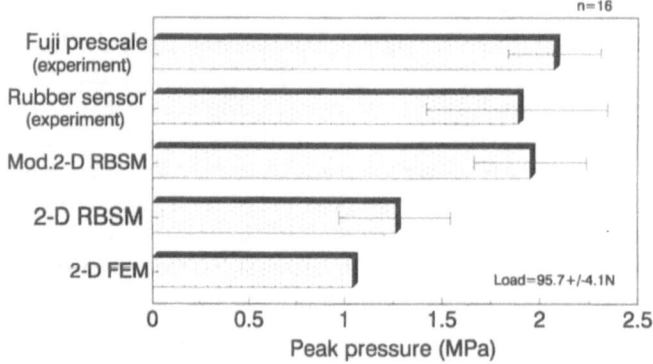

Fig. 5. Estimated peak pressure values calculated using the Modified 2D-RBSM method with non-uniform springs across the length of contact line were similar to experimental values

The spring constant which resulted from modified two-dimensional RBSM was an average of 22.6 N/mm/unit area and the elastic modulus was 25.8 MPa. The peak pressure of the joint obtained from analysis of the modified two-dimensional RBSM was an average of 1.95 MPa, and the average displacement was 0.117 mm. These figures agreed with the results of the actual experiment which showed averages of 2 MPa and 0.117 mm, respectively.

Table 1. Cartilage elastic modulus values estimated by different authors.

Year	Investigators	E (MPa)	Materials
1971	Hayes and Mockros	11.6	Bovine
1975	Johnson	12–45	Bivine
1975	Kempson	8.4–15.3	Femoral head
1976	Hori and Mockros	5.6–10.2	Degenerative cartilage
1978	Askew and Mow	12–50	Bovine
1980	Kempson	10.35	Acetabulum
1990	Ide et al.	19.21	MP joint
		25.8	Theoretical[a]

[a] The results of computer simulation utilizing the Modified 2D-RBSM and Displacement Matching
E, elasticity

Discussion

Load Experiment and Computer Simulation

The results of the analysis based on the two-dimensional RBSM using the spring constant estimated by the experiment showed an average joint pressure 0.6 times that of the experimental result, and a change in position 6.7 times greater. The reason for this discrepancy can be explained by the two-dimensions (2D-RBSM) where the joint depth was not considered. It is a matter of taking a three-dimensional shaped joint and putting it into the two-dimensional analytical model and including the unit depth in the calculations. Thus, for the modified two-dimensional RBSM, where joint depth was also considered, the analysis results come close to the experimental results and we were able to thus prove the usefulness of this modified model.

Spring Constant Depends on Shape and Size

The spring constant calculated by modified two-dimensional RBSM based on experimental studies with displacement matching was 22.6 N/mm/unit area, and the elastic modulus was 25.8 MPa. In the past, many authors have reported marked differences in elastic modulus values from 10 to 50 MPa as estimated by different experimental studies (Table 1) [6, 7]; the elastic modulus value calculated in this experiment is also within the above-mentioned limits (10–50 MPa). It is concluded that this spring constant and elastic modulus of the combined stiffness of the catilage and the subchondral bone based on these experimental studies with displacement matching can be used for computer simulation in human joints to determine the articulating contact pressure.

References

1. Kawai T (1977) A new discrete model for analysis of solid mechanics problems. Seisan Kenkyu 29:208–210
2. Garcia-Elias M, An KN, Cooney WP, Linscheid RL, Chao EYS (1989) Transverse stability of the carpus. J Orthop Res 7:738–743

3. Horii E (1990) Effect on force transmission across the carpus in procedures used to treat Kienbock's disease. J Hand Surg [Am] 15:393–400
4. Ide T, Akamatsu N, Hamada Y (1987) Non-liner stress analysis of various kinds of cementless total hip systems. Jpn J Artif Organs 16:1569–1572
5. Ide T, Akamatsu N, Hamada Y (1988) Biomechanical study regarding cementless total hip system with three falanges. Jpn Biomech 9:39–43
6. Askew MJ, Mow VC (1978) The biomechanical function of the cartilage fibril ultrastructure of articular cartilage. J Biomech Eng 100:105–115
7. Kempson GE (1980) The mechanical properties of articular cartilage. In: The joints and synovial fluid. Academic, New York, pp 177–238

Interfacial Reactions to Bioactive and Non-Bioactive Biomaterials

HIRONOBU OONISHI[1]

Summary. Biomaterials which create chemical and mechanical bonds with tissue include (a) non-porous materials with or without a hydroxyapatite coating, (b) porous titanium alloy beads with or without a hydroxyapatite coating, (c) α-tricalcium phosphate and tetracalciumphosphate bioactive bone cement and polymethylmethacrylate (PMMA) cement, and (d) interface bioactive bone cement made by interposing hydroxyapatite granules between polymethylmethacrylate cement and the bone were used in animal experiments and clinical applications. The common problem with cementless fixation is that some patients complain of slight pain on weight-bearing because a complete initial fixation is not obtained and micromovement of the component may occur. Porous metal with hydroxyapatite coating is found to be better than that without coating for producing earlier and stronger fixation, and problems with fatigue and peeling of hydroxyapatite from the base metal are eliminated. As hydroxyapatite bonds chemically to the bone, pain on weight-bearing due to micromovement should never occur. In order to obtain long-term and stable fixation for severe bony atrophy, bioactive bone cement or interface bioactive bone cement (interposing hydroxyapatite at the bone interface) is desirable.

Key words. Interface — Bioactive — Biomaterials — Hydroxyapatite — Porous metal — Bone cement — α-TCP (alpha-tricalcium phosphate) — TeCP (tetracalcium phosphate)

Introduction

Biomaterials were employed to create chemical [1–4] and mechanical bonds [5–8] in animal experiments and clinical applications. The following materials were used:

---

[1] Department of Orthopaedic Surgery, Artificial Joint Section and Biomaterial Research Laboratory, Osaka-Minami National Hospital, 677-2 Kido, Kawachinagano, Osaka, 586 Japan

308   H. Oonishi

1. Non-porous titanium and alumina (Al_2O_3) with and without hydroxyapatite (HAp) coating [3, 4]
2. Porous titanium alloy beads with and without HAp coating [4, 9–11]
3. Bioactive bone cement and non-bioactive polymethyl-methacrylate (PMMA) bone cement [12]
4. Interface bioactive bone cement by interposing HAp granules between PMMA cement and the bone [13]

Several of the problems of mechanical and chemical bonds will be discussed from the point of view of the histological changes and the bonding strength at the interface with the bone.

Interface Between Bone and Non-Porous Implants

A comparison was done of bone formation on titanium or alumina with and without HAp Coating.

Hydroxyapatite Coating on Titanium by Means of Plasma Spray

HAp and a mixture of HAp and alumina were plasma-sprayed with a plasma jet sprayer onto the titanium surfaces which had been roughed by shot blasting.

The experimental samples consisted of (A) materials sprayed with 100% HAp, (B) those sprayed with a mixture of 80% HAp and 20% alumina, and (C) pure titanium shot blasted as a control.

Materials

The titanium surface processed by shot blasting was from a few to several hundred micrometers in roughness and the HAp surface several tens to several hundred micrometers. No change and resolution due to spraying were found on the phase of HAp. As the mixing ratio of alumina increased, the strength of the membrane became greater.

Animal Experiments and Histology

Experimental samples A, B and C, as above, were implanted into 20 tibiae of beagles. After 2 weeks, reflexion electron microscopic images indicated that the space between the bone and HAp was filled with homogeneous substance. From the coal-HE-stained preparations, and elemental analyses, that space also seemed to be filled with bone mineral. There was no soft tissue observed. In the control group, bone invasion into the rough surfaces had already started, but soft tissue was always present as well. After 4 and 6 weeks, the bone had extensively and completely bonded with HAp (Fig. 1).

Comparisons were made between 100% HAp and the mixture of 80% HAp and 20% Al_2O_3. Throughout the course of time, no histological difference was

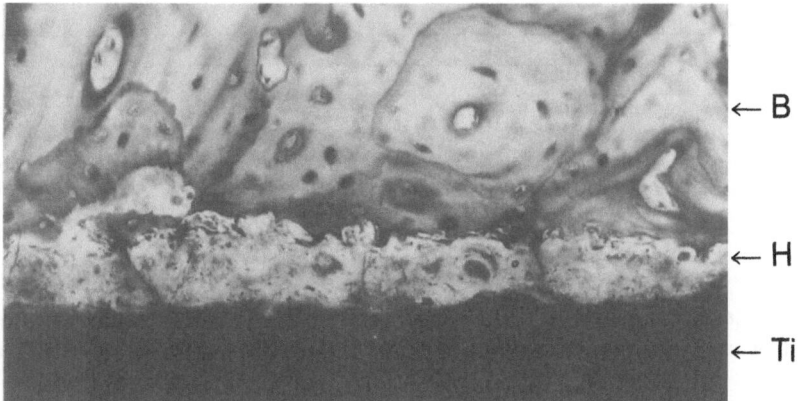

←B

←H

←Ti

Fig. 1. Light microscopic image of titanium plates coated with hydroxyapatite by plasma-spray implanted into the greater trochanters and tibiae of beagles, 4 weeks after implantation (methylene blue stained). *B*, bone; *H*, HAp coating; *Ti*, titanium plate

found. In the control group, bone invasion into the rough surfaces gradually increased, but no difference was noticed after 4 weeks and after 6 weeks.

Adhering Strength to the Bone

Experimental samples A, B and C, as above, were implanted into six tibiae of beagles.

The push-out test was performed 2, 4, 6 and 12 weeks after the implantations. Two weeks after implantation, there was no substantial difference between samples A and C. The adhering strength ratios of A and B were almost coincident with the mixing ratio of HAp to alumina. The stainless steel (SUS 316L) with smooth surfaces, as a control, was very low in adhering strength. For implants A, B and C about 80% total adherence strength was acquired 6 weeks after implantation. In our experiment on porous metal, there was the same tendency (Fig. 6).

After the pushing-out test, implants A and B failed within the HAp layers. Sample C failed between the bone and connective tissues.

Hydroxyapatite Coating on Alumina by Means of the Spatter Method

Method

Three implant test pieces were prepared: (1) a cylinder of Fe-Cr-Al alloy (4 mm in diameter and 12 mm in height) the surface of which had α-Al_2O_3 dispersed by heat treatment (Fe-Cr-Al + Al_2O_3), (2) the same implant test piece over which a coating of HAp was applied by the spatter method (Fe-Cr-Al + Al_2O_3 + HAp), and (3) SUS 316L. The tibial cortical bone of a goat was drilled with the same diameter as that of the test pieces, and each set of three

Table 1. Implant test pieces

	Test pieces		
	$Fe-Cr-Al + Al_2O_3$	$Fe-Cr-Al + Al_2O_3 + HAP$	SUS 316 L
Adhering force to bone (shearing force)	0.9 MPa	5.0 MPa	0.3 MPa
Interface between bone and implant	very thin fibrous membrane	direct bond	thin fibrous membrane

test pieces was implanted into both tibiae of two goats. After 3 months, the goats were sacrificed and the following tests and observations were carried out.

Results and Discussion

The results of the pushing-out test indicated that the adhering force of implants 1, 2 and 3 above to the bone were on average approximately 0.9 MPa, 5.0 MPa, and 0.3 MPa, respectively.

Very thin fibrous membrane was seen between bone and implant 1, however, bone tissues directly touched the implant 2 even after HAp had disappeared (Table 1).

Interface Between Bone and Porous Implants

A comparison was done of bone growth into porous Ti alloy beads with and without HAp coating.

Under Unloaded Conditions

Methods and Results

Histology. Ti-6Al-4V beads with diameters ranging from 300–400 µm (with a mean of 350 µm) and pore sizes ranging from 150–250 µm (with a mean of 200 µm) were prepared. A sample was prepared by coating a titanium alloy bese material with two layers of the beads. In addition, a second group of beads were made rough by sand-blasting and were coated with HAp by plasma-spraying. The two kind of samples were implanted on the bilateral femoral condyles of mature rabbits.

When the beads were coated with HAp, 7–10 days after implantation, the speed of bone invasion was faster than when HAp was absent. Namely, the depth of bone invasion was comparable to that observed in the sample without HAp coating 4 weeks after implantation. Three weeks after implantation, the amount of bone invading even the substrate was larger. Observation of the fracture surface revealed that the collagen fiber forming a net was bound to HAp. Bone formation followed after that and the bone tissue was also bound to HAp. Comparison between beads with and without HAp coating revealed that when the beads were coated with HAp, the invading bone was bound to almost all the HAp coating the beads and, on the whole, the amount of invading bone was extremely large.

Bonding Strength to Bone. Cylindrical samples were implanted in the tibial cortical bone of five mature goats. Each sample was measured for bonding strength (i.e., shear interface strength) by the use of Instron Universal Tester at a push-out rate of 0.1 mm/min cross-head speed.

The bonding strength in the presence of an HAp coating was about four times higher 2 weeks after implantation, about three times higher 4 weeks after implantation, and about twice as high 6 weeks after implantation as that found in the absence of the HAp coating. Twelve weeks after implantation, the bonding strengths with and without HAp more closely approximated each other. Thus, a larger amount of bone invades faster in the presence of HAp coating than in its absence, and as a result, a firmer anchorage is obtained faster (Fig. 6).

Under Loaded Conditions

Methods and Results

Artificial tibial bones coated with titanium alloy beads both with and without HAp coating, were prepared. These were implanted into 12 tibiae of beagles after tibial resection. Histological observations were observed 13 and 26 weeks after the implantations.

Since the amount of bone ingrowth into pores was larger in stems having a porous coating when implanted in the cortical bone than in that implanted in the cancellous bone, the fixability of stem was more stable and firmer in the cortical bone, and the incidence of stem loosening was also lower in the cortical bone.

When the porous coating was coated with HAp as well, new bone tissue had access earlier to the surface of beads in the superficial layer of stem compared to the porous coating without HAp coating. Furthermore, the new bone tissue entered the pores earlier, in larger amounts, and could reach the deepest region. Thus, the stem could be bound to the bone within an early postoperative period (Fig. 2a).

Bone ingrowth was accomplished within 3 months after operation.

When the porous coating had no HAp coating, new bone tissue slowly accessed the surface of the beads surrounding the stem and made no direct contact with the surface of the beads. Therefore, there was a high possibility of stem micromovement at the time of weight-bearing. As a result, not only new bone tissue was prevented from deeply entering the pores, but also the spaces between the new bone tissue and beads was wider (Fig. 2b). This further increased the likelihood of stem micromovement because definitive and firm initial fixation under load could not be established in an early postoperative

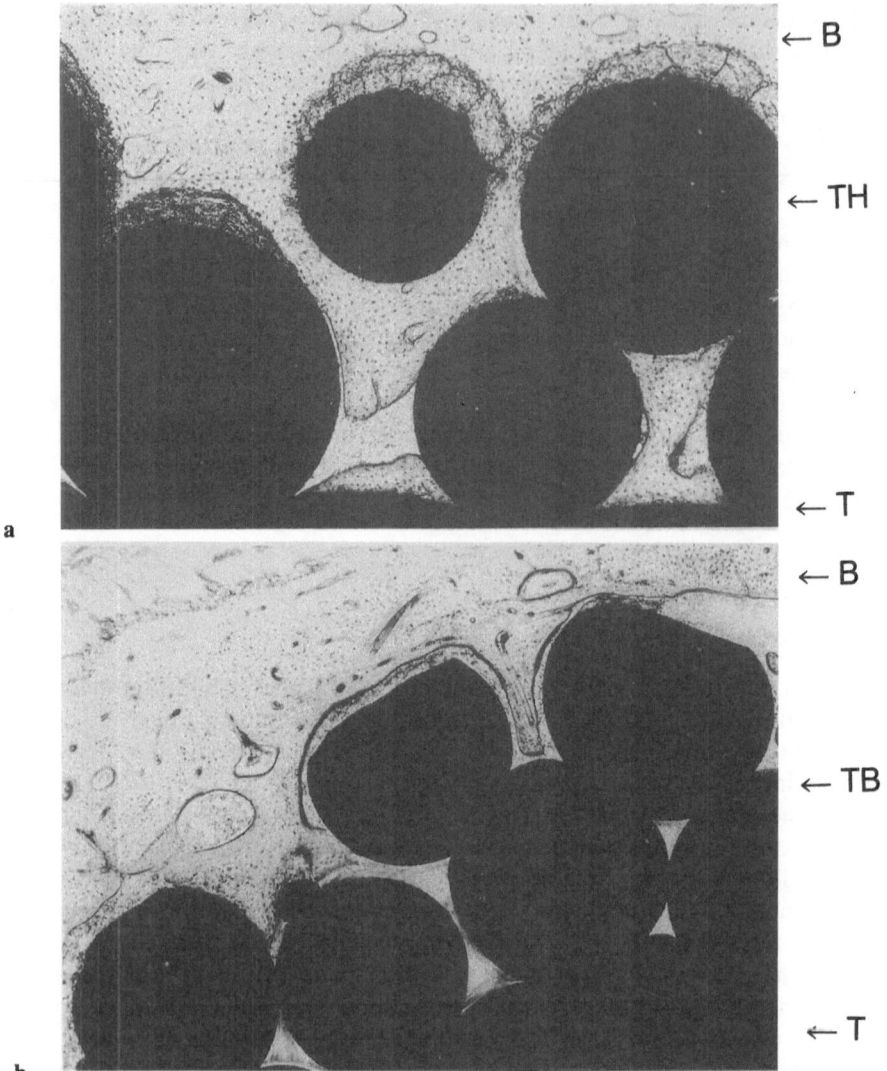

Fig. 2a,b. Artificial tibial bones coated with titanium alloy beads **a** with HAp coating and **b** without HAp coating were implanted into the tibiae of beagles after tibial resection. These light microscopic images were taken 13 weeks after implantation under loaded conditions. *B*, bone; *TH*, titanium alloy beads with HAp coating; *TB*, titanium alloy beads without HAp coating; *T*, titanium alloy cylinder as an artificial tibial bone

period. This condition may clinically cause thigh pain. Thus, it is concluded that the addition of an HAp coating to a porous coating is very useful to prevent not only the occurrence of stem loosening, especially under load, but the occurrence of thigh pain as well.

Clinical Cases

For two revision stems, titanium alloy beads coated with HAp were used. Six months after replacement, a bone-ingrowth was observed not only on the area coated with beads and HAp but also on the area coated with only HAp without titanium alloy beads around the stem.

Interface Between Bone and Bone Cement

Bone cement is useful to obtain stable fixation and to relieve all patients after surgery from pain on weight-bearing due to micromotion for a relatively long period of time. However, in order to obtain the effects for longer periods, even for severe bony atrophy, bioactive bone coment is desirable.

Although various kinds of bioactive bone cement will be developed in the future, they can be roughly classified into (a) all-bioactive bone cement, (b) surface bioactive bone cement, and (c) interface bioactive bone cement based on their functional aspects.

All-bioactive bone cement is completely replaced by bone from its contact surface within a short period.

Surface bioactive bone cement bonds physicochemically to the bone, resulting in direct bonding; however, only the cement surface in contact with the bone is replaced by bone in the fixation between bone and cement.

Interface bioactive bone cement refers to when a bioactive ceramic is interposed between bone and PMMA bone cement. We have employed a method of placing one to three layers of fine HAp granules 100–300 μm in diameter between bone and PMMA bone cement in clinical cases.

Materials, Methods and Results

All-Bioactive Bone Cement

We used a mixture of TeCP (tetracalciumphosphate) and DCPD (dicalcium phosphate dihydrate) as the all-bioactive bone cement. Drilled holes in the femoral condyle of rabbits were filled with the bone cement. The animals were sacrificed at 1, 2, 3, 6, 12 and 24 weeks after surgery to obtain hard tissue specimens.

Bone cement was found to be replaced by bone from its contact surface to a depth of 20–30 μm after 1 week, 700–1000 μm after 3 weeks and close to the center of the cement after 6 weeks, although a relatively large amount of TeCP was sporadically observed. The remaining TeCP decreased markedly in quantity after 12 weeks (Fig. 3).

TeCP bone cement can be effectively used to fill in spaces that appear between bone and a prosthetic component when a cementless artificial joint is inserted because TeCP bone cement is replaced by bone in an early period after surgery.

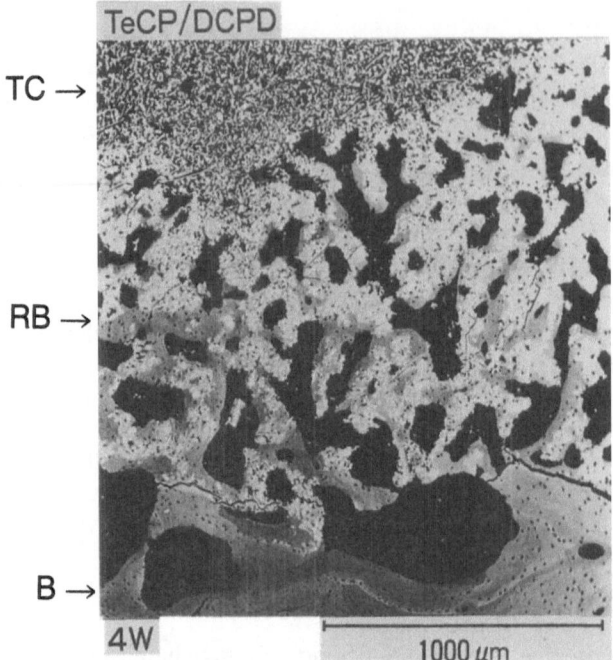

Fig. 3. Reflection electron microscopic image taken four weeks after implantation of all-bioactive bone cement. Drilled-holes in the femoral condyles of rabbits were filled with a mixture of TeCP and DCPD as the all-bioactive bone cement. *TC*, TeCP (tetra-calciumphosphate) as the all-bioactive bone cement; *RB*, TeCP being replaced by bone tissues; *B*, bone

Surface Bioactive Bone Cement

The surface bioactive bone cements we used were (a) alpha-tricalcium phosphate (α-TCP) kneaded with an organic acid (citric acid and malonic acid), and (b) α-TCP kneaded with an organic acid (citric acid).

Powdered α-TCP and a solution of 34.6% (w/w) citric acid and 5.4% (w/w) malonic acid were used as a curing solution.

Through X-ray diffraction, the formation of calcium phosphate dihydrate and HAp was seen on the surface of the hardened cement after immersion in the phosphate-buffered saline (PBS).

Compression strength after one day was similar to that of the PMMA bone cement (80–100 MPa), but became stronger than the PMA cement after 1 week (130 MPa).

The α-TCP began to partially bond to the bone after 1 week, and changed to HAp a depth of about 500 μm from its contact surface after 6 weeks, and about 1000 μm after 6 months, resulting in reasonably wide area of solid bonding with bone (Fig. 4).

The complete change of α-TCP to HAp was confirmed by micro X-ray diffraction analysis.

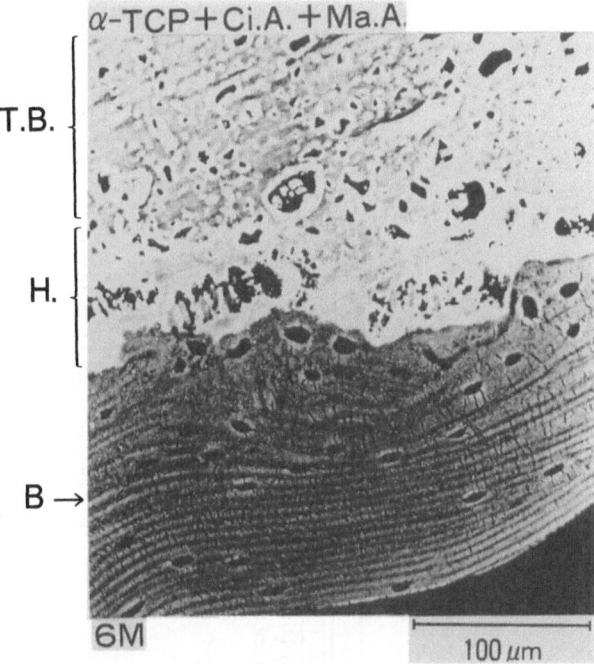

Fig. 4. Reflection electron microscopic image taken 6 months after implantation of surface bioactive bone cement. Drilled-holes in the femoral condyles of rabbits were filled with α-TCP (α-tricalciumphosphate) kneaded with citric acid and malonic acid as the surface bioactive bone cement. *TB*, α-TCP bone cement; *H*, HAp changed from α-TCP surface bioactive bone cement; *B*, bone

A 45% citric acid solution was used for hardening. A 36 wt% polyacrylic acid solution was used as a control hardening solution.

In the animal experiments, the following two methods were employed (Table 2): 1. During surgery, α-TCP was kneaded with either polyacrylic acid (PAA) or citric acid. Then the kneaded cement was filled into the bone and was allowed to harden 2. Cement was hardened into a cylindrical shape and inserted into the bone during surgery.

In both cases, histological pictures were observed and, with the second method, the adhesive strength to bone was also measured. In the first method, a through-hole was made in the femoral condyles of mature rabbits. The α-TCP cements were filled into the through-hole. In the second method, with the above two types of α-TCP cements (PAA and citric acid as hardening solutions) and PMMA, hardened cylinders were prepared which were inserted into a hole drilled in the femoral shafts of six mongrels.

Both in the case in which α-TCP was kneaded with polyacrylic acid during surgery and was filled into the bone, and in the case in which hardened cylinders were implanted, connective tissues were present.

Table 2. Experimental methods

	Kneaded cement		Mardened cement		PMMA
	α–TCP + PAA	α–TCP + Citric Acid	α–TCP + PAA	α–TCP + Citric Acid	
Interface between bone and cement	Connective tissue membrane	direct contact (2 weeks: slight round cell infiltration at 4–6 weeks; bone formation)	connective tissue membrane	direct contact	connective tissue membrane
Bunding force				2 weeks: 0.55~1.5 MPa (mean 1.5) 4 weeks: 1.0~3.5 MPa (mean 2.0) 8 weeks: 2.5~3.3 MPa (mean 2.8)	

TCP, tricalcium phosphate; PAA, polyacrylic acid; PMMA, polymethyl methacrylate

When α-TCP was kneaded with citric acid during surgery and the cement was filled into the bone, slight round cell infiltration was observed 2 weeks later. At 4 and 6 weeks, inflammation gradually disappeared and bone formation gradually increased. Round cell infiltration was also observed in the initial stage because the presence of citric acid during surgery stimulated the bone tissue before the cement hardened. In the case in which the hardened cylinder was implanted, the cylinder started to connect with the bone two weeks later and the connecting strength was 0.55–1.5 MPa (mean of 1.0). After 4 weeks, the connecting strength increased to 1.0–3.5 MPa (mean of 2.0), and after eight weeks, it reached 2.5–3.3 MPa (mean of 2.8).

In long term implantation of 3 years in which a cement consisting of α-TCP kneaded with citric acid was filled into the bone during surgery, the α-TCP cement contacting the bone gradually changed into HAp (also in contact with the bone) to a depth of 50–100 μm. The α-TCP, in the shape of granules, changed into homogeneous HAp. Several $CaCO_3$ crystals, 100–200 μm diameter, were also found in the TCP cement near the bone.

Interface Bioactive Bone Cement

We have developed a method of making bioactive bone cement only at the interface by interposing fine HAp granules between the bone and the PMMA cement. This method involves the placement of one to two layers of fine, porous HAp granules of 100–300 μm in diameter between the bone and cement. The author calls this the interface bioactive bone cementation technique (IBBC).

Experimental Studies.

Tissue Reaction Test. Holes were made in both femoral condyles of 24 mature rabbits, and interface bioactive bone cementation was carried out. New bone tissues had entered a majority of the spaces of the first layer of HAp granules 1–2 weeks after surgery, and a majority of the spaces in the second layer at 4 weeks. All of the spaces were filled with new bone tissue after 6 weeks (Fig. 5).

Bond Strength Test. The IBBC technique was carried out in femoral condyles of 20 mature rabbits. Pushing-out tests were then performed.

At 2–6 weeks, the fracture of the interface from the pushing-out tests occurred at the HAp granule layer. At 12–24 weeks, fracture occurred at the HAp granule layer, the bone around this layer, and at the interface between the PMMA and the HAp. Therefore, this IBBC technique is considered to be a functional gradient material. Both the bond strength of bone cement to HAp and that of the HAp granule layer to bone were adequate in these experiments (Fig. 6).

Clinical Studies. The IBBC technique has been used in total hip replacements and total knee replacements since 1984 [13].

Surgical Procedures. A number of small holes and rough surfaces were made on the surface of an acetabulum.

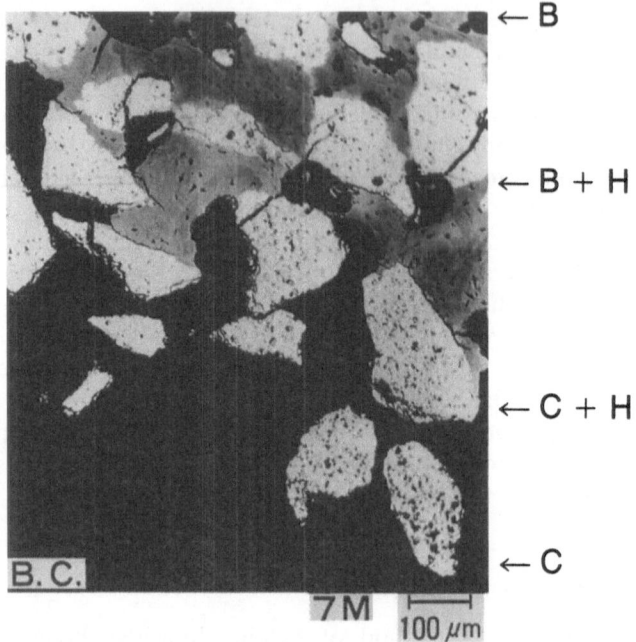

Fig. 5. Reflection electron microscopic image taken seven months after implantation of interface bioactive bone cement was implanted into drilled-holes made in the femoral condyles of rabbits. *B*, bone; *B + H*, bone with HAp; *C + H*, PMMA cement with HAp; *C*, PMMA cement

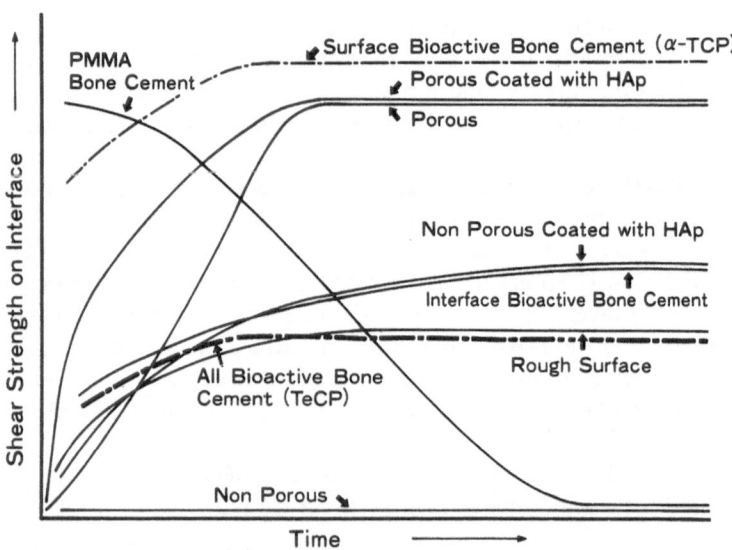

Fig. 6. Comparisons of adhesive strength of several biomaterials to bone over time

In the cemented fixation at bleeding areas during surgery, complete hemostasis had to be performed before cementing. About 2–4 g of fine granules of HAp (100–300 µm diameter) were laid in one to two layers. The same technique was applied to the inner surface of the femur from the proximal end to the distal end, as far as possible, using a special instrument.

The IBBC technique has been applied in over 500 cases.

Clinical Results. Although the configuration of the interface was slightly indistinct in some cases immediately after surgery, the interface showed a distinct configration on X-ray after several months. X-ray findings at the interface after a total hip replacement performed in a non-bloodless area have so far shown no difference after 5 years from those seen after a total knee replacement performed in a bloodless area.

Also, as the HAp bonded chemically to the bone, there is no pain with weight-bearing due to micromotion. The patients were pain free after surgery.

The author is therefore confident that the IBBC technique combines both the advantages of conventional PMMA bone cement fixation and cementless fixation with an HAp coating on the components.

Discussion

Previously, when cement fixation was used, the prostheses appeared to be very secure immediately after surgery, and pain was relieved immediately following surgery in almost all patients. The common problem with cementless fixation is that about 20%–40% of patients complain of slight pain on weight-bearing for a certain period of time after surgery. The cause of these problems is that complete initial fixation is not obtained immediately after surgery and micromotion of the component may occur. However, taking into account long-term results at the inter-face, it may not be desirable to use the PMMA cement. Lately, porous metal has been used to allow early strong fixation with the bone. However, HAp coating on the porous metal surface is considered better for producing earlier and stronger fixation. Furthermore, problems with fatigue and peeling of the HAp from the base metal when HAp is coated on a flat surface have not been solved. However, these problems are eliminated when HAp is coated on the surface of porous metal. After the HAp bonds chemically to the bone, there is no pain with weight-bearing due to micromotion.

At present, long-term and stable fixation cannot be obtained when bone atrophy is severe even when PMMA bone cement is used. Therefore, a bioactive bone cement, which hardens during surgery within ten minutes is essential.

TeCP bone cement can be used to fill in the spaces which appear between bone and component when a cementless artificial joint is inserted into the bone because the TeCP bone cement is replaced by bone shortly after surgery. This procedure secures firm initial fixation, not only giving a possibility of early weight-bearing walking, but preventing the occurrence of micromotion and loosening as well. Furthermore, a large portion of the circumference of the component is allowed to be in contact with the bone.

α-TCP kneaded with citric acid and malonic acid is expected to become common in clinical use.

The IBBC technique (developed by the author) has been used in over 500 cases of total hip and knee joint replacements since 1984, and no radiographic clear zones have appeared between the bone and the PMMA bone cement to date. With either a cementless or a PMMA bone cement fixation, there is always a connective tissue membrane at the bone interface. As a reaction to high-density polyethylene (HDP) wear particles, the bone tissue may become necrotic, through the connective tissue membrane, at the bone interface. However, with the IBBC technique in which the bone is physicochemically bonded to the bioactive materials and there is no connective tissue membrane at the interface, the abovementioned phenomenon does not occur. Therefore, the appearance of loosening at the interface can be delayed.

References

1. Jarcho M, Kay JF, Gummaer KI, Doremus RH, Drobeck HP (1977) Tissue cellular and subcellular events at a bone-ceramic hydroxyapatite interface. J Bioeng 1: 79–89
2. Ducheyne P, Hench LL, Kagan IIA, Martens M (1980) Effect of hydroxyapatite impregnation on skeletal bonding of porous coated implants. J Biomaterials Res 14:225–37
3. Oonishi H, Kohda A, Ishimaru H, Tsuji E (1986) Comparisons of biological fixation to the bone of titanium coated with hydroxyapatite reinforced with alumina. Orthopaedic Ceramic Implants (Japan) 6:73–80
4. Oonishi H (1990) Mechanical and chemical bonding of artificial joints. Clin Materials 5:217–33
5. Galante J, Rostoker W, Lueck R, Ray R (1971) Sintered fibre metal composites as a basis for attachment of implants to bone. J Bone Joint Surg [Am] 53:101–14
6. Cameron HU, Macnab I, Pilliar RM (1978) A porous metal system for joint surgery. Int J Artif Organs 1:104–9
7. Bobyn JD, Pilliar RM, Cameron HU, Weatherly GC (1980) The optimum pore size for the fixation of porous-surfaced metal implants by the ingrowth of bone. Clin Orthop 150:263–70
8. Pilliar RM (1983) Powder metal-made orthopaedic implants with porous surface for fixation by tissue ingrowth. Clin Orthop 176:42–51
9. Oonishi H, Yamamoto M, Ishimaru H, Tsuji E (1986) Comparison of bone ingrowth into Ti-6Al-4V beads coated and uncoated with hydroxyapatite. Orthopaedic Ceramic Implants (Japan) 6:53–64
10. Oonishi H, Kushitani S, Aono M, Ukon Y, Yamamoto M, Ishimaru H, Tsuji E (1989) The effect of hydroxyapatite coating on bone growth into porous titanium alloy implants. J Bone Joint Surg [Br] 71:213–16
11. Oonishi H, Yamamoto M, Ishimaru H, Tsuji E, Kushitani S, Aono M, Ukon Y (1989) Comparisons of bone ingrowth into porous Ti-6Al-4V beads coated and

uncoated with hydroxyapatite. In: Oonishi H, Aoki H, Sawai K (eds) Bioceramics, vol. 1. Ishiyaku Euro America, Tokyo, pp 400–5

12. Oonishi H, Aoki H, Tsuji E (1990) Biological and physical change of α-TCP bioactive bone cement three years after implantation. In: Heimke G (ed) Bioceramics, vol. 2. German Ceramic Society, pp 341–38

13. Oonishi H, Kushtani S, Aono M (1990) Experimental and clinical results of interface bioactive bone cement. In: Heime G (ed) Bioceramics, vol. 2. German Ceramic Society, 410–17

Keywords Index